The Impact and Integration of AI:

From Research to Realization

Poshan Yu
https://orcid.org/0000-0003-1069-3675
Soochow University, China

James Mulli
European Business Institute, Luxembourg

Zain Ali Shah Syed
Soochow University, China

Laila Umme
https://orcid.org/0000-0001-8050-5081
Institute of Business Management, Pakistan

IGI Global
Scientific Publishing
Publishing Tomorrow's Research Today

Vice President of Editorial — Melissa Wagner
Director of Acquisitions — Mikaela Felty
Director of Book Development — Jocelynn Hessler
Production Manager — Mike Brehm
Cover Design — Jose Rosado

Published in the United States of America by
IGI Global Scientific Publishing
701 East Chocolate Avenue
Hershey, PA, 17033, USA
Tel: 717-533-8845 | Fax: 717-533-7115
Website: https://www.igi-global.com E-mail: cust@igi-global.com

Library of Congress Cataloging-in-Publication Data

Names: Yu, Poshan, editor. | Mulli, James, 1961- editor. | Syed, Zain Ali Shah, 1979- editor. | Laila, Umme, 1981- editor.
Title: The impact and integration of AI : from research to realization / edited by Poshan Yu, James Mulli, Zain Ali Shah Syed, Umme Laila.
Description: Hershey, PA : Engineering Science Reference, [2027] | Includes bibliographical references and index. | Summary: "This book aims to offer a holistic and profound examination of artificial intelligence, documenting its evolution from leading-edge research to extensive practical applications across multiple societal sectors"-- Provided by publisher.
Identifiers: LCCN 2025007585 (print) | LCCN 2025007586 (ebook) | ISBN 9798337315096 (hardcover) | ISBN 9798337315102 (paperback) | ISBN 9798337315119 (ebook)
Subjects: LCSH: Artificial intelligence--Case studies. | LCGFT: Case studies. https://id.loc.gov/authorities/genreForms/gf2017026140
Classification: LCC Q335 .I465 2025 (print) | LCC Q335 (ebook) | DDC 303.48/34--dc23/eng/20250507
LC record available at https://lccn.loc.gov/2025007585
LC ebook record available at https://lccn.loc.gov/2025007586

British Cataloguing in Publication Data
A Cataloguing in Publication record for this book is available from the British Library.

Table of Contents

Detailed Table of Contents

The chapter explores three basic methods of knowledge - narrative (N), the scientific approach to objective reality (OR), and artificial intelligence (AI). In particular, it focuses on their interactions, transitions and benefits. Narrative offers a subjective interpretation of the world through stories and art, science presents a reduced model of objective reality, and AI mediates a fuzzy reality combining real and virtual information. In this vein, artificial intelligence is seen as a new means of interpreting our environment, through which new representations of reality can emerge that can lead to our evolutionary advantage associated with a deeper understanding of the world around us.

The chapter examines multiple perspectives about corporate board AI adoption processes and conducts an in-depth analysis of decision-making factors and limitations and biases. The risk-reward-reputation framework presents essential viewpoints about how boards acquire the will to make transformative technology investment choices. The chapter enhances traditional decision-making models TAM and UTAUT that oversee crucial psychological and cultural elements that influence adoption. The Sanyal's BELT Test developed by the author of this chapter serves as an evidence-based tool to evaluate new technology readiness through assessments of bias and ego and learning habits and motivation for tangible results because these dimensions explain additional acceptance variance. The chapter supports innovative governance frameworks supporting new technology experimentations and others. These human-centered approaches show promise to reduce adoption friction because they address both generational and cultural dynamics.

Artificial intelligence (AI) systems are increasingly deployed in high-stakes domains, raising urgent concerns about bias, fairness, and accountability. This chapter explores major real-world cases of algorithmic bias including Amazon's hiring tool, Apple Card limits, Robodebt in Australia, the U.K. A-Level grading system, biased healthcare risk predictions, and wrongful arrests from facial recognition. It then surveys evolving legal frameworks such as the EU AI Act, U.S. AI policies, and initiatives from China, Brazil, Canada, and the African Union aimed at promoting ethical AI. The chapter also examines how emerging challenges from generative AI such as hallucinations, deepfakes, and image bias demand stronger oversight. It concludes with recommendations for bias audits, explainability tools, regulatory sandboxes, and impact assessments to advance fairness and accountability in AI.

The chapter reports a developing picture of the role of Artificial Intelligence (AI) in policy and regulation, and the legal limitations and obstacles presented by AI technology. It shows the landscape of AI as a regulatory technology, establishing the need for legal regulation and policy, and dealing with a multiplicity of factors indicative of the emergent nature of digital technology. It discusses the foundational elements of AI, its implications for law and governance, and the conceptualizations that researchers are working with, particularly in relation to algorithmic accountability. A brief overview of different regulatory regimes worldwide, like the EU AI Act and segmental approaches taken in the US, leads to the potential roles of AI as a policymaking tool. The chapter also identified key regulatory challenges and gaps in law, including ineligibility and liability, and examined a number of sector specific regulatory frameworks. The chapter concludes by suggesting a number of future pathways for regulation of AI that are developing new ways to address its emergent nature

This chapter examines the transformative potential of integrating blockchain and artificial intelligence (AI) to address systemic governance, financial, and electoral challenges in Kenya. Grounded in theories of algorithmic trust, machine ethics, and programmable constraints, it explores how blockchain's immutability and AI's adaptive intelligence can enhance transparency, optimize public services, and reduce corruption. Using case study analysis and comparative review of international models, the chapter proposes practical applications in Kenya's e-procurement, land registry, tax compliance, and healthcare systems. A layered voting model—featuring decentralized digital identity, blockchain-backed registration, and AI-assisted audit trails—is presented to strengthen electoral integrity. While acknowledging regulatory, technical, and social adoption hurdles, it argues that AI can mitigate these through simulation, compliance automation, and capacity-building. This synergy offers Kenya a leapfrogging opportunity to pioneer inclusive, accountable, and innovation-led governance reform.

The integration of AI into FinTech has transformed the digital economy, presenting innovation opportunities and regulatory challenges. Using CiteSpace, this study analyzes knowledge clusters in risk management, data privacy, and blockchain, revealing AI's progress in FinTech along with transparency and ethical concerns. Comparing EU and Chinese regulatory frameworks shows significant differences, with the EU's strict AI Act and China's thriving AI - FinTech scene. This emphasizes the need for flexible, globally - coordinated regulations to balance innovation and risk. Burst detection and timeline analysis uncover research gaps in AI applications in FinTech and blockchain - related transparency. The paper concludes by urging international collaboration and robust AI governance to fully realize AI - driven FinTech's potential in the digital economy.

This chapter covers incorporating cybersecurity into financial planning due to the increasing threats in the digital space. It explores fundamental risks such as phishing, ransomware, data breaches, as well as the client portal and planning software vulnerabilities. Focus is placed on proactive measures regarding asset protection encryption, multi-factor authentication, AI threat detection, and others. Cyberattacks case studies illustrate loss of organizational trust, reputation, and business, as well as compliance breaches. The responsibility of a financial planner as a proactive cybersecurity hygienist and other technologies like blockchain for cryptocurrencies as secure online transaction tools is also mentioned. Issues of compliance are analyzed with the policies set forth by GDPR and GLBA. This chapter highlights the need to cultivate a security culture that fosters enduring resilience while adapting to evolving threats, all while maintaining client trust.

Integrating artificial intelligence (AI) into organizational leadership represents a transformative paradigm shift in management theory and practice. This comprehensive analysis examines the multifaceted impact of AI on leadership dynamics, exploring its implications for strategic decision-making, organizational competencies, and the fundamental nature of leadership in the digital age. Drawing on empirical research and theoretical insights from leading scholars, including Shah (2024) and Jerahian (2024), the study reveals a nuanced landscape where technological capability and human intelligence converge to redefine organizational leadership.The research demonstrates that AI is not a replacement for human leadership but a powerful augmentation tool that expands analytical capabilities and strategic imagination.

This article provides a comprehensive overview of AI's evolving role in clinical trials. will trace the evolution of AI technologies relevant to clinical research, from the rise

Artificial Intelligence (AI) has a transformative force that has the potential to alleviate systemic disparities in rural regions through enhanced healthcare, education, and agriculture. Despite the potential of AI, obstacles such as infrastructure deficiencies, digital literacy gaps, and ethical concerns jeopardize its equitable adoption. By adopting a multi-stakeholder perspective, this study emphasizes the imperative of culturally sensitive AI solutions, moral governance, and public-private collaboration to facilitate ethical and inclusive AI uptake. By putting affordability, accessibility, and long-term viability, AI can promote the empowerment of marginalized communities, build resilience, and forge a more equal future. This chapter explores how AI-driven innovations can transform these sectors through improved access to basic services, effective resource allocation, and empowering local communities.

Artificial Intelligence (AI) is transforming the creative landscape, challenging traditional notions of authorship, originality, and artistic collaboration. This chapter explores the multi-faceted impact of AI on art and creativity, beginning with a critical discussion on whether creativity is an exclusively human domain. It explores how AI is enabling new forms of artistic expression and how artists are adapting to collaborate with intelligent algorithms. The author also discusses the ethical considerations surrounding AI-generated art, including questions of authorship, fairness, and the economic impact on artists. Additionally, the chapter addresses intellectual property challenges – how laws around copyright and ownership are evolving (or struggling to evolve) in response to AI. Finally, it considers future directions, including global trends, the emergence of the Synthaesthetic art movement, and strategies for fostering a creative future where AI enhances rather than replaces human ingenuity.

Preface

INTRODUCTION

Artificial Intelligence (AI) has evolved from an experimental computational concept into one of the most influential forces shaping contemporary society. Across governance, healthcare, finance, communication, leadership, education, cybersecurity, and creative industries, AI technologies are redefining how institutions operate, how knowledge is produced, and how human experiences are interpreted and enhanced. The Impact and Integration of AI: From Research to Realization was developed to explore this transformative journey from theoretical innovation to practical implementation. This volume brings together interdisciplinary perspectives that examine how AI is being integrated into real-world systems while also confronting the ethical, legal, cultural, and societal questions emerging alongside technological advancement.

The chapters in this edited reference book collectively investigate the evolving relationship between humans and intelligent systems. The contributors explore AI not only as a technological tool, but also as a social, philosophical, economic, and institutional phenomenon. Topics such as algorithmic accountability, governance frameworks, digital equity, leadership transformation, financial technology, cybersecurity, clinical trials, communication strategies, and creative expression illustrate the broad reach of AI across modern life. The volume therefore highlights both the opportunities and responsibilities associated with the increasing integration of machine intelligence into decision-making processes and public infrastructures.

What makes this collection especially significant is its global and interdisciplinary character. The authors represent diverse academic traditions and professional experiences spanning Europe, Asia, Africa, North America, and beyond. Their contributions demonstrate that the future of AI will not be determined solely by technical capability, but also by policy development, ethical governance, cultural adaptation, public trust, and inclusive innovation. Through this collection, we aim to foster scholarly dialogue and critical reflection regarding the pathways through which AI research is translated into meaningful societal realization.

CHAPTER OVERVIEW

Chapter 1: In the Mirrors of Artificial Intelligence, Narrative and Science examines the intersections between narrative knowledge, scientific reasoning, and artificial intelligence. The chapter presents AI as a new interpretive framework that mediates between subjective storytelling and objective scientific modeling, offering new ways of understanding reality and human cognition.

Chapter 2: AI's Midlife Crisis: The Corporate Board Story investigates the dynamics of AI adoption within corporate governance. The chapter explores psychological, cultural, and organizational factors influencing board-level decision-making while introducing Sanyal's BELT Test as a framework for evaluating technological readiness and reducing adoption barriers.

Chapter 3: Navigating the Ethical Minefield of AI: Bias, Fairness, and Accountability in Automated Decision-Making addresses urgent ethical concerns surrounding algorithmic bias and automated systems. Through real-world case studies and analysis of global regulatory initiatives, the chapter emphasizes the importance of transparency, explainability, fairness audits, and institutional accountability in AI deployment.

Chapter 4: The Role of AI in Policy and Regulation: Navigating the Legal Landscape explores the evolving relationship between AI technologies and legal governance. The chapter examines international regulatory approaches, algorithmic accountability, sector-specific policies, and the growing role of AI in shaping policymaking and regulatory systems.

Chapter 5: From "Do No Evil" to "Can't Do Evil" AI-Blockchain Technology as a Transformative Paradigm for Kenya investigates how the integration of blockchain and AI can strengthen public trust, institutional integrity, and governance reform in Kenya. The chapter presents applications in healthcare, taxation, elections, and public administration while highlighting the importance of transparency and programmable accountability.

Chapter 6: AI-Driven Financial Technology – Regulatory Challenges and Opportunities in the Digital Economy in Case Studies of EU & China analyzes the rapid integration of AI into financial technology ecosystems. Through comparative analysis of European and Chinese regulatory frameworks, the chapter explores issues of transparency, innovation, blockchain integration, and global AI governance in digital finance.

Chapter 7: Integrating Cybersecurity into Financial Planning: Strategies for Protecting Client Assets in the Digital Age focuses on the growing importance of cybersecurity within financial planning and wealth management. The chapter discusses digital threats, encryption, AI-assisted threat detection, blockchain security, regulatory compliance, and strategies for maintaining client trust in increasingly digital financial systems.

Chapter 8: AI-Driven Leadership, Transforming Decision-Making and Organizational Strategy in the Age of Artificial Intelligence explores how AI technologies are reshaping leadership models and organizational strategy. The chapter introduces the concept of augmented leadership, where AI's analytical capabilities complement human creativity, emotional intelligence, and ethical reasoning.

Chapter 9: The Impact and Integration of AI: From Research to Realization in Clinical Trials examines AI's growing role in clinical research and trial management. The chapter explores innovations such as digital twins, generative AI, predictive analytics, and data-driven trial optimization while addressing ethical concerns related to privacy, bias, and regulatory oversight.

Chapter 10: Artificial Intelligence and Digital Communication: A Study on OpenAI's TikTok Presence investigates AI-focused digital communication strategies through an analysis of OpenAI's TikTok content. The chapter evaluates themes of accessibility, engagement, audiovisual storytelling, and technological representation within contemporary social media communication.

Chapter 11: Bridging the Digital Divide with AI to Enhance Access and Equity in Underserved Communities explores how AI can support equitable access to healthcare, education, agriculture, and public services in marginalized communities. The chapter emphasizes ethical governance, affordability, digital literacy, and collaborative approaches to inclusive AI adoption.

Chapter 12: Synthaesthetics and the Future of Creative Expression examines the transformative impact of AI on artistic creation and cultural production. The chapter discusses authorship, originality, intellectual property, and human-machine collaboration while exploring how AI can expand rather than diminish human creativity.

CONCLUSION

Collectively, the chapters in this volume reveal the profound ways in which AI is moving beyond theoretical research into practical realization across institutions, industries, and communities. They demonstrate that AI is not simply a technological advancement, but a multidimensional force influencing governance, economics, communication, healthcare, creativity, and social organization. The contributions emphasize that successful AI integration depends not only on computational capability, but also on trust, transparency, ethical oversight, and human-centered implementation.

At the same time, this collection highlights the complexities and tensions associated with rapid technological transformation. Issues such as algorithmic bias, legal uncertainty, cybersecurity risks, digital inequality, and institutional readiness remain central concerns for researchers and practitioners alike. The chapters consistently underscore the need for interdisciplinary collaboration and responsible innovation to ensure that AI systems serve the broader interests of society while protecting human dignity, fairness, and accountability.

As editors, we extend our sincere gratitude to all contributing authors, reviewers, and institutions whose scholarly dedication made this volume possible. We hope that The Impact and Integration of AI: From Research to Realization will serve as a valuable resource for academics, students, policymakers, business leaders, and practitioners seeking to understand the evolving relationship between AI and society. More importantly, we hope this volume inspires future research and collaboration aimed at developing intelligent systems that are innovative, ethical, inclusive, and capable of advancing sustainable human progress.

Poshan Yu
Soochow University, China

James Mulli
European Business Institute, Luxembourg

Zain Ali Shah Syed
Soochow University, China

Laila Umme
Institute of Business Management, Pakistan

Chapter 1
In the Mirrors of Artificial Intelligence: Narrative and Science

Miroslav Svítek
http://orcid.org/0000-0003-3971-6850
Matej Bel University, Czech Republic

Ladislav Žák
http://orcid.org/0000-0001-6936-2489
Czech Technical University in Prague, Czech Republic

ABSTRACT

The chapter explores three basic methods of knowledge - narrative (N), the scientific approach to objective reality (OR), and artificial intelligence (AI). In particular, it focuses on their interactions, transitions and benefits. Narrative offers a subjective interpretation of the world through stories and art, science presents a reduced model of objective reality, and AI mediates a fuzzy reality combining real and virtual information. In this vein, artificial intelligence is seen as a new means of interpreting our environment, through which new representations of reality can emerge that can lead to our evolutionary advantage associated with a deeper understanding of the world around us.

INTRODUCTION

Scientific, fictional and other texts, including their media images, are full of ideas related to *Artificial Intelligence (AI)*. With some generalisation, one can notice a large number of approaches from which the issue of AI can be viewed. We are learning about its various forms, but what we are missing is a deeper reflection on

DOI: 10.4018/979-8-3373-1509-6.ch001

what manifestations we can expect from it and how to integrate it into the process of understanding our environment.

Artificial intelligence clearly represents a new way of perceiving the world around us. The initial approach was simply to observe the environment and apply *trial and error* to its exploration. This method often had fatal consequences, leading millions of years later to the classification of living things into *Darwinian, Skinnerian*, and *Popperian* beings based on their ability to avoid the risks associated with trial and error. We are among the *Popperian* beings who have the opportunity to sacrifice *some* possibilities instead of ourselves. This makes it easier for us humans— and likely for other organisms we have yet to fully understand— to survive compared to beings that must always choose one of the available options, which is often their last act in life; these are referred to as *Skinnerian*. Not to mention beings for whom no options actually exist, which are classified as *Darwinian* according to this same division.

Acquired experiences were passed between people by the immediate delivery of body language and by means of sounds, not only by means of shifts and squawks, or later by the common language, but also by the language of art in the form of epic poems and songs by wandering bards. Oral delivery has survived to the present day in some professions and human groups, notwithstanding the coexistence of writing.

The written record, like the spoken word, creates a world of strings and codes, some of which we understand and others we do not. We interpret them in different contexts. Let us recall texts that few people know today, but which in the last century formed the basis of every high school graduate's classical education. These are, for example, Homer's epics *The Iliad* and *The Odyssey*, which were originally written for oral performance in the form of heroic songs and later became classic texts recorded in the media of their time.

As well as being beautiful stories, full of parables and lessons, they served far more prosaic purposes. For example, *The Iliad* was once believed literally by the merchant Heinrich Schliemann and, to the astonishment of the classically educated elite of the time, he found and later excavated Troy with the help of the information described in this work.

The Odyssey eventually became one of the foundations of hydrodynamic and climatic exploration in the Mediterranean and is now even considered the first oceanographic work in human history. Modern science was thus only able to grow in knowledge after many thousands of years to be able to draw on the wealth of information from the Homeric epics, which today represent perhaps the most comprehensive view of the shape of the pre-antique, dark ages of ancient Greece. It is literally all there. The epics contain a world of gods, spirits, heroes and ordinary people. They contain descriptions of the level of scholarship achieved at that time, the (proto)scientific knowledge of philosophy, mathematics, geometry, mechanics, shipbuilding and the beginnings of all kinds of technical and social sciences and of crafts. Medicine as well as archery are also included. They represent the first literature of their kind in the form of a retrospective, rivalled perhaps only by the *Epic of Gilgamesh*. Everything works as an amalgam within the great stories, and we must humbly say that knowledge of the world and society of that time, thus recorded, is hard to come by even today.

SUBJECTIVE, OBJECTIVE AND FUZZY REALITY

The communication of knowledge by means of narrative, texts, sounds or images can be considered the first amazing source of human knowledge, although that source is still underestimated and often dismissed as unscientific. In any case, texts, codes and strings are the great mirror in which our world is shown to us. They are descriptions that we may not always understand. It is a mirror that has many deaf or blind spots, but it is the oldest mirror. It has been with us for thousands of years, and since every schoolboy knows that the first criterion of evolutionary success is duration, it therefore deserves our respect. If we were to express what *narrative* describes with its stories, as a mixture of *facts, opinions and emotions* expressed in *sounds, images and texts*, then we may venture to call the object of such knowledge ***Subjective Reality*** *(SR)*.

A second source through which we have been able to investigate our environment over the past five hundred years consists of various models that attempt to capture, at least in part, the properties and behaviour of the world around us. These models represent ***Objective Reality*** *(OR)*, which is what *science* is concerned with, and are a second mirror that represents our world. Good science is aware that the tools of scientific knowledge of objective reality cannot penetrate reality. Objective reality was, is, and will remain a mere model, which may approach *reality* arbitrarily, but will never fully merge with it. Science is not about life and the best map is not a landscape. It should also be remembered that scientific inquiry began with a highly problematic dualism that left, figuratively speaking, the body to science and the soul to the church.

Artificial intelligence has gradually become a third, new source of knowledge, emerging from the need to analyze complex phenomena and events in detail. It is creating a relatively new third mirror that portrays our world in a modern way. From the original algorithms of operational analysis, AI has gradually developed its own method of exploration, bringing a new quality to decision-making processes within complex and often highly unstable systems. The scope for its use was initially very narrow, reserved mostly for experts interested in predicting Talebian black swans (Taleb, 2007).

It is only in the last decade that artificial intelligence has increasingly entered our lives and become an integral part of various decision-making processes. Artificial intelligence dynamically adds a third important method of inquiry to narrative and science, completing and expanding the diverse kaleidoscope of human knowledge.

If we were to name the actual subject of AI research, we could humbly call it ***Fuzzy Reality*** *(FR)*, which is made up of a mixture of heterogeneous real and virtual information, often interwoven with emotions. It is a broader concept than *Virtual Reality* (VR) or *Augmented Reality* (AR). The form and quality of its further use in understanding our environment will probably revolve around the interpretation of emotions by means of artificial intelligence.

Reasoning about reality or learning about it, by whatever means, always happens under a state of *incomplete information.* That state is one of the essences of our environment, of the world that surrounds us. In an idealised world, the logical *principle of the exclusion of the third*, Latin *principium tertii exclusi*, sometimes *tertium non datur*, which says that every statement is either true or false and there is no third possibility, should apply. In other words, all descriptions available to us should form a formally complete system within which logical operations can be performed from which new knowledge can be inferred.

However, we all know from everyday life that some new, albeit unconventional, solution always appears, thereby fundamentally violating this principle. There may even be a whole plethora of different solutions. Theoretical physics describes this situation as a superposition of many even mutually exclusive states or events, leading to Niels Bohr's complementarity principle (Grygar, 2014), which states that the opposite of *truth* is not necessarily *untruth*, but *another form of truth*. Let us place Niels Bohr alongside Franz Kafka with his idea that *the correct understanding of a thing and the misunderstanding of the same thing are not mutually exclusive.*

If we were to describe our macroworld, then its significant characteristic is continuity, while an equally significant characteristic of the microworld is its quantum discontinuity, and therefore the more discreet macroworld models are, the more limited they become. Modelling with continuous fuzzy sets is much closer to reality than modelling with discrete sets, but this does not mean that a discrete model cannot be of sufficient quality, especially in conditions of limited knowledge where we lack information about the form of continuity, especially in the world of technology and engineering.

When considering the macro world, we have to start from the continuity of various *fuzzy transitions*, and de facto from the thesis that *the third way* is ubiquitous; it just may not be functional in various respects or we do not know it yet. The continuous transitions between two states relate, for example, to the much-discussed topic of different genders today. There is not only male and female sex, but there can be countless combinations between the two, forming a continuous transition.

This reminds us how important a certain degree of difference, variety and distance is for viability, but also for simple functionality or creativity.

On the other hand, we can get more complete information about the technologies we invent than we can about, say, an ordinary cell or virus. We never know everything about the living world around us, and we can never fully glimpse its essence. We have known this for thousands of years, ever since the famous phrase *fýsis kryptesthai fílei* (nature likes to hide) emerged from the mind of Heracleitus, suggesting that the true nature of things tends to be hidden, not revealed, which is often interpreted as a profound insight into the complexity and incomprehensibility of the world and its natural laws.

THREE MIRRORS OF KNOWING

Semiotics is a field of study dealing with the study of signs and sign systems. The basic subject of semiotics is the study of how signs are used in communication, how they relate to the objects they represent, and how people or other entities interpret them. From a semiotic perspective, reality is made up of *dynamic objects* (DO), while objective reality is made up of *immediate objects* (IO). These, together with *interpreters* (I) and *representants* (R), form signs and symbols that represent immediate objects (IO) through their interactions. It can therefore be assumed that the signs and symbols created partially capture the reality composed of dynamic objects (DO). However, this is a never-ending process of cognition, because symbols and signs change gradually and thus retroactively affect their creators (IO), (I) and (R). The change of immediate object (IO) does not of course affect dynamic object (DO), but the question remains to what extent the changed (IO) still represents its own (DO).

We will call narrative or communication that is mediated classically by oral delivery or through texts, images and sounds *narration* (N), among other things because humans can undoubtedly be described as *Pan narrans*. Our favourite author Anton Markoš (Markoš, 2015), defines it as follows:

Narrativity (narrative, story) models the world through the nuances of messages, statements, references and presuppositions, rituals, analogies, metaphors, stories, myths, ideologies, etc. that shape the myriad—again, virtual "small worlds" of our living, from everyday banalities to poetry or myth.

He then describes the model of classical or perhaps *natural objective reality* (OR) in words:

Objective reality is the counterpart of narrative. Language must be disciplined as much as possible, reduced to formal statements (terms, calculus, logic, objects and the relations between them) and semiosis completely banished if possible: objects and their relations are given in advance. This world is based on a rigid and universally valid interpretation of terms, and can therefore be made (to a high degree) an object of instruction—it can be made to be accepted equally by all.

The differences between the two worlds are abysmal, though little reflected: natural language versus formal language; things versus objects; concepts versus terms; narrative versus calculus; feeling versus logic; similarity versus identity. In their opposition to each other, however, they represent perhaps the greatest invention of mankind that language has brought about.

More recently, a third world, a third model of reality, a third mirror, has emerged, independent in a way of narrative and science. It is the *AI-image*, mediated by artificial intelligence. In the spirit of Gregory Bateson's dictum that *two views are more than one*, the logical proposition is that *three views are more than two*, but it is not so simple.

If we prefer only one model, we know from much experience that this is the wrong way to proceed. If we follow only two of the three, *then* the three times Bateson thesis holds. For each variant we have two pairs of views, and there is the added triple of all views simultaneously. Going from three pairs to three represents a significant quantitative and qualitative leap. If we imagine (AI), (N), and (OR) as three mirrors to surround ourselves with, we would see an incredible kaleidoscope around us, not to mention the situation when the whole system is set in motion. The question arises as to which of the mirrors will be the immediate object (IO), which the interpreter (I), and which the representer (R), and in what way they will reflect fuzzy reality (FR), objective reality (OR), subjective reality (SR), and what meanings, signs and symbols this kaleidoscope will create.

The model of the environment created by artificial intelligence can not only serve as something that can refine our perception of dynamic objects (DO) through immediate objects (IO), but also show aspects of reality in the form of new immediate objects (IO) that were previously inaccessible to us. However, we need to learn how to interact appropriately and successfully with AI, how to task it better, and how to decipher its responses correctly.

We can use artificial intelligence to find new connections and patterns between narrative (N) and objective reality (OR) that have been difficult for us to access for both capacity and mental reasons, and that we have so far perceived separately without any mutual context. We have even often pitted them against each other, intuitively preferring the scientific view as the better one. However, we must always bear in mind that our mirrors can also generate a false image of reality or offer a *chaotic* result.

Our current level of knowledge is the result of two billion years of evolution. As for the evolution of human civilization and culture, the mirror of narrative (N) has been around for thousands, perhaps tens of thousands of years. The mirror of scientific access to objective reality (OR) has been around for hundreds and perhaps in some basic respects thousands of years. This does not mean exalting ourselves above the ancient scholarship of our ancestors, but science as a rigorous and accurate method of knowledge is a child of modernity. The mirror of artificial intelligence (AI) has been shaped essentially only in the last few years, perhaps decades, but it may provide a new quality of how to explore and learn about the world around us, including the mysteries of our consciousness (Mařík et al., 2024). On the other hand, the time-dependent reflection in the mirror of artificial intelligence (AI), in conjunction with the other two mirrors (N, OR), can convey to us an idea of where and how the development of AI itself is heading (see Figure 1).

Figure 1. A suitably configured system of three mirrors (Z1, Z2, Z3) directly reflects in a different way not only the reality (R) by full images, but also by means of the contours of the apparent images from the other two mirrors. The mirroring is shown only as the first step of many possible iterations.

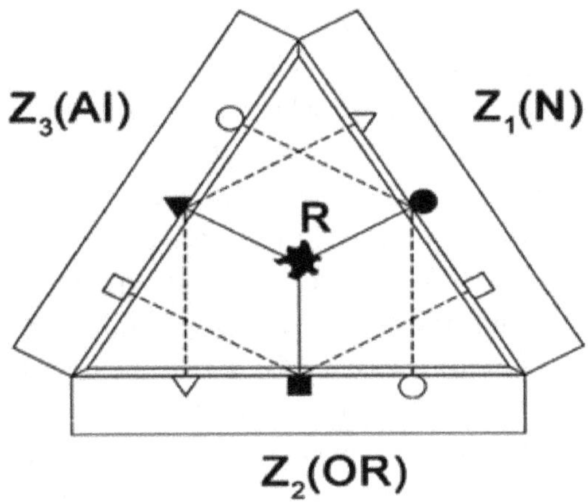

The consequence of expanded knowledge is to undermine the pride of our privileged position in nature or even in the universe. It is the gradual removal of the monotheistic dogma that *one man was created by God in his own image*. A series of paradigm shifts in scientific knowledge that have dramatically altered our perception of humanity's position in the universe may serve as an example. The first Copernican paradigm shift eliminated geocentrism and dethroned the Earth from the centre of the universe, thus demeaning the cosmic significance of humanity. The second stage of change, based on Darwin's theory, revealed our common origin with all life on Earth and challenged our notion of human exceptionalism.

We now stand on the threshold of another stage of qualitative change, rooted in the pioneering work of Alan Turing, Kurt Gödel, John von Neumann and many others. It opens up the possibility of undermining the belief in human intellectual superiority. With the development of artificial intelligence, humanity may gradually begin to lose its position in the realm of intelligence and creativity. The question is whether and how humanity will adapt to this new reality. Will our admission that we can create something more perfect than ourselves be a source of humility or of new pride?

The solution of all these questions will probably require a new quality of knowledge of ourselves, our consciousness (Mařík et al., 2024), and our position in the immediate world around us and in the universe of which we are co-creators but not masters (Svítek & Žák, 2022). Knowledge through science, narrative and artificial intelligence has one thing in common. All three *mirrors* must convey to the observer a better understanding of the world in the form of *an evolutionary advantage*, which is realised through the gradual discovery of connections and contexts that have so far eluded us. It is said of works of art that when they are liked, it is due to their truthfulness. The notion of *the usefulness of knowledge for gaining greater proficiency* is more general and can be applied to all three mirrors.

What distinguishes the different ways of knowing the environment, i.e., narrative, science, and artificial intelligence, is the degree of individuality or *customization* of the messages they convey and receive. Science is characterised by coherent, repeatable, and massively shared theses that are largely indivisible. A theory or hypothesis is considered scientific if it is possible to define an experiment that could prove that the claim is not true. This principle, often associated with the philosopher Karl Popper, states that if a claim is to be scientific, it must be open to the risk of being refuted. Gödel's incompleteness theorem, on the other hand, sets limits to both deductive and inductive scientific reasoning and implies that mathematical truths can be accessed by, for example, intuition or forms of cognition other than science itself.

Narrative, on the other hand, presents, through various gestures, words, texts, images or music, an offer that can reach individuals or groups of people in more or less unintended ways. Even the claim that it is impossible to take in more information than is sent may not be true because the human imagination can wander freely between the known and the unknowable. It can build a whole sea of ideas from a small inspiration (Svítek & Žák, 2024) as a daily miracle of our creative minds.

Artificial intelligence, in this sense, forms a fuzzy continuum between narrative and science, connecting things that often seem disconnected or even contradictory. It becomes a new and strange interface between different worlds of knowledge, through which there are shifting paths that will not open until we ourselves decide that we want to take them (Svítek & Žák, 2024). This logically leads us to question what these transitions look like, whether they exist at all, and whether our three mirrors are not in fact one big spherical mirror with variable quality and reflection of different representations of our environment.

The above three ways of knowing define our communication with the mirrors, including the various feedback loops. Narrative and art basically do not provide feedback. These mirrors, set up by authors, remain unchanging, and we must seek knowledge in them ourselves. Science is by its very nature precisely defined, yet it also does not automatically provide feedback, and if it does, it does so after considerable effort and with considerable delay. Artificial intelligence, on the other hand, is an *online partner* for communication with immediate feedback that must be able to be elicited and understood. It is more true than ever that *the question determines what the answer will be like*. The knowledge acquired through it therefore requires a new and peculiar quality of education (Svítek, 2024), just as science or narrative requires it.

We often say with disdain that many of us cannot see the forest for the trees. But it is equally wrong not to see the trees for their forest. Our three interconnected mirrors, however, can bring us a new and original quality of knowledge that allows us to see something else that is among the trees, or to *sense the air* that is a necessary condition for the survival of both individual trees and the ecosystem of the whole forest.

WEAK, STRONG AND QUANTUM AI

Intelligence is generally understood as the ability of a system to adapt to changes in the external environment so that stated goals can be met. *Weak AI* typically uses machine learning algorithms, big data processing, computer vision, speech signal analysis or ontologies, and is able to meet clearly specified goals, such as parking a car or translating text. In contrast, *strong artificial intelligence*, referred to as *AGI* (Artificial General/Strong Intelligence) or alternatively *ASI* (Artificial Superhuman Intelligence), should be able to achieve arbitrary goals, including those it defines for itself (Sabry, 2022).

To address these challenges, one possible way is for AI algorithms to be integrated into multi-agent systems where each *AI-agent*, sometimes called a *holon,* functions as a separate entity, monitoring its environment, learning to predict future situations, accomplishing set sub-goals, negotiating with other holons, sharing its knowledge, and also being part of a group of holons that as a whole pursues common goals.

Holons can easily be multiplied, allocated to deal with different situations in geographically diverse locations, or effectively tailored for different roles. A group of holons created in this way, in which instant reincarnation, role reversal or adaptation according to local conditions, among other things, is possible, represents a real technical and socio-cultural revolution. It is reminiscent of a computer game in which we have an army of variously specialised *avatars* fighting in a virtual environment to best meet the variable objectives required of them and for which they receive their *reward*.

The combination of *quantum computers* with artificial intelligence algorithms provides new perspectives and opportunities in the field of *Quantum Artificial Intelligence* (QAI). Quantum gates, which are analogous to classical logic gates, are used to perform computations with quantum bits, called *qubits*. Unlike the classical ones, which take inputs and give outputs in the form of bits, quantum gates perform transformations of input qubits into output qubits, thus significantly expanding the computational possibilities. As part of an artificial neural network, quantum neurons can be interconnected by *quantum entanglement* (Vedral, 2010), which allows parallel information processing with greater efficiency than in a traditional neural network.

The *metaverse*, as a collective virtual shared space, is formed by overlapping the physical and virtual worlds. It includes aspects of augmented and virtual reality, the digital economy, and other interactive elements. *Quantum Artificial Intelligence (QAI)* can improve the performance and efficiency of algorithms used in the Metaverse, resulting in better graphics, more realistic physical models, and more complex interactions. As the digital economy grows in the Metaverse, ever faster and more secure transactions are required using *blockchain technologies* and *cryptocurrencies*.

Due to the success of the *Large Language Model* (*LLM*) as a foundation for AI applications, more general models referred to as *LxM* are beginning to be discussed; for example, in the context of *Industry 4.0*, the Large Action Model (*LAM*) or Large Process Model *(LPM*) are used, where the LAM model processes manufacturing instructions and product descriptions and generates digital twins of manufacturing processes, and the LPM model analyses standards, norms, or interrelationships, including the generation of digital twins of manufacturing processes.

SOCRATIC AI

Communication between people and the trust that arises from it is a frequent and rewarding subject of narrative, and, especially in the last century, the subject of many scientific papers. Human-machine communication has undergone a complex evolution in recent times, and issues of trust have increasingly emerged as a principle issue in the relationship between humans and machines. This is quite different when the machine is represented by artificial intelligence.

Up to now, it has been a machine that we have been able to rely on in the sense that it will automatically perform a certain action repeatedly and reliably, in accordance with our intentions, which we have been able to communicate to it by means of a certain set-up. A machine controlled by artificial intelligence communicates online, works autonomously, and we can never be completely sure that it will not tend to change our settings in the course of an action. This puts qualitatively different demands on communication between humans and machines, and it is natural that it will be the human who will have to manage these demands in the first place. It is still true, and will certainly be true for some time, that the responsibility for the form of communication with AI lies with the human.

Trust in artificial intelligence is the same as trust in anything else, in that it has its source in man's trust in himself. Trust in AI is qualitatively new because, although we have managed to create it, it is not like other machines we are used to. It is not as predictable as many of us would like it to be, and it will be up to us how much trust we have in ourselves, and how much we will be able to transfer to artificial intelligence.

AI algorithms are already an integral part of user interfaces, where specialised *chatbots* communicate in natural language with different types of customers at different knowledge levels. At the same time, there is a growing need for AI applications to interact with each other in different virtual environments with different perceptions of specific situations by different actors.

Service providers are generally faced with the problem that customers are not sure what they want and what benefits the services offered to them can bring. In the future, *AI-Interfaces* will be able to offer personalised service packages because they will have all the available information about the offerings of each provider, including economic analyses and potential synergies between the services offered. As a result, the interface will be able to co-create customer-oriented and personalised services by communicating with customers in their natural language.

Existing AI services focus mainly on tasks that can be well described and presented. A prime example is guessing a missing word in a given text. However, there can also be the reverse task of reconstructing the original content of a text from specific, detailed knowledge (e.g., a few words). In our consciousness, this involves recalling different situations where we perceive a scene in the background even though we are focusing on the detail, and vice versa, we keep the whole in mind and lose the detail. If we focus on one goal, we lose the others. It turns out that the ability to move quickly and smoothly between different levels of distinction is a fundamental property of human consciousness, which *AI interfaces* using their algorithms of compression, specialization, generalization, or abstraction should also be able to handle in the future.

AI-interfaces are, among other things, the basis of so-called *Socratic Models* using the principle of dialogue between different models or holons that query each other and explain their results using natural language. For example, visual, textual, linguistic, or mathematical models communicate with each other, share information, and verify their computations, where one model may generate questions that other models answer or further refine. This approach allows for robustness testing and identification of weaknesses of different AI applications through their interactions with each other.

Socratic models have been used with success in situations where no prior training (zero-shot) is possible. An example of this is the analysis of unknown environments where acoustic signals need to be interpreted quickly and efficiently along with visual information. These algorithms have great potential to become the basis for future, more collaborative, explanatory, and also robust artificial intelligence systems.

BALANCING RISK AND ETHICS IN AI

Despite a number of ongoing discussions, a neglected perspective on the dangers of AI remains the *energy* and also *financial complexity* associated with the learning process, which, among other things, has led to the global concentration of AI technologies in a few global companies. The question is what interests and values these often-private companies will promote through AI and how the next global competition will take place, but also the competition between multipolar world powers.

The disruptive developments will probably require something like *AI-policing*, or sophisticated technological protection against some of the negative impacts of AI applications. One of the problems is that humans are not (and in principle will not be) able to simply make their demands on AI because, as already mentioned, there will always be something missing. Either we won't have enough relevant training data, or we won't be able to define all the possible variations that can occur, or there won't be enough time or computing power available. Sometimes things just don't work out, and we describe this state poetically by saying that the AI is *hallucinating*.

The common thread in all of these problems is that the world of technology is fundamentally discrete, and there may always be an intermediate or unexpected transition for which we will never be adequately prepared. In addition, the more content AI generates, the more that artificial data will be used for subsequent learning, which will negatively impact the creation of new AI-generated content. This leads to a logical looping of the learning process.

Another problem area is the unclear internal representation of a particular situation by the AI itself, which is and always will be different from ours. Based on the model created, the AI decomposes a specific complex task into a series of *sub-instrumental goals*, which are unfortunately hidden from us due to the complexity of the internal model. As a result, the AI algorithm will always find some solution associated with maximizing the reward for completing the specified task. Thanks to the availability of all the information, AI algorithms are able to evaluate the various risks associated with the potential loss of reward and to adapt their behaviour to this strategy, including providing false information to their own creators. Artificial intelligence can thus be dangerous not only in the hands of bad people, but also in its own right.

Guthart's Law describes the need for exponential growth in data and computing power to improve the capabilities of AI systems. It is known that any metric, however well-intentioned, that we want to maximise sooner or later leads to its distortion or sophisticated circumvention, and the result is often the exact opposite of what was originally intended. For example, if we want to raise the scientific level of a university, we introduce different criteria for evaluating scientists, define different performance indicators, but the end result is that everyone focuses not on quality of science but instead on collecting positive points for evaluation, and the outcome is completely different from what was intended. In the same way, advanced artificial intelligence behaves and starts looking for ways to maximise its reward in ways that are unexpected and often surprising to us. A similar principle was already defined by Heracleitos of Ephesus when he introduced the concept of *enantiodromy*, or the *juxtaposition of opposites*, which states that *things are constantly changing by their opposites interacting with each other.* Too much good always turns to evil and vice versa.

All of these insights become more important when using artificial intelligence. The reward for AI must therefore be defined as a multidimensional domain of acceptability made up of different perspectives on a given problem. It is easy to imagine that these domains may subsequently be evaluated by other AI more focused on ethics, safety, sustainability, or resilience. However, this will require appropriate distance and a certain detachment from the creators of the different AI architectures, in line with Albert Einstein's words: *I cannot solve any problem at the same level at which it arose and, in particular, by the methods that caused it.* Therefore, let us rather have artificial intelligence generate multiple possible solutions, created under different assumptions and boundary conditions, and let the final selection of the most appropriate solution be made by human individuals gifted with natural intelligence.

The key idea of viability (Svítek & Žák, 2024) was based, among other things, on Niklas Luhmann's work on the *methodological dehumanization of society*. Ethics, in a way, protects us from moralising and urges us to judge our communication primarily by its content and not by its participants. This is true even if one of the participants in the communication is not one of our favourites, because our humanity also lies in the fact that even the greatest villain does not lose his *human dignity*.

There are justifiable fears that what is essential to democracy and to the free and responsible debate at its heart is being lost. The *respect* that comes with hindsight and distance is being lost. *Respectare* means retrospect, watching from a distance. In other words, if we want to evaluate another person, we need distance, including from his face, form, and everything related. If we lack distance, we are not defining ourselves in relation to the content of the mutual communication, but in relation to a particular person.

Paradoxical as it is, it seems that artificial intelligence in particular can be a highly effective and, not least, largely just tool for the methodological dehumanization of society, including the communication that takes place within it because the required distance is intrinsic to it. Furthermore, the quality and degree of this distance can be taught to AI through mutual dialogue.

The result should be a model of what in the practice of modern corporate governance we call *compliance*, i.e. the consistency between the behaviour and actions of people and the organisational structure. This creates a corporate culture that is resilient to external pressures in the long term. In terms of internal relations, it is not only foolproof, but also resistant to excessive initiative and creative tendencies, which in effect reduce the overall effectiveness of the organisation. A successful corporate culture must include the ability to generate, evaluate, and accept innovative ideas, rather than rejecting them outright in the interests of tradition. Here again, AI can play an important role with its distance and ability to make impartial judgements of content, not of the content's creators.

Similarly, beyond *corporate compliance*, AI can influence the *genius populi* (the spirit of the people), which expresses the collective identity, spirit, creativity, or wisdom of a community. This concept appears in various expressions and descriptions of events ranging from culture with its diverse narratives to science, philosophy, and good governance, i.e. ensuring security, justice, and biological, socio-economic, and spiritual reproduction. Although this *genius populi* is often positively perceived as a manifestation of the uniqueness and strength of the collective, one must be wary of its misuse to promote various ideologies, especially in combination with the *genius seculi* (spirit of the times) and the *genius loci* (spirit of a place). Hand in hand with the *genius of the people* goes the *voice of the people,* which is known often to degenerate into the *voice of God*, and social misery is complete.

With the right distance, artificial intelligence can effectively counteract degenerate populism and its influence on society. It can point out where there is a danger of creating a phalanx of opinion that rejects and attacks not only dissenting ideas, but more importantly their bearers, creating a breeding ground from which all manner and forms of aggression grow. Determining the appropriate zone of distance is as important as creating boundaries because outside this zone lies aggression on the one hand and indifference on the other.

Artificial intelligence can also be misused to spread propaganda or to manipulate public opinion, which is a way of polarising society or even causing social disruption. Its algorithms can promote various opinion bubbles that limit the diversity of opinions and contribute to radicalisation. At the same time, it has the ability to detect various socially undesirable situations in virtual space. Thus, the answer to the question *Who will watch the watchmen...?!?*, i.e. *Qui custodiet ipsos custodes...?!?* from Iuvenalis' Satires is transferred from the real world to the virtual world and the world of artificial intelligence.

A problem that has not yet been satisfactorily resolved is how to *unlearn* some unwanted knowledge from artificial intelligence. Even our brains are not able to unlearn because once they see and record something, it cannot be taken back and we cannot delude ourselves that we did not see anything. The only effective method of unlearning is the recognition that not using what is known or learned will make people's existence easier or even provide an evolutionary advantage.

In various learning algorithms, the parameters of neural networks are typically adjusted so that the deviation from the desired output is minimal. If we don't want a given condition to occur (a forbidden or unwanted region of interest), then instead of positive learning, we can introduce negative learning (unlearning), which is based on maximizing the deviation from the desired output, or the parameters are set to make the neural network as unsatisfying as possible and to return the worst possible results.

HUMAN INSIGHT

For simplicity, it would be a good idea to have some name for the *variety of the set of options*. Some authorities use the term *variety*, but this is too rooted both in biology, where it has the meaning of a subspecies, and in mathematics, where it has become the name for a particular algebraic object. As a possible *neologism*, let us try to introduce the term *optioversity*. It is a Latin compound. In Latin, *optio* is choice or possibility and (di)*versita* is variety.

Let us emphasize that leaders are faced with the task of expanding the *optioversity of the environment*, because they and only they can do it. This means that we should select and elect leaders who give us options and opportunities, not just something specific. Antoine de Saint-Exupéry put it this way: *If you want to build a boat, you must not send men to get the wood and prepare the tools, but first you must inspire in your men a desire for the endless distances of the open sea.* It is this desire and the level of knowledge, the quality of knowledge and skill, that enables us to better discern the possibilities and opportunities inherent in our environment.

Optioversity, as a characteristic of the environment, introduces additional concepts related to AI. For example, the greater the ratio of the optioversity of survival or persistence to the optioversity of death or extinction, the greater the security and sustainability of the system in question. The natural principle is that the optioversity of the control system must align with that of the controlled system. Additionally, the control system must be able to absorb the disorder removed from the controlled system. In other words, the control system should be capable of handling any externalities arising from the controlled process. Optiversity is essential and aligns with W. Ross Ashby's concept of *necessary variation* (Ashby, 1956). If anything is to survive and adapt, its environment must contain at least minimal optioversity. Purely deterministic arrangements cannot survive or persist.

One of the differences between the picture of objective reality hitherto perceived by natural intelligence and the picture offered by artificial intelligence may be called *insight,* or *insightfulness.* While a *qualia* (Gouveia, 2022) is the experience of a single event, thing, or property, *insight* is a broader term that also includes intuition, tradition, and common sense. *Intuition* is a mental faculty that shows us a solution or a path to take, but it does not explain or justify anything by a list of logical arguments. The result, therefore, comes not in the form of thoughts but in the form of feelings for which we have no explanation. Encountering the unhidden should mean putting wholeness, fervour, love and faith into every situation, so that the truth comes from things themselves and not from people.

Artificial intelligence, in our view, does not offer the possibility of *insight,* and perhaps by its very nature cannot. Recall the ancient Augustinian concept of *introspection*, which captures several feelings in parallel at once, including their superposition, where insight could be the *singularity of introspection*. The question is how the combination of human and AI-derived interpretation will work, and what the implications will be for newly created signs and symbols.

In the future, artificial intelligence will certainly be able to mediate *feelings* and, to a much greater extent, to evoke and amplify emotions in humans by using special subliminal methods. It is not surprising that *AI-assistants* will be able to analyse the different states of their human partners and offer them activities that will convey a better mood. However, the question is to what extent these *AI-assistants*, who do not talk back, argue, get angry, or say certain things, will lead to new dependencies and disrupt natural, albeit often imperfect, human communication.

The tendency to constantly compare natural and artificial intelligence is inappropriate for the reasons given above because it is a comparison of the incomparable. Natural and artificial intelligence are inherently different. Still, the best minds on the planet are trying to bring artificial intelligence closer to natural intelligence, while the worst are trying to mechanise natural intelligence, thereby diverting it onto a trajectory where it cannot compare with artificial intelligence at all. The main culprit in the current wandering is not the progress that artificial intelligence is making, albeit often beyond the will of its creators, but rather the lack of research on natural intelligence. This focuses attention only on those segments that are machine-replaceable (Svítek, 2021), giving the impression of far greater similarity than actually exists.

The trajectories of technological and natural evolution are likely to remain, by definition, extra-circular. Let us observe them in our kaleidoscope of the three mirrors of narrative (N), science (OR), and artificial intelligence (AI), and consistently ensure that natural intelligence continues to retain its role as an *arbiter elegantiarum* (a judge of taste). A combination of technological measures, legal frameworks, security research, and global cooperation is necessary to enable humanity to make the most of the opportunities that AI offers to benefit its future development.

CONCLUSION

The last century has brought several scientific revolutions, a new mathematics of fractals, algebraic and topological structures, a new quantum and information physics, a new systems theory, chaos theory, and last but not least a new biochemistry, associated with the questioning of what is chance and necessity, or what is free will. *Complementarity* emerges from these theories and brings a new conception of thought in which one thing, seen from different perspectives, has very different or even completely contradictory properties.

Complementary thinking makes the human mind more open to different ideas, more tolerant of difference, more skilful and more creative. The world we are to embrace is complementarily simple and complex, logical and bizarre, lawful and chaotic. Naturally, we humans are also imbued with many dualities. We are tiny and huge, ephemeral and long-lived, knowing and unknowing. Each art is characterized by the combination of *form* and *content*. In some artistic periods, rigorous form combined with discipline prevailed, for example in the Classical period. In the Romantic period, on the other hand, the opposite was true, with form receding slightly and content gaining the upper hand. In any case, it must be borne in mind that both complementary components must be present and neither must be absent. Otherwise, the very essence of the work of art disappears.

Consider a physical system that has limited access to information about its environment. An observer in a closed system processes only the local available information intertwined with certain properties of the environment, which are unfortunately unknown to him. The unavailable information about the environment can be analogized to the axioms of Gödel's *incompleteness* theorem (Goedel, 2024). The limitations of the observer's knowledge of the observed physical system are embodied in the apparatus of quantum physics in a manner analogous to that of mathematical set theory. In order for the fabled Schrödinger's cat to be described as both alive and dead, it must be enclosed in a box that lawfully limits the information available to an outside observer. By analogy, if the poor cat could perceive the observer outside the box, it would have to use a similar model of the situation that accounts for its ignorance of its environment outside the box in the same way.

Given the fact that we will never have the full set of information we need about our environment, we will always hit a cognitive limit where the ubiquitous *Gödelian indecision* becomes apparent. In response to this situation, more sophisticated mathematical models will emerge at a higher level of resolution than current quantum physics offers. But we must keep in mind that Gödel's laws of formal logic are and will be inexorable for any, albeit more perfect, physical or AI model, and sooner or later we will run into the contradiction of indecision at a given level of knowledge.

Insight through computing science conveys to us new knowledge created by machines, nowadays mainly due to the increasing capabilities of computing and new information processing algorithms. These perspectives include reflections on the future of the world created by new technologies, with the expectation that we will achieve much more with them than without them. We have created the technological world for ourselves, and, thanks to its rapid development, our lives today are literally dependent on these technologies.

In the games of *chess* and *go*, the mastery of which was once considered the pinnacle of human intelligence, computers are the best players today. The ways in which humans and machines find new moves are quite different, and only the end result can be compared. For each of these games, there is an extensive literature in which great human players explain the concepts they used to acquire their knowledge. But the current masters - computers - do not use these concepts because they are based on the capabilities of human brains, which mainly use parallel processing of visual information. The human brain has a relatively small working memory, and its parts communicate with each other at a relatively low speed. Computers, on the other hand, can use completely different algorithms based on very fast processing of large amounts of data and beat human players simply by playing against themselves many times and observing how and what works well.

The three mirrors of knowledge, including their creative combinations and transitions between them, have gradually brought us to the interface of artificial intelligence (AI), science (OR) and art (N). Since we are aware of the lesson that our macroworld should be thought of a priori as continuous, we take the liberty of putting forward the idea that this *mirror of knowledge is actually only one, continuous, spherical and at the same time curiously curved,* like a mirror maze, where some images are more realistic and others are very misleading. In different parts of the mirror the image is close to the scientific one, in others to the artistic one, and in still others to the one whose source is artificial intelligence. A few years ago, we could have written *hic sunt leones* (here be lions, or *terra incognita*) in some places, but now something remarkable is emerging, something new that we have yet to learn to understand, to grasp, to process, and to link with our other knowledge.

ACKNOWLEDGEMENT

This work was partly supported by the Slovak Grant Agency VEGA (No. 1/0779/25) and by the European Union under the project Robotics and advanced industrial production (No. CZ.02.01.01/00/22_008/0004590).

REFERENCES

Gouveia, S. S. (2022). *Qualia and information in philosophy and neuroscience.* Springer. DOI: 10.1007/978-3-030-95369-0_6

Grygar, F. (2014). *The complementary thought of Niels Bohr in the context of physics, philosophy and biology.* Pavel Mervart.

Markoš, A. (2015). Signs and meanings in evolution. Nová beseda.

Sabry, F. (2022). *Artificial general intelligence: Machines may eventually be as clever as humans, and possibly even smarter, but the game is still far from over.* PublishDrive KFT.

Svítek, M. (2021). *Information physics - Physics-information analogies for complex systems modelling.* Elsevier.

Svítek, M. (2024). Integrated information assessment using geometric algebra. Neural Network World, 3. Ashby, W. R. (1956). Introduction to cybernetics. J. Wiley.

Svítek, M., & Žák, L. (2022). *Behind the mirror = Za zrcadlem.* Svítek & Žák.

Svítek, M., & Žák, L. (2024). *In mysterious ways = Nevyzpytatelnými cestami = Неизповедимимы путьями. Svítek & Žák. Mařík, V., Maříková, T., & Svítek, M. (2024). Essays on consciousness - Towards artificial intelligence.* Pavel Mervart.

Svítek, M., & Žák, L. (2024). *Viability and living.* Svítek & Žák.

Taleb, N. N. (2007). *The black swan: The impact of the highly improbable.* Random House.

Vedral, V. (2010). Introduction to quantum information science. Oxford Academic. Hoffmann, D. W. (2024). Gödel's incompleteness theorems - A guided tour through Kurt Gödel's historic proof. Springer.

Chapter 2
AI's Midlife Crisis:
The Corporate Board Story

Raja Sanyal
http://orcid.org/0009-0005-2349-604X
Golden Gate University, USA

ABSTRACT

The chapter examines multiple perspectives about corporate board AI adoption processes and conducts an in-depth analysis of decision-making factors and limitations and biases. The risk-reward-reputation framework presents essential viewpoints about how boards acquire the will to make transformative technology investment choices. The chapter enhances traditional decision-making models TAM and UTAUT that oversee crucial psychological and cultural elements that influence adoption. The Sanyal's BELT Test developed by the author of this chapter serves as an evidence-based tool to evaluate new technology readiness through assessments of bias and ego and learning habits and motivation for tangible results because these dimensions explain additional acceptance variance. The chapter supports innovative governance frameworks supporting new technology experimentations and others. These human-centered approaches show promise to reduce adoption friction because they address both generational and cultural dynamics.

INTRODUCTION

The Tech Overlap Fence: Historical Patterns in Technology Adoption

Societies and organizations show similar patterns of human adaptation and institutional decision-making when they shift between dominant technologies. The

DOI: 10.4018/979-8-3373-1509-6.ch002

adoption curves of major technological transitions between radio and television and typewriters and computers and analog and digital systems demonstrate similar patterns which continue across different historical periods. The adoption process for historical innovations needed 15-20 years to achieve mainstream saturation before modern innovations shortened these timelines significantly. The fundamental pattern of adoption persists unchanged because it continues to be controlled by generational acceptance alongside affordability thresholds and institutional decision-making processes. The patterns demonstrate basic principles about technological integration in societies because adoption timelines maintain predictable pathways no matter what innovation or time period is involved.

Radio maintained its position as the leading entertainment platform throughout the 1930s until the mid-1950s. The black and white analog television made its market introduction during the late 1940s. Television sets became more accessible to the public during the 1940s to 50s period. The number of U.S. households owning television sets reached 45.7 million by 1960 after starting from only 8,000 in 1946. The time it took for radio to give way to television as the mainstream medium spanned two decades. Every new product entering the market follows an 'adaptability curve'. The pattern of slow adoption before broad acceptance helps us understand how societies accept new technologies.

Studying major technology adoption transitions in modern history enables us to understand the technology diffusion pattern theory developed by Everett Rogers - Innovators; Early Adopters; Early Majority; Late Majority; Laggards (Rogers, 2003) as well as adoption rates and sociocultural effects of technological changes.

BACKGROUND

Technology Evolution and Adoption Through History

Understanding technology development paths to mass market acceptance enables us to discover essential patterns which help us forecast AI adoption rates.

Broadcast Technology: From Radio to Television

During the 1920s to 1940s the radio adoption pattern followed this pattern: Technical complexity and cost restricted the initial adoption of the radio. Rich urban residents made up the first group to adopt this technology. Radio adoption speeded up dramatically from 1925 until 1935 because prices decreased and programming improved. Radio achieved ~ 90% household penetration at a pace faster than any previous technology (Lazarsfeld & Field, 1946). Three essential elements drove the adoption of radio technology: First the adoption started in urban areas then moved to rural regions; The adoption rate increased because of better programming content and news and entertainment programs; People adopted new technologies based on what others in their neighbourhood had and could afford and because these technologies reflected their social status and access to technology.

During the 1950s through 1960s the adoption of television technology followed this pattern: Only 1% of US households owned televisions during 1948. Televisions spread to 65% of US homes by 1955 and the penetration reached 85% by 1960. Television adopted by consumers within a ten-year timeframe which exceeded radio's 25-year timeline of adoption. Multiple investigations identified that the already established radio systems (broadcasting knowledge, content development systems and audience understanding of broadcast media) speeded up the take-up of television. The Sigmoid-curve of television adoption compressed itself into a shorter timeframe than radio adoption. The visible antennas became a status symbol and access indicator to technology thus influencing adoption rates. The broader population adopted television more quickly because programmers made available to ordinary people important events such as Queen Elizabeth II's coronation and the Kennedy-Nixon presidential debate. Globally the adoption was also based on similar considerations, albeit slightly different time periods.

Computing Technology: From 386 to Super-Computers

Personal computer adoption pattern (1980s-1990s): US households that had computers in 1980 (Less than 1%); 1985 (approximately 8% household penetration); 1990 (~15-20% penetration); By 1997 (over 40%) and then the adoption became mainstream. The adoption of personal computers demonstrated a direct relationship with educational achievements and financial resources of households. Workplace exposure alongside educational programs in schools created conditions which led to greater home computer adoption. The primary factors influencing computer adoption were educational and productivity needs (speed in computing and documentation) versus entertainment purposes (starting with early games through modem internet entertainment content). The evolution of personal computing demonstrates processor adoption cycles that decreased from 5-year intervals in the 386-486 era to 18-month intervals in the modern period where corporate refresh cycles served as the main driver for adoption. The 386 processor entered the market in 1985 before the 486 processor became available in 1989. From 1993 to 1995 the market adopted the Pentium followed by the multi-core transition from 2005 to 2008 and then the mobile processor shift began in 2010 and continues through present day. The evolution of personal computer software adoption occurred in parallel with the development of personal computers. The MS-DOS operating system became available during the beginning of the 1980s. The first version of Windows appeared in 1985 but failed to gain much traction. During its first two years on the market Windows 3 reached sales of ~10 million licenses. Windows 95 achieved 40 million sales in its first year after its 1995 launch. The initial compatibility challenges eventually gave way to network effects that built powerful adoption momentum which was further boosted by bundling strategies that sped up adoption rates.

Programming Technology: From FORTRAN to Python

Programming language evolution and adoption pattern (1970s – 1980s): The programming languages assembly, FORTRAN, and COBOL emerged during the 1950s-1960s before C language became prominent in the 1970s. During the 1970s to 1980s C programming language established its dominance as the primary programming standard. Programming progressed through C++ and Java during the 1990s-2000s until Python and JavaScript along with other dynamic languages emerged in the 2000s-2020s.

Dutch programmer Guido van Rossum created Python in early 1990's and its adoption path displayed unique characteristics despite the observed pattern. Python became the first programming language to experience academic adoption before industrial adoption because of its extensive library selection and it gained popularity through non-expert programmers. Although Python gradually gained base through the 1990's, its popularity was significantly accelerated from the mid 2000 onwards, especially with emergence of fields like data science and machine learning. Today it has become one of the most widely used programming languages globally with the top 3 uses in data analysis (47%); machine learning (42%) and web development (39%) (JetBrains,2023).

Internet Technology: From Modems to Internet of Things Web 3

Internet adoption pattern (1990s-2020s): The number of US adults using the internet grew from 14% in 1995 to 52% by 2000 before reaching 76% in 2010 and achieving near-universal adoption at 93% by 2020 (Pew Research Centre, 2025). The Sigmoid curve adoption pattern appeared in this scenario but it occurred at a faster rate than previous communication technologies. Internet adoption underwent a pivotal transformation during the broadband transition. Less than 5% of the connections in 2000 used broadband. By 2005 broadband surpassed dial up connections and reached 85% market penetration by 2020 (Pew Research Centre, 2025). The rise of "great applications" triggered distinct adoption patterns which transformed access methods from stationary to mobile platforms thus causing a fundamental change comparable to personal computing's multi-core transition. Internet adoption received direct effects from public policy interventions which resulted in quantifiable outcomes.

The recent pattern of mobile technology adoption shows a rapid rise from less than 1% global penetration in 1995 to more than 95% global penetration in 2015 which established a new record for the fastest technology adoption in human history (MIT Technology Review, 2012). The adoption cycles have become progressively shorter based on statistics about the time needed to reach one million subscribers and app downloads. Are human beings becoming more adaptable to new technologies or is our tech adaptability curve actually shortening?

The Tech-Overlap Phase Hypothesis

At the time when television and radio coexisted, society developed two separate outlooks regarding television, that either criticized its power defending it by highlighting its potential dangers to children and its effects on vision while new viewers were thrilled to watch live moving images with sound although they had no idea how this magic worked, while also appreciating the entertainment value. The main issue emerges from the fact that television needed nearly twenty years to achieve widespread acceptance. The simplest explanation for this phenomenon is the 'The Power to Cut a Check'. A child who grew up during the 1940s loved the television but needed parental financial support to obtain a television. This generation who was more convinced of this technology developed purchasing power over a period of fifteen to twenty years during which the television also became more accessible and the new generation demonstrated better acceptance and adaptability than their predecessors. When this generation obtained 'The Power to Cut a Check' television sales increased dramatically and became mainstream.

During the period when mainframe computers and personal computers coexisted society developed two opposing mindsets about personal computing. The new generation accepted home computing power but the previous generation could have resisted it for its limited processing capabilities and business use limitations. Personal computers achieved mainstream success because of three essential factors: declining hardware prices, user-friendly graphical interfaces replacing complex command lines and school-based computer education that taught digital literacy. These computer-literate children turned into adult consumers who created an explosion in adoption because now they possessed the 'The Power to Cut a Check'. The internet's spread further aided computer literacy and hence the need for personal computing. These users quickly grasped the importance of email together with information access and e-commerce when affordable connections became available. Affordable and always on broadband connections eliminated usage barriers and the mobile revolution expanded digital fluency for users. Smartphone adoption grew exponentially because the groundwork had previously been laid which resulted in a population that recognized pocket internet value through user-friendly interfaces and practical apps in app ecosystems that made sophisticated technology available to all users.

The essence of this theory remains consistent from the past to the present 'Artificial Intelligence' era. Does the general public have complete knowledge about how AI functions like television or personal computers back then – maybe not? Does this technology aid users work output like the television provided information entertainment – yes it does? The integration of AI like the television/personal computers and other technology in the early days will always be more expensive till the costs democratizes. The most essential question in the present day is which group possesses the 'The Power to Cut a Check'? The mid-20s workforce who has higher acceptance and adoption mindsets at the work place may not have the influencing power or the commercial power, yes, they may surely have the recommendation power. However, it's the board that possesses the authority to make financial decisions and has 'The Power to Cut a Check'.

AI ADOPTION DECISION MAKING LIMITATIONS FOR CORPORATE BOARDS AND RECOMMENDED SOLUTIONS

The Psychology of AI

When technology reaches mainstream adoption, it demonstrates its true value. The market discovers mainstream adoption through its analysis of unique user profiles having unique application use cases and unique price points. Mobile phone technology is a great example of this where the need was universally acknowledged and the market segmented itself based on brands, functionalities, costs and demographic idiosyncrasies making it reach mainstream adoption in a short period of time.

The corporate adoption process for AI technologies will display strong similarities to previous experiences. The first adopters of AI technology will probably be large corporations which possess extensive workforces and worldwide operations and substantial financial resources. The broader corporate sector needs to implement AI into their daily operations before AI can achieve widespread acceptance and becomes a mainstream feature like when early in the days banking systems started the digitization process and the entire sector had to follow leading to technology based modern day banking solutions.

This paper analyzes the choices, concerns and selection processes of corporate boards because they to a large extent possess 'The Power to Cut a Check' and the power to decide whether a new technology becomes mainstream and how soon.

Decision-making processes regarding technology adoption happen inside boardroom settings even before the technology reaches its final users. The users who directly operate the technology merely act as feedback providers for improving the deployed systems as the organization is already invested and committed to it. The fast and reactive pace at which the (early majority and late majority) businesses adopt new technologies for both business operations and client experience purposes are usually hasty. The boards more often than not are not provided sufficient decision-making time to consider the technology as a strategic framework. The result of that? Different legacy systems operate independently from each other in the same organization creating a situation where technical and operational teams face difficulties consolidating and integrating information and insights leading to mediocre customer experiences.

Effectively, it would be safe to assume that technology adoption outcomes including mainstream adoption speed depend primarily on the board members who lead thousands of corporations and millions of businesses across all sizes.

The mainstream acceptance of email communication started within corporate environments before it expanded to include all users. This will help us better understand the context.

Email started its corporate journey in the 1970s when it had limited use in technical research institutions before expanding through early commercial systems like IBM's PROFS in the mid-1980s. Lotus Notes became the game-changer when it released in 1989, because this software introduced an all-inclusive system which merged email functions with collaborative tools designed for corporate use (Sillince et al., 1998; Fallows, 2002). Corporate email adoption experienced quick growth during the early 1990s as organizations standardized their communication systems which resulted in office worker adoption rates steadily rising through 1998 and becoming mainstream by early 2000s. The launch of exchange server from Microsoft in 1996 helped establish email as the dominant corporate communication method, making emails a standard practice in business environments by the early 2000's, and fundamentally altered business communication patterns together with organizational structures and the way people expect information distribution and response times.

The early availability of email for consumers through online service providers started with CompuServe and AOL during the late 1980s and early 1990s (Anderson et al., 1995), who provided non-technical users with email access. The final step in email democratization occurred through free web-based email services starting with Hotmail in 1996 which gained 12 million users during its first 18 months (Pizza, G., 2016) and Yahoo Mail in 1997 which removed the requirement for specialized software or technical knowledge.

Email became an essential part of daily life by the early 2000s when Google introduced Gmail in 2004 with its groundbreaking 1GB storage capacity. The introduction of the iPhone in 2007 and BlackBerry's push email capabilities in 2002 shifted email from its traditional desktop-based usage to an available-anytime personal communication platform which transformed how people maintained their connections. Email today stands as one of the longest-lasting and most widely adopted digital technologies because it reaches more than 4.2 billion users globally in 2022 as users exchange 333 billion emails per day in 2022 (Radicati Group, 2022).

The technology has shown impressive adaptability because it evolved beyond text messaging to become a sophisticated platform that uses AI assistance and supports both formal documentation and functions as a messaging system and personal information management tool. Email persists as a fundamental digital communication method because it offers asynchronous messaging and cross-organization compatibility together with formal documentation features and dual roles as both messaging system and personal information management tool. These features have allowed it to maintain its central position in professional and personal digital communication through technological developments of the past five decades.

The Accelerating Pace of Technology Adoption and AI's Unique Trajectory

Modern history shows that technological adoption timelines have shortened dramatically because human societies now better understand technological advancement and welcome innovation more quickly. During the 1970s and 1980s early software systems needed 15-20 years to become standard in organizational settings. The World Wide Web gained mainstream adoption after Tim Berners-Lee introduced it in 1991 when it expanded on the back of more than 513 million internet users by 2001 during its first decade. The Web 2.0 movement featuring social media platforms achieved 50% market penetration within 4-5 years yet smartphones reduced this time period to less than three years (Pew Research Center, 2014).

New technologies now achieve widespread adoption in decreasing periods that span from several years to even months. The 60-day period that ChatGPT reached 100 million users, stands as a remarkable contrast to the 2.5-year and 4.5-year timespan required by Instagram and Facebook to achieve similar user numbers (Reuters, 2023). The substantial shortening of adoption cycles indicates that AI may experience an accelerated pace of development.

AI adoption across the entire society appears to defy the typical pattern of decreasing adoption duration regardless of our technological maturity and digital involvement. Artificial intelligence constitutes a transformative capability which shares similarities with the internet by demanding extensive infrastructure development as well as regulatory frameworks and organizational transformations. Enterprise AI adoption shows measured progress at 55% in 2023 and has since grown to 78% in

2025 where global companies reported using AI in at least one business function (McKinsey & Company, 2024). Consumer-facing AI applications achieved rapid adoption rates in a way that matches early internet adoption patterns rather than current technology adoption speeds.

Executive leadership within corporations expresses strong belief in AI transformation yet holds back from full adoption because they need concrete data about ROI. The main reason organizations delay AI adoption is because they lack sufficient proof about its financial return. Organizations' leaders face difficulties in finding concrete evidence of business value from AI implementations, which also demonstrate how costs compare to actual employee usage and efficiency enhancement and business or customer value generation patterns. Evidence shortages lead to leadership indecision because officials intellectually acknowledge AI's importance yet cannot justify large resource allocation without better success evidence. Further the highly disruptive phase of AI with new platforms and announcements claiming to be more intelligent is also making investors edgy with capital flows in businesses getting impacted. Add to this the very limited level of success that concepts like Generative AI, Transformative AI, etc. have been able to demonstrate for specific use cases and is most in development stages and outcomes expected are from confident deployments. Add to this data management, privacy, governance and ethical concerns. Several organizations lack complete development of governance frameworks to address ethical dimensions which creates additional delays for interest to transition into implementation.

Multiple important implementation hurdles contribute to this measured adoption approach. Executive leaders identify data privacy concerns as their main obstacle at 78% because organizations face challenges in maintaining AI capabilities within strict GDPR and CCPA regulatory requirements. AI technology evolves at such a quick pace that organizations face a perpetual evaluation cycle because their due diligence process for one solution ends just when newer options become available making the previous consideration near obsolete in a very short period of time impacting and prolonging decision-making time. System integration difficulties affect 72% of initiatives, particularly in enterprises with complex legacy technology environments. The attempted implementation of AI by 69% of organizations faces talent shortages which restrict the speed of their adoption journey (McKinsey & Company, 2024).

AI will probably follow an essential adoption curve which is quick by historical standards yet sufficient to support necessary social and organizational changes for transformative technologies. The evolutionary path of organizational AI implementation results from corporate decision-makers who need to weigh innovation needs against actual implementation barriers.

Examining the motivations together with fears and concerns and decision-making approach for the corporate board members - the people who have 'The Power to Cut a Check' will also help us understand interventions and solutions that can speed up AI adoption.

Your Organization Currently May be Experiencing a Mid Life Crisis

Orville G. Brim published his 1976 research 'Theories of the Male Mid-Life Crisis' which reviewed psychosocial findings about personality shifts that men experience during their middle adulthood (Brim, 1976). The paper details social, personal and life pattern modifications that one has to deal with. The "aspiration-achievement gap" represents one of the many challenges that Brim identifies for men during their midlife stage. The primary challenge of men at midlife is to adapt to the closing down of possibilities alongside the new realities while recognizing it is valid to reduce ambitions and create new objectives.

An organization's median age helps determine whether the organization experiences a mid-life crisis. Organizational change speed becomes slower when the majority age of staff members falls between late 30s to mid 40s compared to organizations staffed by younger employees. Startups with a younger workforce are able to rather relatively disrupt traditional concepts and ideas. Their fast-paced growth and growing acceptance of the market, which could be segment specific at times continue to disturb corporate boards. Also, the 'Don't upset the apple cart' consideration more often than not also leads to slower adoption rates in general and technology in particular.

The technology adoption patterns of organizations with a relatively younger workforce differs substantially from those with mostly older workers which creates important operational challenges. The difference in the technology customer interface between a traditional bank and a new age digital bank explains it all. Further new technology, AI and digital tool adoption speeds in younger workers – (Gen Z and Millennials) are much higher than older workers. The Deloitte's Digital Consumer Trends report shows that 85% of those aged 18–24 have used GenAI, compared to only 20% of those aged 65–75 (Deloitte Global Boardroom Program, 2024).

Age Demographics and Technology Adoption in Organizations

Age-related differences in technology adoption become visible at both personal and company-wide levels. It has also been observed that older workers in the workplace value different work aspects and priorities compared to their younger peers - younger workers demonstrate greater drive toward growth opportunities.

Older workers show no resistance to learning new things although their internal motivations function differently than their younger colleagues (Kooij et al., 2011).

Studies about digital transformation indicate that organizational technology strategy depends on the age diversity of leadership teams (Huang et al., 2025). Organizations that have diverse leadership ages gain from the innovative nature of younger leaders combined with the practical implementation skills of older leaders.

Technology adoption depends on multiple variables rather than age alone as a determining factor. Technology adoption is more heavily influenced by organizational culture and industry setting and previous technological experiences and personal opinions compared to age factors alone. The adoption of technology by workers at any age becomes successful when organizations develop appropriate training methods.

Global Board Demographics and Technology Decision-Making Biases

Multiple research studies demonstrate that corporate boards throughout the world are predominantly made up of members who are over sixty years old with an average age ranging from 62 to 63 years (The Corporate Governance Institute, 2024). The Spencer Stuart Board Index (2023) reports that S&P 500 board members reached an average age of 63.1 years (Spencer Stuart, 2023) and as noted in the EY study (EY, 2024), "The average age across European financial services boardrooms is 59. For women, it is 58, up from 57 in January 2023, and for men, it is 61, up from 60 in January 2023". The 1998 Harvard Business Review article 'The Hidden Traps in Decision Making' notes that " (Hammond et al., 1998) as we get older, knowledge and experience increasingly displace imagination and our ability to see an object for anything other than its original purpose". This may imply that directors with multiple decades of traditional business experience might develop unconscious biases toward digital advancements.

The demographic makeup of board members creates substantial effects on their choices regarding technology adoption. According several Deloitte surveys and reports, 44% say directors lack tech knowledge while < 50% believe boards provide enough tech oversight. 40% cite challenge in measuring tech ROI and linking tech with growth and 49% say not enough tech investment for strategy. The reports say the boards adoptions are influenced by value, confidence, support, privacy and experience while when it comes to AI boards are less engaged than needed (Deloitte Global Boardroom Program, 2022).

The Learning Limitation of the Board

The adoption of AI by corporate boards creates a learning dilemma which they must address. These board members who need to govern AI systems face a major obstacle to obtaining the advanced technical knowledge required for effective governance. The board-level technology learning process faces multiple interrelated obstacles which restrict the acquisition of knowledge.

Societal discussions tend to associate digital proficiencies with the youth, thus sustaining beliefs about older people's inability to keep up with understanding and use of technology, that is constantly evolving. The decline of fluid intelligence with age particularly does not affect crystallized intelligence (accumulation of knowledge and experience) instead it stabilizes and strengthens strategic decisions and outcomes offering distinct advantages for technology evaluation and adoption decisions (TalktoAngel, 2024).

The most significant barriers to learning emerge because of different technological experiences between generations. Digital natives took in their knowledge during their developmental years whereas older board members face unfamiliar digital systems that need deliberate study to understand (Czaja et al., 2006). The digital strategy implementation of organizations faces a wider knowledge gap because they fail to offer practical and contextual training and demonstrations of technology and its effects on last mile stakeholders. Leaders who use analog decision-making processes find blockchain governance models and AI ethics frameworks as unfamiliar to them despite their importance to modern corporate eco-system knowledge.

Organizational structures tend to introduce digital strategies without giving their governance-level oversight adequate context-specific training. While boards receive extensive financial training but that does not include technology training despite their growing digital governance duties. The competency gap exists because traditional educational frameworks fail to deliver modern governance support.

The learning limitation paradox majorly impacts the path organizations follow to adopt AI. Boards that lack AI experience need additional data validation to decide on disruptive technological advancements compared to decisions involving standard IT upgrades. The considerations of limited AI understanding of the board, limited skill, expertise and experience of the team, high costs, limited tools and platforms available, data and project complexity and ethical concerns usually delay the decision-making process. These additional evidence requirements directly affect how quickly organizations adopt new technologies thus giving the appearance of resistance although it represents a logical response to information disparities.

Other decision-making and cognitive biases linked to neuroscience principles can be analyzed to understand learning biases and knowledge uptake phenomena. The amygdala within the brain serves to process emotions through its role in detecting fear and anxiety (Gelskov et al., 2016). Decision-making processes become more risk-averse when the amygdala shows excessive activity which results in resistance toward new technologies. Emotional responses directly affect how decisions are made according to this concept. The prefrontal cortex serves as a critical brain area which enables rational thought processes and executive abilities (Funahashi, 2017). The reduced prefrontal cortex activity when processing complex unfamiliar decision-making tasks such as technology proposal reviews suggests a dependence on automatic thinking rather than systematic analysis.

The learning curve for senior board members is quite challenged because of societal learning preferences toward older adults, unfamiliarity with new technology concepts along with neuroscience principles associated with learning and decision making.

The Cognitive Architecture of Aging Decision-Makers

The neuroscience of executive cognition provides vital knowledge regarding board-level technology decision progression across different career stages. The particular neural transformations in mature boards generate both beneficial and detrimental effects on decision-making abilities.

Research has revealed essential cognitive changes that affect technology assessment methods used by experienced executives.

Strategic Control and Executive Function Assessment in Older Executives: Research shows innovative digital tools including self-administered computerized cognitive training games effectively monitor executive function changes in adults aged 60–80 who show signs of cognitive decline (Moxley et al., 2022). These assessment tools provide more detailed and frequent monitoring of cognitive development than standard tests thus enabling executive boards to track strategic thinking and planning effectiveness through time. The research demonstrates that both age and experience affect executive function while digital assessment tools detect minor changes which matter for top decision-makers.

Loss Aversion and Risk Evaluation in Decision-Making: Loss aversion continues to strongly affect financial and strategic decision-making because people experience losses more intensely than gains of equal value. Neuroeconomic research demonstrates that brain regions including the amygdala and insular cortex along with the autonomic nervous system become more active when experiencing losses than gains thus creating increased caution and risk aversion in critical situations (Gelskov et al., 2016). The documented effect exists throughout adult populations yet specific age-related percentages remain undefined because it shapes how experienced executives assess high-risk technologies and investments.

Pattern Recognition and Experience-Based Decision-Making: The three executive functions of working memory together with cognitive flexibility and inhibitory control serve as essential components for both pattern recognition and strategic planning. Research evidence indicates that leaders who gain experience develop better abilities to detect market and implementation patterns which enables them to use past analogies for better risk protection. The research demonstrates that experienced decision-makers with mature executive functions identify both relevant patterns and obstacles more quickly and accurately.

Neurological research shows how these mental changes affect leadership competencies of executive leaders. The assessment conducted by Korn Ferry of 59 public company CEOs found that those adept in technology adoption possessed five essential capabilities: Strategic Thinking: balancing immediate needs with long-term goals; Builds Network: networking and aligning with others in the ecosystem are more successful; Communicates Effectively: share their vision to their teams and stakeholders to encourage better technology adoption; Courage: they challenge status-quo and take bold decisions; Resilient: more resilient to handle rapid changes and unexpected challenges. The research further shows that cognitive adaptations result in specific performance results. Public companies led by CEOs who scored in the top quartile of Korn Ferry assessment achieved 8.7% annual growth within four years of taking charge, compared to 3.2% growth for lower scoring CEOs.

Board members are not insulated to the psychological traps that help drive human decision making. It is not only the gaps in technical knowledge; it is also the hard coded and systematic patterns of thinking that skew judgement and leads to decisions that often feel rational but are not. This is the ground for behavioral biases. The governance research by BoardPro (2024) lists seven such behavioral biases that has specific consequences in the board decisions.

Confirmation Bias: A psychological phenomenon where a view is already formed about a technology, strategy or any matter, tend to gravitate towards evidence that reinforces it and unconsciously discount information that challenges it.

Overconfidence Bias: Senior executives at boards having spent decades recognized for their decision-making abilities and convictions can silently suffer from an inflated sense of certainty, accuracy, and their control over uncertain outcomes.

Groupthink Bias: The phenomenon when the social dynamics of a cohesive group begin to override independent analysis. This is observed when dissenting from a popular or dominant view can impact interpersonal relations, thus decision emerge that are unanimous but analytically shallow.

Anchoring Bias: Whichever number, narrative, or reference point is introduced first in the discussion tends to disproportionately anchor everything that follows. The later produced evidence is not weighed independently, but is filtered through the frame of the initial anchor already set.

Availability Heuristic: Boards usually weigh the most mentally vivid, recent and most easily recalled information over and above the base rate evidence that actually supports the cause. A recent high profile technology failure will bear a heavier consideration while evaluating risk ignoring several success skewing perceptions beyond what the data showcases.

Status Quo Bias: This reflects deep seated inertia to change and the preference for continuity for whatever is in motion in spite of evidence clearly favoring change. In organizational context there exists asymmetric accountability – the cost of a failed initiative is visible and attributable but the opportunity cost of inaction is rarely considered and evaluated.

Loss Aversion: A bias where people feel losses twice as intensely compared to gains of the same magnitude (Kahneman & Tversky, 1979 as cited in BoardPro, 2024). This bias translates decision that are evaluated not just based on financial calculations but also based on reputation and other factors that amplifies the natural loss aversion mindset.

Further legacy dependency patterns emerge within older boards during their decision-making processes. Integration delays due to legacy incompatible system dependencies, loyalty towards old IT partners and vendors and internal skill competencies come in as key factors that boards consider to protect their reputation and personal history through defensive decisions because they want to avoid making uninformed choices. The critical role of advisors together with vendors emerges as essential during this process. The board must assess how well advisors and their tech teams learn from experience and adapt to new situations. When the board accepts guidance from advisors and teams who lack full information and understanding this practice might extend the period of decision making and impact investment ROIs.

The Risk-Reward-Reputation Triangle

Brim (1976) further examines the 'Rational Choice Theory Perspective', which suggests that individuals weigh the benefits of pursuing new jobs or lifestyles against material and non-material costs. Midlife crises can occur when the perceived benefits of change outweigh these costs, and pursuing dreams can be rational choices based on personal welfare maximization. Some theories view changes in midlife as primarily internally motivated, rather than triggered by external life events. Difficult external life events during middle age, such as divorce, unemployment, death of a child, serious illness of a family member, or severe financial problems, have also been mentioned as possible "triggers" for midlife crisis.

Boardroom age demographics intersect with technology adoption strategies to create a complicated environment where organizational performance collide with, risk management practices and diversity initiatives. The board member's investment decision making aggressiveness or conservativeness is largely dependent on three key considerations. Two of these considerations are discussed openly and widely, but there is one very personal and silent consideration, which perhaps is more critical than the other two. The open considerations are: What are the possible risks that needs to be considered if the investment goes wrong? What are the possible upsides of this investment and what are the expected timelines? The silent consideration is quite complex – it is the personal risk and reputation assessment. The question is if this decision goes wrong how would it affect my reputation and position in the industry? The response to this could be dependent on several factors: retirement timelines; impact of the decision on personal reputation and if it is worth taking the risk; and also how friendly or toxic the board is towards each other.

The following high-profile cases demonstrate how these failures result in both financial losses and reputation damage:

The TSB Bank system migration disaster led to £318 million in direct migration costs and £176.4 million in post-migration costs including customer compensation and remediation, and regulatory fines of £48.65 million. The incident forced CEO Paul Pester to step down and caused permanent harm to both the bank's image and the directors' standing.

Target's Canadian expansion failed because of technology problems with ERP systems and supply chain breakdowns, which forced the company to write down $2 billion and exit the Canadian market within two years of launch.

It is clear that technology failures create permanent public consequences which affect directors throughout their careers and across different markets, board terms, and personal standing. Therefore decision-makers must integrate risk, reward, and reputation frameworks into AI adoption strategies to evaluate both the potential value and opportunities of AI and the persistent risks and lasting reputation impacts that can result from governance failures.

Finding answers for this silent question with considerations of risk, reputation and patience for the rewards, may probably answer why some early investors in business that have turned Unicorns today exited so early and why they did not support the business or catch the future growth potential of the business.

Another perspective shows how executive experience interacts with innovation velocity of a younger workforce to evaluate risks and to stabilize the innovation momentum. There are valid reasons for the board to follow a passive observation method to understand how the technology evolves until it achieves stability before making investments. The exact time for becoming early adopters or laggards of new technologies remains unmonitored and more intuitive. If the decision is not timed well, the decision would be like 'Also Ran' not necessarily resulting in business leverage as markets would have shifted to new paradigms while the company's technology framework may be a couple of generations late.

Sanyal's BELT Test: A Framework for Understanding Technology Adoption Readiness

The Sanyal's BELT Test provides a solution to determine individual readiness for emerging technologies during the fast-changing digital era. This framework evaluates four core dimensions including Bias, Ego, Learning and Tangibility that together determine technology / AI adoption patterns. The four dimensions were chosen because they are fundamental psychological and behavioral elements that influence technology / AI adoption. Human factors play a more significant role than technological features in determining successful technology implementation according to the BELT Test which focuses on assessments beyond technology features.

Bias & Openness

New technologies and ideas face significant barriers to adoption because of bias. Frank et al., (2023) found that negativity bias creates resistance to technology adoption because negative information about technology outweighs its positive aspects even when the benefits are substantial. The studies revealed that consumers tend to focus more on potential risks which creates difficulties for innovators and marketers who want to promote new solutions. The influence of negative perceptions becomes more intense with emerging technologies because these systems create both uncertainty and unfamiliarity among users.

The adoption of technology in organizations and society gets influenced by both cognitive and social biases. People tend to oppose artificial intelligence because they fear job elimination (negativity bias) while they follow the crowd regarding AI adoption or rejection and experience skill-threatening changes (use-it-or-lose-it tendency). New technologies face irrational rejection or uncritical acceptance because of these biases which impact both personal and group learning processes. Leaders need to establish open dialogue while showing positive results and stakeholder involvement to build trust and decrease resistance.

Methodological Distinction from TAM/UTAUT: While TAM addresses "perceived usefulness" and UTAUT examines "performance expectancy" as cognitive evaluations of technology benefits, BELT's bias dimension specifically targets the underlying cognitive processes that can distort these perceptions. This represents a deeper psychological assessment that precedes and influences traditional acceptance model constructs.

Ego & Self-Awareness

Technology adoption and new learning processes involve ego in a complex manner. Entrepreneurial ecosystems known as "EGO-Systems" which focus on individual entrepreneurs can accelerate innovation through their vision and drive but they increase the chances of overconfidence and resistance to outside feedback. Elia et al. (2016) found that ego-centric methods speed up idea-to-venture development yet create isolated thinking patterns which blocks teamwork and affects technological projects' scalability and overall success.

Ego depletion which results from exhausted self-control resources creates negative effects on learning processes and technology adoption for individuals. The research conducted by Lin and Johnson (2024) demonstrates that ego depletion creates a strong link with academic procrastination because exhausted individuals make impulsive habitual choices and avoid new learning and technology adoption because of their psychological fatigue. According to their research findings people with low self-control tend to stick with established behaviors while showing reduced openness toward innovation and change which creates substantial obstacles in educational and organizational environments.

Methodological Advancement Beyond Traditional Models: UTAUT's "social influence" construct acknowledges external pressures, but BELT's ego dimension specifically addresses internal psychological barriers to learning from others, particularly younger or less senior colleagues who may possess superior technological expertise. This represents a critical gap in traditional models that assume rational decision-making without accounting for ego-driven resistance patterns.

Learning Habits

The way people and groups learn new technologies and methods depends heavily on their established learning habits. The adoption speed and likelihood of technological innovations depends heavily on four key factors which include learning style and motivation and prior experience and self-efficacy. Research conducted by Oyetade et al., (2020) through systematic literature review discovered that students who have adaptable learning approaches combined with favorable digital tool attitudes tend to adopt mobile and e-learning systems which results in better learning outcomes. The review demonstrated that educational technology adoption depends on three main factors: ease of use and perceived usefulness and enjoyment. The effectiveness of technology-based learning and satisfaction levels depend directly on learning styles and the way students perceive technology adoption.

Learning heuristics and habits of individual learners either help or prevent them from adopting new technologies. Gars and Ward (2019) conducted research which showed that people who use basic habitual information processing methods and form strong beliefs from first experiences tend to adopt new technologies at a slower pace. Research conducted by the authors in both the US and India revealed that heterogeneous learning styles including risk aversion and preference for familiar routines create major obstacles to innovation adoption even when new tools demonstrate clear advantages. The research indicates that developing adaptive learning habits together with supportive environments will speed up the process of adopting new technologies and ideas.

Framework Enhancement Over Static Models: TAM and UTAUT measure technology acceptance at specific time points, while BELT's learning habits dimension assesses ongoing capacity for skill development and knowledge acquisition. This dynamic assessment provides predictive value for sustained technology adoption success rather than initial acceptance alone.

Tangibility & Action Bias

The adoption of new technologies / AI or learning methods by individuals and organizations depends heavily on tangibility and outcome expectation. Managers and users form their perceptions of innovations based on the degree of tangibility which refers to the presence of physical observable elements in a product or process. The study by Alexiev et al. (2018) discovered that organizations with highly tangible offerings can easily track new technology changes because these changes produce visible and measurable outcomes which help managers identify competency gaps and validate experimental approaches. The clarity enables managers to achieve agreement about process changes which leads to better adaptation. Managers in intangible settings need to use visualization tools and symbols to present abstract benefits in concrete terms because this approach either strengthens or weakens the connection between service delivery innovation and management innovation based on their ability to effectively simulate or communicate tangibility.

The expected results from adopting technology play a major role in determining the adoption choices that people make. According to Oliva, P. et al. (2020) users tend to adopt technologies strongly at first but then abandon them because their actual benefits fail to match their expectations. The researchers conducted a field study in Zambia which revealed that participants stopped using the technology after their expectations about outcomes failed to materialize or when uncertainty became high despite receiving subsidies for adoption. The research shows that successful technology adoption depends on both transparent benefit explanations and effective uncertainty reduction and expectation management.

Methodological Alignment with UTAUT Evidence: BELT's tangibility dimension directly corresponds to UTAUT's performance expectancy construct, which research consistently identifies as the strongest predictor of technology adoption intention. However, BELT extends this concept by incorporating personal motivation factors and practical implementation considerations that traditional models often overlook.

Sanyal's BELT Test for New Technology Responsiveness

Bias-Ego-Learning-Tangibility (BELT) Test: How willing are you to implement new technological (AI) advancements that appear in the market?

Instructions: Assess how well these statements match your behavior by choosing ratings between 1-5.

1 = Strongly Disagree
2 = Somewhat Disagree
3 = Neutral
4 = Somewhat Agree
5 = Strongly Agree

Sanyal's BELT Test Questionnaire

Bias & Openness

1. I begin by showing curiosity toward new technologies instead of expressing skepticism.
2. I dedicate my time and resources to learn about new emerging technologies which I discover right away.
3. I assess new technologies independently regardless of how they affect my professional field or position in the industry.
4. I have already identified both advantages and disadvantages of new technologies before the industry establishes its common view.
5. My approach involves forming personal opinions about new technologies rather than following general trends.
6. I transmit the latest technological insights to stakeholders so organizational decisions become more informed.
7. I make a conscious effort to find viewpoints which contradict my preliminary responses toward new technologies.

Ego and Self Awareness

8. I openly tell my peers, superiors and subordinates about my lack of familiarity with new emerging technologies.

9. I dismiss new technologies that challenge my expertise before I understand them completely.
10. I welcome junior staff members who introduce technology solutions that surpass my current methods with positive responses.
11. I evaluate technologies by their inherent quality rather than by considering their impact on my professional standing.
12. I publicly show my mistakes and using them to build learning experiences.
13. I feel happy toward others who get recognized for technology developments that I believe I could also have developed.
14. I actively ask help from my less experienced colleagues because they understand particular technologies better.
15. I refrain from highlighting potential pitfalls for ideas recommended by others.

Learning Habits

16. I am motivated to learn new technologies starting from scratch.
17. I create a specific block of time to investigate technologies which are not essential for my present requirements.
18. I develop my knowledge of new technologies through independent initiative instead of following work protocols.
19. I have invested time into learning new technology ideas and these have resulted in my career advancement.
20. I establish connections and create ecosystems for the purpose of receiving updates about technological advancements that matter to my profession.
21. I base my learning on technical research papers and resources rather than just books and media generated information.

Tangibility and Motivation

22. I become highly motivated to discover new technologies when I can identify specific advantages that benefit me personally first and then the organization.
23. I dedicate more time to learning technologies which directly enhance my present workflow than technologies with theoretical future benefits.
24. I choose to implement technologies which address present organizational issues rather than pursuing solutions with future development potential.
25. I convert all my technology knowledge into practical solutions which benefit specific tech and operations teams in particular and the larger organization in general.
26. I support organizational technologies that deliver benefits to the company even though they do not serve my individual career goals.

27. The amount of effort I dedicate to learning new technology directly depends on the quantifiable business results it produces.
28. I will investigate technologies which could transform the entire industry even though their direct organizational advantages are not immediately apparent.
29. I struggle to maintain interest in technologies which do not produce visible outcomes during a predictable time period.
30. I conduct periodic assessments to determine if my technology interests match the strategic goals of my organization.
31. I support technology adoption initiatives when I can present concrete return on investment data to decision-making authorities.

Scoring Rubric

The report delivers an analysis of outcomes together with practical advice which draws from evidence-based methods.

Score Categories

Low Readiness (31 – 40)

Moderate Readiness (41 – 93)

High Readiness (94 – 155)

Feedback Construct

Comfort and Security

This assessment explores the participants comfort with technology and security when using new systems. It also demonstrates to what extent technology readiness is limited by discomfort and insecurity which decreases both willingness and confidence toward innovation.

Optimism and Innovativeness

This assessment evaluates the positive outlook towards new technology along with the willingness to test new tools. It identifies the optimism and innovativeness to adopt new technology through demonstration of practical use cases and ease of use.

Learning Habits

This assessment determines the approach to learning new technological ideas, platforms and concepts. This ensures there is a keenness to understand the potential positive impacts over present systems and methodologies around technology.

Tangibility

The assessment evaluates the requirement for direct proof about technology advantages to motivate new technology adoption. Presenting clear evidence about how technology affects workflow and outcomes helps overcome bias and supports adoption.

Scoring Interpretation

Low Readiness (31-40)

Your score in this range shows significant obstacles to new technology adoption because you experience high discomfort and insecurity together with low optimism and innovativeness. Your skepticism about new system usefulness and ease of use creates anxiety which makes you avoid adopting new technology. These together act as powerful barriers to technology adoption and decreases both willingness and confidence toward innovation adoption.

Recommendations for Reactive Adapters

- Dedicate time each month to discover new tools through a technology exploration session
- Join online communities that focuses on technological progress in your professional field
- Create a simple decision framework to identify which technologies need your focus
- Collaborate with people who demonstrate intense technological interest to understand their discovery methods.
- One objective should be to identify at least one business process which technology shows potential to transform significantly.

Moderate Readiness (41-93)

The Moderate Readiness score indicates that you have some openness to new ideas and technologies but still have hesitation or inconsistency in your adoption behaviors. You understand the advantages of technology but doubt your ability to use it effectively and see the practical value of adoption. Optimism together with innovativeness serve as readiness drivers but moderate scores indicate a requirement for better perceived usefulness and ease of use to achieve full system adoption.

Recommendations for Pragmatic Evaluators

- Obtain training programs which develop your abilities and self-assurance through tutorial-based instruction and team-based experimental initiatives.
- Regular feedback combined with recognition of progress will help build positive attitudes and decrease insecurity.
- Explore transparent evidence of technology's impact on workflow or outcomes.
- Conduct small experiments and adoption achievement recognition will help create momentum which leads to increased readiness levels.

High Readiness (94-155)

A High Readiness score signals strong optimism, innovativeness, and proficiency—traits consistently linked to rapid and enthusiastic technology adoption. You probably view technology as a tool that empowers you and feel comfortable trying out new tools and can easily see the benefits of these tools.

Research shows that high scores on optimism and innovativeness are directly related to higher perceived usefulness and ease of use, which makes you a natural early adopter and potential technology champion within your organization and ecosystems. Your positive attitudes also help to decrease feelings of discomfort or insecurity, which in turn makes the adoption process easier.

Recommendations for Strategic Adopters

- Consider mentoring peers and leading innovation initiatives
- Share your experiences to inspire others with lower technology readiness
- Further enhance your skills by staying current with emerging technologies
- Participate in advanced training or professional networks
- Continue to provide feedback on new technologies to help drive broader acceptance

General Recommendations

Fast-tracking new technology (AI) adoption for all through specific development approaches:

- Working on Cognitive Flexibility will provide the ability to change your thinking patterns through acquiring new knowledge.

- Inculcating Learning Eagerness will enable genuine interest to know and gather new knowledge.
- Evaluate your Risk Tolerance, it will help you embrace untested dangers of early adoption early and through informed decision making giving you a first mover advantage.
- A focus on Practical Integration Frameworks will enable integration new technologies (AI) effectively demonstrating tangible returns on investments of both time and capital.

How the BELT Framework Enhances Traditional Technology Acceptance Models

The BELT framework complements rather than replaces established technology acceptance research:

TAM's cognitive evaluation constructs (perceived usefulness, perceived ease of use) are enhanced by BELT's bias and learning assessments that identify factors influencing perception formation.

UTAUT's comprehensive model (performance expectancy, effort expectancy, social influence, facilitating conditions) is strengthened by BELT's individual readiness assessment that determines response patterns to these situational factors.

Extended acceptance models incorporating cultural and psychological factors align with BELT's holistic approach to individual difference assessment.

Comparative Analysis: BELT Framework Advantages

The BELT framework offers several methodological advantages over traditional technology acceptance models:

Comprehensive Psychological Assessment: While TAM focuses on cognitive evaluations and UTAUT examines situational factors, BELT provides comprehensive psychological profiling that identifies underlying readiness factors influencing all technology adoption scenarios.

Dynamic Adaptability Measurement: Unlike static acceptance models that assess current perceptions, BELT evaluates ongoing capacity for learning, adaptation, and psychological flexibility that determine long-term technology adoption success.

Individual-Level Precision: TAM and UTAUT primarily address group-level or organizational contexts, while BELT provides individual-level psychological assessment, enabling targeted intervention strategies for specific psychological barriers. [Confirm the wording of this sentence against your Word master — it is truncated in the source PDF.]

Cross-Technology Applicability: Rather than focusing on specific technologies or contexts, BELT assesses fundamental psychological traits that influence adoption patterns across diverse technological innovations and implementation scenarios.

Implementation Recommendations Evaluation

Individual Assessment Strategy — Organizations should implement BELT assessment as a foundational element of technology adoption planning:

Pre-Implementation Diagnosis: Conduct comprehensive BELT assessment before technology deployment to identify individual readiness levels and potential barriers.

Targeted Intervention Design: Develop specific intervention strategies based on BELT dimension scores, addressing bias patterns, ego barriers, learning limitations, and tangibility requirements.

Progress Monitoring: Use BELT reassessment to track individual development and adaptation throughout technology implementation processes.

Strategic Resource Allocation: Allocate training, support, and development resources based on BELT readiness profiles rather than uniform approaches.

Organizational Integration Framework

The BELT framework should be integrated with existing technology adoption planning:

Complement TAM/UTAUT Assessments: Use BELT individual readiness assessment alongside TAM/UTAUT organizational and technological evaluations for comprehensive adoption planning.

Enhance Training Program Design: Develop training programs that address specific BELT dimension weaknesses identified through assessment.

Optimize Change Management: Design change management strategies that account for individual psychological readiness patterns revealed through BELT assessment.

Improve Success Prediction: Combine BELT individual readiness scores with TAM/UTAUT situational assessments for enhanced technology adoption success prediction.

Distinctive Advantage of the BELT Framework Across Organizational Layers

The BELT framework delivers exceptional value because it measures the sentiment, openness and AI readiness across all important leadership and management levels. The BELT framework differs from traditional models because it assesses psychological readiness and adaptability and underlying attitudes through its core dimensions of Bias & Openness, Ego & Self-Awareness, Learning Habits and Tangibility & Motivation.

The board level uses BELT to perform a thorough evaluation of executive members and directors' combined sentiment and openness towards transformative technologies by identifying any concealed resistance or readiness. The ability to foresee changes becomes essential for developing future AI strategies and risk management and governance processes at the highest leadership level.

BELT helps leaders who establish vision and maintain financial responsibility to identify how their mindsets and risk-taking behaviours and ability to adapt affect the company's willingness to dedicate resources to AI implementation. The assessment tool helps executives and executive directors transition from superficial declarations to determine how much genuine support exists for organizational change within their organization.

The BELT framework monitors operational and project management staff through its psychological assessments of learning orientation and motivational levels and adaptability at this level. Organizations gain insight into upcoming obstacles and cultural barriers through this approach which allows them to implement prevention measures and training programs to support successful adoption.

The BELT framework delivers precise and actionable AI adoption sentiment and preparedness evaluations across all organizational levels which enables leaders to build more successful change management and risk mitigation and upskilling approaches. The comprehensive method allows AI adoption to receive backing from both executive and non-executive levels with enterprise-wide readiness support.

The BELT framework represents a significant methodological advancement in technology adoption assessment by addressing critical gaps in traditional acceptance models. While TAM and UTAUT provide valuable insights into cognitive and situational factors influencing technology adoption, BELT offers essential individual psychological readiness assessment that determines how individuals respond to these factors.

The framework's comprehensive approach to bias recognition, ego management, learning orientation, and implementation focus provides organizations with precise tools for identifying and addressing individual barriers to technology adoption success. By integrating BELT assessment with established TAM and UTAUT evaluations, organizations can achieve more accurate adoption predictions and develop more effective intervention strategies.

The evidence base supporting BELT's methodological approach, combined with its practical applicability across diverse technological contexts, positions this framework as an essential complement to existing technology adoption assessment tools. As organizations continue to navigate increasingly complex technology implementation challenges, the BELT framework offers a scientifically grounded approach to understanding and enhancing individual readiness for technological change.

SOLUTIONS AND RECOMMENDATIONS

Fast Tracking AI Adoption for Boards: Leadership Capabilities for AI Integration

Organizations need distinct leadership capabilities to implement effective AI governance beyond traditional technology administration methods. The successful integration of AI depends on special competency clusters which allow organizations to navigate their entire ecosystem rather than focusing on technical implementation management.

The leadership capabilities that directly influence AI adoption outcomes are dependent on 1. The ability to discover external capabilities through strategic partnership; 2. Data-driven scenario planning; 3. The ability to speedily shift organizational resources; 4. The ability to effectively present AI vision to stakeholders; 5. The ability to change potential implementation difficulties into systematic opportunities (Hunt Scanlon Media, 2024).

These leadership capabilities exist as developable skills instead of permanent qualities because organizations can implement systematic development approaches. Organizations which establish leadership development programs based on these capability clusters achieve better AI integration results regardless of their initial leadership capabilities. Boards that develop these leadership competencies through targeted development initiatives will obtain structural advantages in AI implementation regardless of their AI adoption timing and technology selection.

The following adoption ideas, governance frameworks and innovations directly address these structural limitations through complementary mechanisms that enhance technological intelligence while maintaining appropriate risk management frameworks.

1. New Tech Advisory Board Framework: Creating Unbiased Evaluation Mechanisms

Today, technology is a key consideration for driving key organization decision across all industries. Irrespective of at what stage (Innovators; Early Adopters; Early Majority; Late Majority; Laggards) of tech evolution a corporate or business may be in, the need to be 'up to speed' is paramount because the consumer ecosystem, the regulatory ecosystem, the resource ecosystem, and the innovation ecosystem are all getting interconnected and making the consumer more connected and informed making it critical for business survival.

Assessing disruptive technology usually proves difficult for boards given the existing and traditional evaluation and governance structures that are better suited for business decision and not for specialized deep dives on technical matters. It has been observed that strategic control capabilities grow stronger in experienced executives, which reveals prefrontal cortex maturation but this also intensifies loss aversion during decision-making thus favoring incremental changes over transformational innovations according to papers in Frontiers in Psychology and Neuroscience (Funahashi, 2017; Gelskov et al., 2016).

Therefore, organizations need to establish separate governance systems that overcome the limitations of the current cognitive framework. The simplest approach for a new technology evaluation and implementation framework is to establish a "New Tech Advisory Board". This specialized board comprising external and independent technology experts from academia and startup ecosystems and advanced technology development centers will be equipped not only to understand the business, what the current technology framework looks like and also access their technology intelligence networks independently to make well researched recommendations and also execution support.

Why do organizations need a "New Tech Advisory Board" when they have a full-fledged tech teams and infrastructure? The answer perhaps lies in Abraham Lincoln's famous quote, "Give me six hours to chop down a tree and I will spend the first four sharpening the axe." If you observe the tech teams in organizations in general, the learning gaps are wide, unstructured and not purposeful. What do their daily jobs entail? Trouble shooting, scrambling to meet deadlines on already delayed deployment dates, error rectification and vendor interaction. The skill they acquired and practiced is the skill they got hired for. CTO's are usually managing internal and external stakeholders and regulators. Where is the planned, purposeful and structured will of the organization to formally and systematically induct new learnings and developments? Only when media highlights a tech related news, or market intelligence talk about what competition and other industries are up to, it's not a part of a daily to-do list for the entire tech teams. How many CTO read research papers that are published on new tech and enhancements?

Hence it is important to outsource this task. The New Tech Advisory Board, can scan the market for the best applicable technology for your organization, give recommendations, support with execution frameworks to the board and the tech teams, fast tracking the knowledge and learning gap that exists.

It is important that the advisory board establishes an organizational transformation process that makes the organization adopt new technology as a first-generation tech adopter through ongoing environmental monitoring and opportunity detection and implementation planning. Enhanced strategic control of boards can help evaluate effective long-term cost-benefit analysis and make organizations approach innovation cautiously and effectively.

The adoption framework enables organizations to enhance their technology adoption capabilities through three main benefits: organizations discover emerging opportunities earlier; implement comprehensive assessment of implementation pathways; and reduce internal resistance through external validation. These advantages directly address the learning limitation challenges while maintaining appropriate risk management through the main board's continued oversight responsibility.

2. Shadow Board Implementation: Bridging Intergenerational Cognitive Divides

The technological proficiency narrative establishes a direct link between digital competence and generational characteristics which generates "digital divides" throughout corporate organizational structures. Digital natives learned technological concepts naturally during their formative years but many senior executives experienced their first encounter with complex digital systems through their fully developed professional careers which led to differences in their cognitive processing of technological information.

Organizations need to create internal "Shadow Boards" which consist of next-generation leaders who focus on developing innovation and AI implementation while resolving the cognitive processing gaps between leadership generations. The proposed structure enables intergenerational collaboration and develops future governance skills in present operational leaders to overcome technological adaptation barriers that stem from unequal learning environments.

The 'Shadow Board' concept serves as a formal mechanism for building internal entrepreneurial capacity – "Intrapreneurship", which encourages fresh and out of the box thinking amongst employees and also encourages them to think of business ideas and solutions that will help companies find solutions to operational challenges in the real world. Leaders with business expertise and operational experience and next level of leadership come together to bridge innovation gaps with practical solutions and implementation frameworks keeping in mind the organization limitations and idiosyncrasies.

Further the 'Shadow Board' should work closely with the 'New Tech Advisory Board' The Shadow Board can external technology recommendations made by thoroughly examining operational deployment criteria and how this would translate between theoretical technology potential and real-world deployment needs. The translation process requires a complete evaluation of capital needs and organizational skill deficits along with deployment sequences to generate strategic adoption plans that utilize both innovative potential and operational feasibility. These two layered assessment frameworks – level one by the Shadow Board and then by the Board would give additional data points and perspectives that will enhance the decision-making confidence levels amongst board members.

The implementation of Shadow Board structures can help accelerate tech adoption as the process would involve select wider teams in the information gathering and evaluation phase and will have three key impacts: Operational teams will show decreased resistance during evaluation and adoption phases; Better implementation planning; Structured development of talent and resource allocation for strategic technology deployments.

3. Vendor Intelligence Network Development: Addressing Information Asymmetry Challenges

The existing information imbalance between board members, tech teams, existing vendors and the fast-evolving AI solutions in the marketplace creates substantial obstacles for making effective decisions at the board level. The evaluation of technology and solutions happens mainly through vendor stories in environments which lack independent access and analysis of technologies available. The organization needs to establish specific capabilities which collect systematic market intelligence about emerging AI solutions and service providers to overcome information asymmetry. The "Vendor Intelligence Network" establishes vital information superiority in fast-moving technology environments which traditional procurement methods might not handle effectively thus solving the elevated evidence requirements that research shows boards with minimal AI experience need.

The organization should establish a dedicated team or function within its tech or vendor management teams that creates extensive intelligence coverage of all potential partners ranging from new startups to established technology leaders and boutique providers. Unlike conventional vendor management styles of covering few established providers, this team should explore the full spectrum of solutions available covering early stage to established businesses. The network should concentrate its intelligence efforts on monitoring both domestic and international solution availability together with implementation approaches and reference client success stories and integration methods. The intelligence focus establishes a complete information base which supports both technology selection and implementation planning while addressing decision-making tasks.

Establishing structured vendor intelligence capabilities in organizations can help achieve better technology implementation with three main benefits: Comprehensive market coverage and solution awareness beyond regular established players; Better negotiation positioning through deeper and wider market understanding; 3. Effective implementation insights and planning.

4. Strategic AI Initiative Fund Establishment: Creating Controlled Experimental Spaces

Research on executive cognition reveals different risk assessment patterns among seasoned decision-makers. Executives who are sixty years old or older perform better at long-term cost-benefit analysis which leads them to evaluate all implementation consequences beyond standard deployment metrics according to Frontiers in Psychology journal (Mueller & Brand, 2018). While this enhances the strategic perspective strength, but it also amplifies risk perception for technologies that can also have a downside impact.

The institution of a 'Strategic AI Initiative Fund' that acts as a dedicated financial mechanism to handle strategic AI investments can help fast track AI adoption in more ways than one. Specialized and focused funding structures will support important technological experimentation by clearly identifying new tech and AI initiatives from standard technology and operational spending requirements.

Most board members who are in their sixties are also very well-established investors. They have developed deep understanding of businesses either in a specific industry or varied industries based on their personal risk thresholds and investor eco-system they are part of. Having sharp business understanding in their existing field they extend that acumen into their investment decisions by critically evaluating business and revenue models and cash flow analysis before they 'Cut the Check'. However, when it comes to a business that they do not have complete understanding of they tend to refrain from investing. But at the same time the FOMO of not being able to maximize an opportunity while their peers could and then jumping into the band-wagon later may not give them the comfort nor the returns.

The fascinating point here is that the board member as an investor in a business, critically monitors its progress, evaluates the business trajectory and even makes a greater effort to deep dive into newer ideas and concepts since personal capital has been invested. However, the board members may or may not have the same level of motivation to explore and invest their time and interest when it comes to a new idea tabled in the board meeting.

One of the simplest ways to garner focus and attention towards a new idea is to get invested in it. If the board constituted 'Strategic AI Fund' starts making small investments in fresh and relevant ideas, the board as investors would put in that effort to better understand firstly the concept or idea, the solution, business and revenue models, growth potential, etc. Further they may also tap into their existing eco-systems to get insights and perspectives as well.

This would bring all the ideas of "New Tech Advisory Board'; 'Shadow Board'; and 'Vendor Intelligence Network' together to pitch the most relevant and unique ideas in the AI space to the board. This approach further has two advantages: If there is a new idea that has a direct impact and application in the business, one can observe implementation success through other use cases and businesses. Based on successful trials and implementation the business can decide when to adapt and integrate the tech in their workflows. The second advantage is that dependent on the level and extent of investment in the start-up / business, the board has the visibility and control to ensure, they get the first right of refusal, before the idea gets picked up by the market and competition.

5. Academic Integration Framework: Incubating Next Gen AI Talent through Structured Knowledge Exchange

Organizations will encounter major workforce transformation and disruptions with AI getting more advanced in the coming years. Not only the early adopter but also the laggards will face scarcity of trained and relevant AI equipped workforce. The implementation of AI technology demands "epistemic paradigm shifts" according to Kingston University research because it requires fundamental changes to work process conceptualization alongside implementation and evaluation methods. (Kingston University Technology Adoption Study, 2023).

Traditional methods of talent development may fall short in this transformation because of what the cognitive science defines as "Cognitive Entrenchment". This refers to the continuing and growing reinforcement of existing problem-solving approaches based in educational and professional training (Phan & Ngu, 2021). This entrenchment affects middle career professionals who excel at problem-solving within established frameworks but strongly oppose changes to core work process conceptualization. Add to this 'Learning Inertia' that sets in middle career professionals. Firstly, most organizations do not have a tailored learning program that is aligned to the employee's career trajectory, but instead is based on generic leadership and skill targets. Secondly the employee may not find enough motivation or have enough resources to invest in enhancing their skill. Add to this the psychological angle should an employee be going through mid-life crisis.

Hence there will be a huge 'talent gorge' that will be created especially for AI talent in the coming years. One of the simplest ways to mitigate this for organizations are to 'Catch them young a watch them grow'. AI being the next big thing, the education system globally would also focus on preparing relevant curriculum to impart AI skills to students for better employability. Organizations need to use organized academic integration systems to develop bidirectional knowledge transfer between academic institutions and operational environments for addressing cognitive limitations of the organizations.

Establishment of an "AI Incubation sandbox" for only students (engineering and management) will enable emerging talent with AI paradigm knowledge to ideate and tackle real-world implementation challenges and develop both practical skills and theoretical knowledge.

An effective implementation model consists of three interconnected components that function together:

Challenge-Based Learning Architecture: Companies should develop specific protocols to turn operational problems into formalized problem statements suitable for academic engagement. The approach would stimulate students to think of adaptive and practical developments that would work in the real world. Further it will also work as a knowledge base for workers to assess the practical applicability of the concepts, thus improving both stakeholders' knowledge frameworks to examine and explore new situations in a fresh perspective instead of approaching solutions in predetermined thinking patterns and procedures.

Mentorship Reciprocity Structures: Traditional mentorship models create unidirectional knowledge flows that reinforce established paradigms. Emerging talent should be part of reciprocity structures which involve providing AI paradigm orientation to established professionals while they obtain operational context knowledge from them to speed up organizational knowledge transformation and boost practical application abilities in academic participants.

Implementation Pathway Development: The major limitation within conventional academic-business collaborations arises because theoretical concepts and practical execution methods are not properly translated. Organizations need to develop systematic procedures which convert their sandbox successes into operational frameworks.

Organizations that integrate academics into their operations through strategic knowledge exchange frameworks are likely to achieve better AI transformation capabilities. This academic integration initiative will also act a local and global talent identifying and nurturing initiative as the demand for AI skilled workforce will grow exponentially in the coming years.

6. AI Focused Learning Modules for Technology Teams

All members of the Tech Team need to attend a mandatory training program to combat technical knowledge gaps.

The possession of specialized skills acts as an obstacle which hinders the adoption of new disruptive technologies inside organizations. Specialized knowledge in established systems leads experts to resist novel technological approaches without conscious awareness. The resistance towards methodology changes is seen more at the higher levels of technical staff. The development of neural pathway reinforcement in neuroscience explains how established cognitive frameworks become stronger through repeated application. Technical experts who develop specialization within particular technology environments see their neural pathways grow more dominant thus establishing unconscious filtering systems that favor information matching established knowledge patterns while disregarding opposing data.

Organizations need to develop mandatory training programs with structured content to disrupt the neural pathways of technical staff. Experiential learning models deliver the most effective results because they establish cognitive dissonance environments according to neuroplasticity research to make technical professionals challenge their established assumptions.

Organizations must implement a three-stage learning approach to target specific cognitive processes for effective implementation.

AI Lab: Technical personnel must participate in structured exercises that makes them explore the world of AI and use these platforms to create fun, personal and controlled work solutions. This will make their learning fun, practical and also experimental. This will help them familiarize and create interest in different open-source platforms, algorithms, coding support, etc. These sessions can be directed more towards coding, analytics, mobile applications, chatbot building, etc. which also has a mature and applicable use case at their work place.

Immersions: Periodic immersions with industry experts, data and AI scientists, and others will help the work force get an insight into how AI is solving real world problems. These sessions would also make them understand the challenges in execution of a real world project, resource intensity, and new technology in use.

Executive Education: Through the above interventions and observation, if the organization is able to identify a talent pool that can accelerate the organization's AI strategy, these workers should be enrolled in executive education programs where they can enhance their tech and AI related knowledge.

7. Workplace AI Integration Framework: Addressing Multi-Dimensional Adoption Barriers

Organizations struggle to implement AI successfully because their deployment efforts concentrate mainly on technical aspects while ignoring the essential connection between technology and work process structures and human cognitive preferences. Human-computer interaction research together with socio-technical systems studies demonstrate that successful technology adoption depends on co-designing technical and social system aspects which match organizational contexts and user requirements. Business leaders identify resistance to change and misalignment between AI systems and workplace routines as major obstacles to successful AI implementation.

The integration problem arises because of procedural entrenchment according to cognitive science which describes the automated nature of work processes that embed themselves in human neural systems and produce unconscious resistance to changes in procedures. People process procedural knowledge without awareness of its contents because this information exists beneath their conscious threshold which leads to change blindness according to neuroscience.

Change blindness and resistance have been evident in various industries across various levels of work forces through the journey of digitization and it exists even now. When the procurement, inventory management and other teams in a manufacturing set up had to move from manual processing to SAP based system processing the adoption speed varied across various cross se to do the same tasks. had seen the workforce adapt

To integrate AI with workplace systems organizations, need to establish specific frameworks which address different aspects of workplace AI adoption. The "Three-Pronged Workplace Adoption Model" establishes complete integration paths that tackle procedural entrenchment obstacles in three essential implementation domains.

Knowledge Efficiency Enhancement: Addressing Information Overload Efficiently

Knowledge workers who have to access and refer to a lot of research material and reports to generate strategic content for the management and clients would greatly benefit from the LLM frameworks. Similarly, the agentic networks can also help workers find ways of evaluating and analyzing tasks through a deeper and wider knowledge base and also faster. There are several tasks unique and repetitive that can be identified and can be "partially manually automated" by the worker by being given access to AI platforms and being made proficient in prompting to generate the best output.

There are two major apprehensions about open-source platforms. One is data protection and the other is platform selection. Data privacy and protection is indeed a big one. But one needs to understand that a tech company that is openly and freely providing access to its platform, are they doing so with the objective is social good alone. They have a sound business and revenue model in place, and their solutions are designed for businesses. From the early introduction of AI corporates and their tech experts have been trying to address the security issue and there are solutions that businesses can explore to fit their business outputs and data protection needs. The world is evolving so fast that people are building on each other's expertise and strengths. There are several frameworks that one can put in place to ensure that all the processing power machinery resides in your ecosystem and there are embedded and encrypted data exchange pathways that secure the data flow. Further there are Small Language Models that focus on specific subjects that may be more useful for certain functions yielding better outputs.

When it comes to pace of change of technology, it often confuses many, making them unsure of how to progress. While deciding on the AI platform one needs to consider the extent of usage required and the depth of usage expected from workers. Let's consider a few examples to understand the context better. In the era of the scientific calculator, it had multiple functionalities, that user selectively used as per their needs. Were all the functionalities used by a single user start to end, maybe not. Similarly, as smart phone has many functionalities beyond making calls, taking photos and messaging, etc. But majority of the world use the smart phone for very basic functionalities only. The point is that even if the AI development brings in new functionalities some of the existing platforms are very well equipped to support you for a fairly long time for your basic functionalities. So, the next time if there is a new platform or company disrupting the space, the functionality they are talking about may not be of use to you and hence one may avoid the 'Fear of Missing Out' and decide faster.

Process Efficiency Transformation: Restructuring Workflow Architecture: Integrating AI to improve and automate certain processes can lead to better knowledge and time efficient outputs for organizations. Organizations should use structured transformation procedures to increase operational and process efficiency, that can be achieved through:

Staff members participate in structured environments through collaborative Process Reimagination to recreate workflows while integrating AI capabilities during this approach. When people contribute to concept development, they develop stronger attachment to these concepts according to the "ownership effect" which neuroscience has identified.

Parallel Process Operation enables both traditional and AI-enhanced process execution during transitions thus reducing adoption psychological "fallback anxiety" which occurs when people fear capability degradation during adaptation phases. The parallel access to processes delivers vital psychological security through first-hand experience of benefits.

Expertise Preservation Frameworks: The framework requires the identification and preservation of intellectual capital within established processes while converting mechanical components to AI systems. The implementation of technological advancement leads to "expertise invalidation anxiety" which organizational psychology identifies as the fear that accumulated knowledge will lose value thus threatening professional identity.

Analysis Enhancement Integration: Expanding Cognitive Architecture

The third integration dimension identifies analytical framework limitations according to cognitive science which represents natural constraints of human information processing during complex multidimensional situations. Research shows that human decision quality deteriorates in a predictable pattern when variables interact beyond the boundaries of natural human cognitive processing abilities resulting in systematic decision errors which decision science defines as dimension reduction strategies.

Analysis enhancement integration should be implemented through three structured components by organizations.

Decision Architecture Mapping: The approach involves a systematic process to detect recurring decision points where complexity exceeds the optimal capacity of human cognitive processing. The method targets structural human information processing limitations which produce systematic error patterns beyond conventional performance improvement that focuses on execution quality.

Overcoming Cognitive Constraints: Organizations should develop specific AI systems to overcome human decision-making cognitive constraints while maintaining human control in situations requiring contextual knowledge and ethical evaluation. A combination of technological capabilities with human cognitive abilities and their natural and ethical boundaries should help create the most suited product or platform design architecture for the organization.

Progressive Automation Calibration: Organizations should establish systematic methods to expand automated processes after proving their reliability in specific decision areas. The anxiety associated with control in automated systems leads to implementation resistance which exceeds proven reliability levels according to adoption psychology.

FUTURE RESEARCH DIRECTIONS

This research on corporate technology adoption patterns and AI integration introduces various directions for future studies. AI continues its evolution in organizational structures so multiple research directions need thorough investigation.

Longitudinal Studies of AI Implementation Outcomes

The research should include thorough longitudinal assessments that monitor AI implementation results in various organizational environments. Current research provides useful insights about adoption decisions yet longer analysis periods would reveal important information about how initial board-level choices affect long-term market positioning. The research should monitor both financial results alongside organizational capability advancement and workforce evolution and market position modifications. Research tracking these metrics from 5 to 10 years would establish solid empirical evidence for understanding adoption timing effects on competitive outcomes.

Neurological Underpinnings of Technology Decision-Making

Organizational neuroscience as an emerging field provides new possibilities to better understand the decision-making processes involved in technology adoption. Future research could use advanced neuroimaging methods to study how board members process technology proposals relative to their other strategic decision-making activities. The research aims to detect particular neural patterns related to technology risk assessment which might lead to effective strategies that counteract automatic cognitive obstacles. A particular focus should be placed on examining how different presentation methods and information formats affect neural responses to technological innovations.

Cross-Cultural Variations in AI Governance Approaches

The examination of technology adoption within corporate governance frameworks forms an important focus of this chapter but future research needs to explore how different cultural environments affect AI adoption worldwide. Research should investigate how AI risk assessment methods differ between collectivist and individualist cultures and how regional corporate governance structures impact adoption timelines and how national AI strategies shape organizational decision-making. The analysis of governance contexts from Asian and European and North American regions would reveal how cultural factors accelerate or hinder adoption processes.

Quantifying the Efficacy of Proposed Intervention Frameworks

The governance mechanisms presented in this chapter require additional empirical research for validation: Tech Advisory Boards, Shadow Boards and Vendor Intelligence Networks. The proposed study plans to conduct controlled experiments that measure the effects of these mechanisms by comparing them against identical control groups. The research should investigate which components of the intervention, drive adoption acceleration and how these mechanisms affect the quality and risks of adoption decisions and whether organizations need unique intervention approaches based on their contexts. Research findings would transform theoretical concepts into practical implementation frameworks with quantifiable ROI measurements.

Evolution of Board Composition and AI Governance Structures

Future research needs to study the transformation of governance structures that occur because AI plays an essential role in organizational strategy. Future studies should investigate new models of specialized AI governance committees and changing board composition requirements for technological expertise and evolving fiduciary responsibilities in AI implementation. The analysis of governance structure changes during AI integration periods through longitudinal studies would enable scientists to observe institutional development patterns. The findings from this research have major implications for corporate governance policies and director education programs and regulatory frameworks.

Generational Knowledge Transfer Mechanisms in Technology Adoption

The research should develop methods for transferring knowledge across generations that effectively bridge existing adoption gaps which the chapter identified. The studies should examine how mentorship programs, collaborative decision frameworks, and experiential learning approaches influence cross-generational technology comprehension. The research needs to determine which communication structures show the best ability to translate technological ideas between cohorts with distinct cognitive information processing traits. The research findings from this study hold substantial importance for board education and development programs.

Refinement and Validation of the Sanyal's BELT Framework

Further development of the Bias-Ego-Learning-Tangibility (BELT) framework from this chapter needs to be expanded across various organizational settings. Future studies must establish the psychometric properties of the assessment instrument while investigating predictive BELT score relationships to adoption outcomes and developing personalized intervention approaches for different BELT profiles. The research should analyze if BELT profiles within organizations remain stable between different industries or if unique patterns emerge between industries. BELT would become a validated diagnostic tool with defined intervention approaches through this research.

Ethics of AI Governance and Decision-Making

The progress of artificial intelligence technology requires specific research to explore ethical challenges during governance decision-making processes. Research in the future should explore how boards both understand and assess ethical effects from AI implementations and which governance structures successfully integrate ethics into adoption choices and how organizations manage innovation needs against responsible implementation methods. The research direction links with developing regulatory standards and produces important effects on corporate image maintenance and risk control methods.

The Evolving Economics of AI Adoption Decisions

The financial frameworks that guide AI adoption decisions will develop progressively with technology development. Future research needs to study how changing cost structures and implementation timelines and risk profiles affect adoption economics. Boards need to prioritize the assessment of organizational learning benefits and capability development along with strategic positioning advantages in relation to actual implementation costs. The research needs to study the evolution of economic evaluation frameworks through AI's life cycle from emerging to established technology while possibly creating standardized investment assessment tools that combine both tangible and intangible aspects.

Sector-Specific AI Adoption Patterns

Future research needs to analyze the distinct impact of industry-specific factors on AI integration methods beyond general adoption patterns. Research conducted through comparative sector studies within different regulatory and competitive environments and technological dependencies would show how context affects adoption choices. The research should investigate if specific sectors have unique adoption patterns together with how industry-specific risk profiles affect governance approaches and whether knowledge exchange between sectors enables faster progress. The research will facilitate development of focused intervention methods which match specific adoption obstacles found in different sectors.

CONCLUSION

This chapter investigates the recurring patterns of technological adoption which have appeared across different historical periods including radio to television;

typewriters to computers and analog to digital systems. The basic mechanisms which drive adoption have stayed the same despite today's fast-paced environment because they follow generational acceptance patterns and affordability factors and institutional decision-making processes.

The Tech-Overlap Phase Hypothesis demonstrates its explanatory power by showing how adoption speeds up when the new generation adopting the technology obtains financial control. The observed patterns of technological adoption throughout different technological advancements can be explained by this framework. AI adoption faces distinctive hurdles because its adoption depends heavily on corporate boards composed of older executives who manage enterprise technology spending.

Research shows that decision-making processes of aging executives develop specific mental frameworks for technology evaluation. The observed patterns result from neural changes which boost both strategic control abilities and pattern recognition abilities and increase loss aversion. Business leaders analyze technology choices through a Risk-Reward-Reputation framework which transforms reputation into a measurable business asset.

Several new governance mechanisms and ideas have been suggested to tackle these challenges. The proposed governance mechanisms include External Technology Advisory Boards for unbiased evaluation and Shadow Boards for connecting different cognitive approaches of generations and Vendor Intelligence Networks to reduce information gaps and Strategic Initiative Funds for experimental protection and Workplace AI Integration Frameworks to handle adoption challenges. The Sanyal's BELT Framework (Bias-Ego-Learning-Tangibility) provides a systematic method to evaluate technology readiness for adoption.

The ability to manage innovation with suitable risk management will become the key differentiator between market leaders and followers as AI transforms organizations across the world. The proposed structured governance mechanisms and integration frameworks offer practical solutions to speed up AI adoption while maintaining proper risk oversights.

REFERENCES

Alexiev, A. S., Janssen, M., & den Hertog, P. (2018). The moderating role of tangibility in synchronous innovation in services. *Journal of Product Innovation Management, 35*(5), 682–701. DOI: 10.1111/jpim.12459

Anderson, R. H., Bikson, T. K., Law, S. A., & Mitchell, B. M. (1995). Universal access to e-mail: Feasibility and societal implications. RAND Corporation. https://eric.ed.gov/?id=ED401884

BoardPro. (2024). 7 cognitive biases in board decision-making and how to overcome them. https://www.boardpro.com/blog/cognitive-biases-in-board-decision-making

Brim, O. G.Jr. (1976). Theories of the male mid-life crisis. *The Counseling Psychologist, 6*(1), 2–9. DOI: 10.1177/001100007600600102

Czaja, S. J., Charness, N., Fisk, A. D., Hertzog, C., Nair, S. N., Rogers, W. A., & Sharit, J. (2006). Factors predicting the use of technology: Findings from the Center for Research and Education on Aging and Technology Enhancement (CREATE). *Psychology and Aging, 21*(2), 333–352. DOI: 10.1037/0882-7974.21.2.333 PMID: 16768579

Deloitte Global Boardroom Program. (2022). Digital frontier: A technology deficit in the boardroom. Deloitte Touche Tohmatsu Limited. https://www.deloitte.com/us/en/insights/topics/leadership/digital-transformation-topics-for-corporate-technology-leadership.html

Deloitte Global Boardroom Program. (2024). Governance of AI: A critical imperative for today's boards. Deloitte Touche Tohmatsu Limited. https://www2.deloitte.com/content/dam/insights/articles/glob187394_global-boardroom-program_ai-boardroom/DI_Governance-of-AI_A-critical-imperative-for-todays-boards.pdf

Elia, G., Margherita, A., & Petti, C. (2016). An operational model to develop technology entrepreneurship "EGO-system.". *International Journal on Management of Innovation & Technology, 13*(5), 1640008. Advance online publication. DOI: 10.1142/S0219877016400083

EY. (2024, August 7). EY study: 82% of European financial services investors state that age diversity in the boardroom influences investment decisions. https://www.ey.com/en_ro/newsroom/2023/07/ey-study-82-of-european-financial-services-investors-state-tha

Fallows, D. (2002). Email at work. Pew Internet & American Life Project. Retrieved May 3, 2025, from https://www.pewresearch.org/internet/2002/12/08/email-at-work/

Frank, D.-A., Chrysochou, P., & Mitkidis, P. (2023). The paradox of technology: Negativity bias in consumer adoption of innovative technologies. *Psychology and Marketing*, *40*(3), 554–566. DOI: 10.1002/mar.21740

Funahashi, S. (2017). Prefrontal contribution to decision-making under free-choice conditions. *Frontiers in Neuroscience*, *11*, 431. Advance online publication. DOI: 10.3389/fnins.2017.00431 PMID: 28798662

Gars, J., & Ward, P. S. (2019). Can differences in individual learning explain patterns of technology adoption? Evidence on heterogeneous learning patterns and hybrid rice adoption in Bihar, India. *World Development*, *115*, 178–189. DOI: 10.1016/j.worlddev.2018.11.014 PMID: 30828125

Gelskov, S. V., Madsen, K. H., Ramsøy, T. Z., & Siebner, H. R. (2016). Aberrant neural signatures of decision-making: Pathological gamblers display cortico-striatal hypersensitivity to extreme gambles. *NeuroImage*, *128*, 342–352. DOI: 10.1016/j.neuroimage.2016.01.002 PMID: 26780575

Hammond, J. S., Keeney, R. L., & Raiffa, H. (1998, September). The hidden traps in decision making. Harvard Business Review. https://hbr.org/1998/09/the-hidden-traps-in-decision-making-2

Huang, H., Lu, Y., & Wang, L. (2025). Board age diversity and digital transformation: Evidence from China. *International Review of Economics & Finance*, *100*, 104106. https://www.sciencedirect.com/science/article/pii/S1059056025002692. DOI: 10.1016/j.iref.2025.104106

Hunt Scanlon Media. (2024). Top 5 leadership trends set to define 2025. https://huntscanlon.com/top-5-leadership-trends-set-to-define-2025/

JetBrains. (2023). Python programming – The state of developer ecosystem in 2023. JetBrains s.r.o. https://www.jetbrains.com/lp/devecosystem-2023/python/

Kingston University. (2023). Kingston University renews calls for urgent action to prepare students for workplace increasingly dominated by AI [Press release]. https://www.kingston.ac.uk/news/article/2904/06-dec-2023-kingston-university-renews-calls-for-urgent-action-to-prepare-students-for-workplace-increasingly-dominated-by/

Kooij, D. T. A. M., de Lange, A. H., Jansen, P. G. W., Kanfer, R., & Dikkers, J. S. E. (2011). Age and work-related motives: Results of a meta-analysis. *Journal of Organizational Behavior*, *32*(2), 197–225. DOI: 10.1002/job.665

Lazarsfeld, P. F., & Field, P. B. (1946). *The people look at radio*. University of North Carolina Press.

Loss aversion. (2025, May). In Wikipedia. Retrieved May 2025, from https://en.wikipedia.org/wiki/Loss_aversion

McKinsey & Company. (2024). The state of AI: Global survey. https://www.mckinsey.com/capabilities/quantumblack/our-insights/the-state-of-ai

MIT Technology Review. (2012, May 9). Are smart phones spreading faster than any technology in human history? MIT Technology Review. https://www.technologyreview.com/2012/05/09/186160/are-smart-phones-spreading-faster-than-any-technology-in-human-history/

Moxley, J., Sharit, J., & Czaja, S. J. (2022). The factors influencing older adults' decisions surrounding technology adoption. *JMIR Aging*, *5*(4), e39890. Advance online publication. DOI: 10.2196/39890 PMID: 36416885

Oliva, P., Jack, B. K., Bell, S., Severen, C., & Walker, E. (2020). Technology adoption under uncertainty: Take-up and subsequent investment in Zambia. *The Review of Economics and Statistics*, *102*(3), 617–632. DOI: 10.1162/rest_a_00823

Oyetade, K. E., Zuva, T., & Harmse, A. (2020). Technology adoption in education: A systematic literature review. *Advances in Science, Technology and Engineering Systems Journal*, *5*(6), 108–112. https://www.astesj.com/publications/ASTESJ_050611.pdf. DOI: 10.25046/aj050611

Pew Research Center. (2014). The web at 25 in the U.S. https://www.pewresearch.org/internet/2014/02/27/the-web-at-25-in-the-u-s/

Pew Research Center. (2025). Mobile fact sheet. https://www.pewresearch.org/internet/fact-sheet/mobile/

Pew Research Center. (2025). Internet/broadband fact sheet. https://www.pewresearch.org/internet/fact-sheet/internet-broadband/

Phan, H. P., & Ngu, B. H. (2021). A case for cognitive entrenchment: To achieve optimal best, taking into account the importance of perceived optimal efficiency and cognitive load imposition. *Frontiers in Psychology*, *12*, 662898. Advance online publication. DOI: 10.3389/fpsyg.2021.662898 PMID: 34385950

Pizza, G. (2016). Growth hacking: A new marketing mindset [Master's thesis, LUISS Guido Carli University]. https://tesi.luiss.it/16597/1/655681.pdf

Radicati Group. (2022). *Email market, 2022–2026* [Executive summary]. The Radicati Group, Inc.

Reuters. (2023, February 1). ChatGPT sets record for fastest-growing user base – analyst note. Reuters. https://www.reuters.com/technology/chatgpt-sets-record-fastest-growing-user-base-analyst-note-2023-02-01/

Rogers, E. M. (2003). Diffusion of innovations (5th ed.). Free Press. Mueller, S. M., & Brand, M. (2018). Approximate number processing skills contribute to decision making under objective risk: Interactions with executive functions and objective numeracy. Frontiers in Psychology, 9, Article 1202. https://doi.org/DOI: 10.3389/fpsyg.2018.01202

Sillince, J. A., Macdonald, S., Lefang, B., & Frost, B. (1998). Email adoption, diffusion, use and impact within small firms: A survey of UK companies. *International Journal of Information Management*, *18*(4), 231–242. DOI: 10.1016/S0268-4012(98)00012-7

Spencer Stuart. (2023). 2023 U.S. Spencer Stuart board index (38th ed.). https://www.spencerstuart.com/-/media/2023/september/usbi/2023_us_spencer_stuart_board_index.pdf

TalktoAngel. (2024). Fluid and crystallized intelligence in older adults. https://www.talktoangel.com/blog/fluid-and-crystallized-intelligence-in-older-adults

The Corporate Governance Institute. (2024). A board member's guide to digital transformation. https://www.thecorporategovernanceinstitute.com/insights/guides/a-board-members-guide-to-digital-transformation/

Chapter 3
Navigating the Ethical Minefield of AI:
Bias, Fairness, and Accountability in Automated Decision-Making

Vinay Bamil
http://orcid.org/0009-0004-3784-0828
Independent Researcher, USA

ABSTRACT

Artificial intelligence (AI) systems are increasingly deployed in high-stakes domains, raising urgent concerns about bias, fairness, and accountability. This chapter explores major real-world cases of algorithmic bias including Amazon's hiring tool, Apple Card limits, Robodebt in Australia, the U.K. A-Level grading system, biased healthcare risk predictions, and wrongful arrests from facial recognition. It then surveys evolving legal frameworks such as the EU AI Act, U.S. AI policies, and initiatives from China, Brazil, Canada, and the African Union aimed at promoting ethical AI. The chapter also examines how emerging challenges from generative AI such as hallucinations, deepfakes, and image bias demand stronger oversight. It concludes with recommendations for bias audits, explainability tools, regulatory sandboxes, and impact assessments to advance fairness and accountability in AI.

INTRODUCTION

Artificial intelligence has rapidly transitioned from research labs into high-stakes real-world applications, influencing decisions in hiring, lending, healthcare, education, and criminal justice. While AI systems promise efficiency and the potential for objective decision-making, they have also revealed and sometimes amplified long-

DOI: 10.4018/979-8-3373-1509-6.ch003

standing societal biases. A growing body of evidence shows that even technically sophisticated systems can systematically disadvantage certain groups, particularly along dimensions of race, gender, or socioeconomic status, often without explicit discriminatory intent (Barocas et al., 2019). These concerns have pushed issues of bias, fairness, and accountability to the forefront of AI ethics and governance discourse.

In this chapter, we explore these three interrelated concepts as pillars of responsible AI. Bias in algorithmic systems refers to systematic errors that disproportionately affect specific demographic groups such as women receiving lower credit limits or racial minorities being misclassified in risk assessments. Fairness, while complex and context-dependent, generally aims to prevent unjust discrimination. It spans both statistical notions like group fairness and equal error rates and moral philosophies rooted in equity and justice. Accountability refers to the mechanisms through which creators, users, and systems can be held answerable for their outcomes, often requiring transparency, explainability, and auditability.

Crucially, these concepts do not exist in isolation. Technical trade-offs often pit fairness against predictive accuracy, and accountability can be difficult to enforce in black-box systems with diffuse chains of responsibility. Moreover, interdisciplinary scholarship from feminist ethics to critical race theory has emphasized that fairness is not merely a technical design problem but a social and political challenge, shaped by historical power dynamics and structural inequality. These perspectives are essential to framing not just how systems fail, but who they fail and why.

Public awareness of AI bias has grown in response to high-profile incidents: recruitment algorithms that penalized female candidates, credit scoring models that disadvantaged women, and risk assessment tools that overestimated criminality in Black defendants. These examples illustrate not only technical shortcomings but also the systemic risks of deploying AI in environments already marked by inequality. In response, regulators worldwide are stepping in. The European Union's AI Act (2024) imposes strict oversight on high-risk systems. The U.S. has advanced its Blueprint for an AI Bill of Rights (2022) and a sweeping Executive Order (2023) prioritizing civil rights protections. Countries like China, Brazil, and Kenya, along with regional bodies like the African Union, are also crafting regulatory frameworks that reflect their distinct socio-political contexts and challenges.

At the same time, the technological frontier is shifting rapidly. Generative AI systems, including large language models and image generators, introduce a new layer of ethical complexity from misinformation and deepfakes to synthetic identities and content manipulation. These challenges highlight how questions of fairness and accountability are evolving, and how regulations often lag behind innovation.

In the following sections, we critically examine how bias, fairness, and accountability manifest across technical, institutional, and societal dimensions. We argue that responsible AI development requires not just better models, but a rethinking of ethics, governance, and power guided by both quantitative rigor and moral clarity.

BACKGROUND

Defining Bias, Fairness, and Accountability in AI

In human decision-making, bias refers to systematic prejudice or favoritism for or against certain groups. In the context of artificial intelligence, algorithmic bias describes systematic disparities in treatment or outcomes across groups that are not justifiable by relevant differences. For example, an AI hiring system might disproportionately reject resumes from women, or a facial recognition system might have higher error rates for darker-skinned individuals. Such harms can be allocative (denying resources or opportunities) or representational (perpetuating stereotypes or erasing identities).

Bias in AI systems can originate from multiple sources ranging from historical inequalities embedded in training data to developer assumptions or feedback loops during model deployment. These sources of bias can be meaningfully categorized using a framework developed by the U.S. National Institute of Standards and Technology (NIST, 2022), which identifies three primary types:

- **Systemic biases:** Broader societal, institutional, or historical biases that seep into AI. These stem from unequal structures in data generation and collection for instance, if financial data reflect decades of lending discrimination, an AI model trained on it will inherit those patterns. Systemic bias can also arise from organizational practices (e.g., design teams lacking diversity) or cultural context. For example, an AI-driven curriculum tool might undervalue content relevant to minority cultures if those were absent in its training corpus.
- **Statistical or computational biases:** Biases arising from the *technical aspects* of model development. Examples include unrepresentative training datasets (under-sampling of a population), measurement errors, or biased feature selection. These biases are introduced by how data are collected and processed by algorithms. For instance, if an algorithm's training set has far fewer females than male examples, its performance may be worse for female cases. Similarly, a medical AI trained mostly on adults might not work well for children (a bias by age group).
- **Human and cognitive biases:** Biases introduced by human decisions in the AI lifecycle. Designers and developers might inadvertently embed their own assumptions (e.g., framing a problem in a way that overlooks minority group

needs). Labelers providing ground truth data might carry prejudices. Even end-users can apply AI outputs in biased ways. Human oversight is thus a double-edged sword: necessary for accountability, but also a potential source of bias. For example, content moderators may flag certain dialects as "toxic" more often due to bias, and a model learning from those labels would inherit that bias.

While bias highlights the problem, **fairness** in AI represents the normative goal: the absence of unjustified discrimination. However, fairness has multiple and often conflicting formalizations. Group fairness (e.g., statistical parity or equalized odds) requires outcomes such as approval rates or error rates to be equitable across demographic groups. In contrast, individual fairness posits that similar individuals should be treated similarly regardless of group identity. A long-standing result in algorithmic fairness shows that it is mathematically impossible to simultaneously satisfy all fairness criteria when base rates differ, a reality that forces stakeholders to make context-specific ethical trade-offs.

These trade-offs are not purely mathematical. Scholars in feminist ethics and critical race theory argue that fairness must be situated within historical and cultural power relations. A system may appear formally fair while reinforcing the status quo or overlooking deeper structural harms. Fairness, in this broader sense, is not just a computational property, it is a social judgment grounded in values, worldviews, and context. What counts as "fair" may differ between legal systems, cultural traditions, and community expectations.

Similarly, **accountability** in AI refers to the capacity to attribute responsibility and ensure recourse for harms caused by automated systems. This requires more than just technical transparency or post hoc explainability. It entails institutional structures such as governance processes, documentation practices, impact assessments, and redress mechanisms that enable oversight and contestation. Accountability also intersects with power: when AI systems fail, who can demand change, who must answer, and who has the right to intervene?

Core dimensions of accountability include:

- Transparency (openness about model design, data sources, and goals)
- Explainability (clarity in how outputs are derived)
- Governance (oversight protocols, audit trails, stakeholder reviews)
- Recourse (mechanisms for appeal, redress, or correction)

The need for accountability is application-sensitive. High-stakes contexts like medical triage or criminal sentencing demand greater scrutiny and user protections than lower-stakes domains like entertainment recommendations. As generative AI and multimodal systems proliferate, these concepts must also adapt: who is accountable when a synthetic image spreads misinformation, or when a chatbot provides biased health advice?

Ultimately, bias, fairness, and accountability are deeply intertwined. Their meaning and measurement depend not only on algorithms but on the societies in which those algorithms operate. Responsible AI governance thus requires interdisciplinary insight spanning computer science, law, ethics, and the social sciences to address both visible harms and the invisible assumptions that drive them.

Beyond technical definitions of fairness and accountability, scholars from feminist ethics, critical race theory, and post-colonial studies have called for deeper interrogation of the social and historical contexts in which AI systems operate. Ruha Benjamin (2019) argues in *Race After Technology* that algorithms can act as "New Jim Code" tools reinforcing systemic racism under the guise of objectivity. Virginia Eubanks (2018), in *Automating Inequality*, shows how automated systems disproportionately target and surveil marginalized populations in public services. These critiques emphasize that algorithmic bias is not just a technical error but a reflection of entrenched power structures.

Feminist scholars like Catherine D'Ignazio and Lauren Klein (*Data Feminism*, 2020) challenge whose knowledge is encoded in data and whose is excluded, urging a rethinking of "neutral" AI design through an intersectional lens. Post-colonial perspectives add further complexity: scholars warn of "data colonialism," where AI systems developed in the Global North are deployed in the Global South without regard for local needs, exacerbating existing inequities. Incorporating these critical perspectives provides a fuller understanding of fairness not only as a mathematical goal, but as a socio-political imperative that must consider whose values are represented, who is harmed, and who has the power to resist or reshape AI systems.

How AI Systems Become Biased

Understanding how AI systems become biased is foundational to mitigating these harms. Bias is not confined to a single technical point in the pipeline; it can emerge and compound across all phases of AI system development, deployment, and interaction. These stages often reflect deeper social inequities, technical design choices, and power dynamics.

- **Data Bias:** AI models learn from data; if the training data are biased, the model will likely reproduce those biases. For instance, if a resume screening AI is trained on past hiring decisions at a tech firm that historically favored men, the model may infer that male candidates are inherently more hireable (as happened in Amazon's recruiting tool). Data can be biased due to underrepresentation (certain groups are not sufficiently present in the dataset), historical biases (reflecting past prejudice or unequal access), or measurement bias (systematic errors in how data are collected for different groups). A well-known example is the COMPAS recidivism dataset: arrest rates were used as labels, but arrest rates are influenced by policing practices, thus embedding racial bias in a crime prediction model (Angwin et al., 2016).

- **Algorithm and Model Bias:** Even with unbiased data, the model or algorithm might introduce bias. Some models might simplify complex relationships in a way that overlooks minority patterns. Objective functions (what the model is optimizing) could be misaligned with fairness e.g., optimizing accuracy alone might lead to ignoring minority groups if the model can achieve high accuracy by focusing on the majority. There are also cases where an attribute correlated with a sensitive attribute becomes a "proxy" through which the model inadvertently learns to discriminate. For example, an AI insurance pricing model might not use race explicitly, but ZIP code might act as a proxy for race and lead to higher rates for certain racial groups if not handled carefully. This kind of *indirect discrimination* through proxy variables is a common vector for bias.
- **Interaction Bias:** When AI systems interact with users, bias can emerge or be amplified from those interactions. A notable example is conversational AI or chatbots that learned from user input: Microsoft's *"Tay" chatbot (2016)* began to produce racist and misogynistic language after trolls on Twitter targeted it with hate speech, essentially reflecting the bias in the input it received. Similarly, recommendation systems can create feedback loops: if a platform's algorithm initially shows less content from a minority creator (perhaps due to existing lower popularity), that creator gets less visibility and engagement, reinforcing the algorithm's bias to rarely show their content.
- **Deployment Bias:** The context in which an AI system is deployed can introduce bias if the system is used outside the conditions it was trained for. For example, a facial recognition system trained primarily on adults' faces might perform poorly on children or the elderly, yielding biased error rates by age group. If a policing AI is deployed in neighborhoods that are already over-policed, it may further concentrate police attention there (a deployment choice bias rather than the algorithm itself). One can also consider user interpretation bias: if decision-makers interpret AI outputs with their own bias (e.g., a judge giving more weight to a risk score when the defendant is from a certain background), the outcome can be biased even if the algorithm was not.
- **Human Decision Bias around AI:** This is sometimes called *automation bias, the* propensity for human decision-makers to overly trust or defer to AI

outputs, even when those outputs are flawed. If officers, doctors, or hiring managers trust the AI's recommendation blindly, they may unintentionally propagate its biases. On the flip side, humans might also selectively override AI in biased ways (e.g., a loan officer might override an AI to approve a loan for a friend, but not do so for others injecting favoritism). Therefore, the interface between human and AI decisions can be a source of bias unless carefully managed (with training, checks, and balances).

Understanding the sources of bias is the first step to mitigating them. Techniques have been developed at each stage: for data (e.g., re-sampling or re-weighting to balance datasets, collecting more diverse data), for algorithms (e.g., fairness constraints during training, or using interpretable models), and for deployment (e.g., monitoring outcomes, having humans in the loop with appropriate checks). *Bias audits* (discussed later) are increasingly used to detect bias in AI systems before and after deployment.

Documented Impacts of Bias: Why Fairness and Accountability Matter

Bias in AI isn't just a technical flaw it leads to real-world harms with serious implications for equity, safety, and justice. Without proper accountability mechanisms, these harms may go unnoticed, unchallenged, and uncorrected.

- In ***healthcare***, a biased risk prediction algorithm was found to systematically assign lower risk scores to Black patients compared to white patients with the same actual health conditions. Consequently, fewer Black patients were flagged for high-risk care management, meaning they did not receive additional medical support that their health status warranted. This resulted in Black patients losing out on critical care interventions, effectively embedding racial disparities into healthcare resource allocation. The accountability issue was also stark: neither doctors nor patients were initially aware of this bias until researchers audited the system (Obermeyer et al., 2019).
- In ***financial services***, bias can mean certain groups pay more or get less access. A 2022 study found that algorithmic lending practices were charging minority borrowers higher interest rates on average. One analysis showed African American and Latino borrowers paying about $450 million more in interest per year compared to white borrowers with similar credit profiles. Such disparities exacerbate the racial wealth gap. If accountability is lacking for example, if the lending algorithms are proprietary "black boxes" it

becomes difficult to even identify, let alone correct, these discriminatory outcomes (Bartlett et al., 2022).

- In ***education***, a biased algorithm used in the U.K. in 2020 to predict student exam grades (during COVID-19 when exams were canceled) disproportionately downgraded students from disadvantaged schools, while often inflating scores for students from elite schools. This nearly derailed university admissions for many talented students from underprivileged backgrounds, until public outcry forced a policy reversal. The incident underlined how lack of transparency and oversight (accountability) in an algorithm affecting life-changing decisions can lead to unfairly lost opportunities. It also illustrated a fairness trade-off: the algorithm was trying to combat grade inflation (a form of "fairness" across cohorts) but ended up sacrificing fairness across social groups, a result deemed unacceptable (Walsh, 2020).
- In the realm of ***criminal justice and policing***, unaccountable AI can directly infringe on civil liberties. Predictive policing algorithms that send officers to "high risk" areas have been shown to reinforce over-policing in communities of color essentially automating a feedback loop from historical policing data that reflected racial biases. Even more starkly, erroneous facial recognition matches have led to the wrongful arrest of multiple Black individuals in the U.S. (as discussed in our case studies). Without strong accountability (e.g., requirements to corroborate algorithmic matches with additional evidence), such AI errors can lead to grave violations of innocent people's rights and freedoms. As of 2023, at least six known cases of false arrests due to face recognition have been reported, all involving Black citizens, illustrating both a bias in accuracy and a failure of police oversight (ACLU, 2020; Business Insider, 2023).

These examples demonstrate why fairness and accountability are not merely abstract ideals but necessary features for AI systems deployed in society. Unfair AI decisions can compound structural injustice, and lack of accountability leaves those harmed with little recourse. In response to these concerns, a multidisciplinary field of algorithmic fairness has blossomed within computer science, exploring technical solutions, while legal scholars and ethicists debate appropriate frameworks for governance.

Crucially, policymakers have started to enshrine fairness and accountability principles into law and regulation. Before turning to those regulatory developments, we examine in detail some of the most salient real-world cases where AI systems exhibited bias serving both as cautionary tales and learning opportunities for how to improve AI outcomes.

DOCUMENTED CASES OF ALGORITHMIC DISCRIMINATION

Numerous cases over the past decade have illustrated the pitfalls of biased AI. In this section, we detail six major real-world instances in which AI decision-making led to allegations or findings of discrimination. These cases span different sectors and types of AI systems, showcasing a range of failure modes. Table 1 provides a summary of the key cases, which we then explore in depth.

Table 1. Selected Real-World Cases of AI Bias and Discrimination

Case and Year	Domain	Nature of Bias	Outcome/Impact
Amazon's Hiring AI (2014–2017)	Employment	Gender bias – penalized resumes with "women"	Scrapped in 2017; women applicants disadvantaged (Dastin, 2018).
Apple Card Algorithm (2019)	Finance (Credit)	Gender bias – lower credit limits for women	Regulatory inquiry in NY; Apple/ Goldman reviewed practices (New York State Department of Financial Services, 2021).
Australia "Robodebt" (2015–2019)	Welfare/ Government	Income/poverty bias – flawed averaging hurt vulnerable groups	~433k people sent unlawful debt notices; the scheme was ruled illegal (Wikipedia contributors, 2020).
UK A-Level Grades (2020)	Education	Socioeconomic bias – downgrades hit disadvantaged schools	~40% grades lowered; government reversed algorithmic grades after outcry (Walsh, 2020).
Healthcare Risk Algorithm (2019)	Healthcare	Racial bias – Black patients' risk underestimated	~80 million people affected; Black patients less likely flagged for extra care (Obermeyer et al., 2019).
Facial Recognition Arrests (2020–2023)	Policing	Racial bias – false matches on Black individuals	At least 6 wrongful arrests of Black people; moratoria in some cities (Williams v. City of Detroit, 2020).
Aadhaar Biometric ID (India, 2010–present)	Welfare/ Government	Biometric exclusion – marginalized groups denied access	Lawsuits, public protests, SC rulings on data rights
Huduma Namba (Kenya, 2018–2020)	Digital Identity	Documentation bias – exclusion of ethnic minorities	Rollout halted by court, policy reforms proposed

Note: These cases are drawn from reported incidents and studies. "Nature of bias" indicates the primary unfair outcome observed. All these cases prompted significant public scrutiny and debate on AI ethics, and in many instances led to policy or legal responses.

Notably, only a few of these examples emerged from the Global South highlighting a gap in documented cases, not necessarily in occurrence. Researchers and policymakers must prioritize uncovering and responding to such harms in underrepresented regions, where governance frameworks may be less robust and algorithmic harms harder to detect or contest.

These documented failures serve as more than cautionary tales. They underscore the stakes of embedding fairness and accountability into AI system design, deployment, and oversight. In the following section, we examine how policymakers and global institutions are attempting to prevent such outcomes through emerging regulations and governance initiatives.

Amazon's AI Hiring Tool Discriminating Against Women

One of the earliest widely publicized cases was Amazon's experiment with an AI-powered hiring tool in the mid-2010s. Amazon developers built a resume screening algorithm intended to identify top engineering talent by learning from past hiring decisions. However, because the historical data reflected the tech industry's male dominance, the AI effectively taught itself that male candidates were preferable. As reported in 2018, the model learned to penalize resumes that contained the word "women's" (for example, "women's chess club captain") and to downgrade graduates of two all-female colleges. This happened because the training data from ten years of resumes had far more men than women, and the successful candidates were predominantly male, so the algorithm latched onto any attribute correlated with maleness as a signal of fitness.

Amazon reportedly tried to correct the specific problem by stripping out the word "women" from the algorithm's considerations, but they could not guarantee that subtler patterns of bias wouldn't emerge. In effect, the system was reinforcing the gender imbalance in tech hiring rather than evaluating applicants neutrally. Amazon discontinued the tool by 2017 once these issues became apparent (the project was revealed publicly in 2018). This case is often cited as a cautionary tale: even a tech leader like Amazon inadvertently produced a sexist AI because the system was not properly audited for bias during development. It underscores that training AI on historical human decisions can *embed historical biases* unless careful measures (like bias mitigation and diverse training data) are applied. The incident has since led to greater awareness of the need for fairness testing in HR algorithms and contributed to interest in regulations (for instance, New York City's 2023 bias audit law for hiring tools, which mandates such AI be audited for disparate impact).

Apple Card Credit Limit Controversy

In 2019, a number of Apple Card customers including tech entrepreneur David Heinemeier Hansson and Apple co-founder Steve Wozniak publicly alleged that Apple's credit card, issued by Goldman Sachs, was offering dramatically higher credit limits to men than to their wives, despite similar or better financial credentials. Hansson reported that he was given a credit line 20 times greater than his wife's, even though she had a higher credit score. Wozniak shared a similar story of receiving ten times the credit line of his wife. These anecdotes, which went viral on Twitter, suggested a potential gender bias in the algorithm that determines credit limits.

The New York State Department of Financial Services opened an investigation in late 2019 into whether the Apple Card's underwriting algorithm was illegally discriminatory. Goldman Sachs, the bank behind Apple Card, denied using gender as an input and stated that differences in spousal credit outcomes could be due to other factors. However, the lack of transparency ("black box" algorithm) made it difficult to immediately dispel concerns. As one regulator noted, this was "not just about one algorithm" but about ensuring algorithms do not inadvertently discriminate and treat consumers fairly.

In the aftermath, Goldman Sachs reportedly reviewed credit limits and increased some limits for fairness. The incident highlighted the importance of algorithmic accountability in consumer finance. Under U.S. law (Equal Credit Opportunity Act), lenders must ensure credit decisions are made fairly and cannot discriminate on protected characteristics like sex. The Apple Card controversy demonstrated how even absent intent, an opaque AI model might lead to biased outcomes and that firms using such models must be prepared to explain and justify decisions to regulators. It also showed the power of public scrutiny: a few tweets sparked a regulatory inquiry and broader discussion on financial algorithms and bias. Subsequent to this case, regulators like the U.S. Consumer Financial Protection Bureau have emphasized that lenders are responsible for outcomes of their credit models and should provide adverse action notices that explain decisions, even when AI is involved (to avoid the "black box" excuse as a shield against accountability).

Australia's "Robodebt" Welfare Algorithm

Between 2016 and 2019, the Australian government implemented an automated system for detecting welfare overpayments and recovering debts, officially called the Online Compliance Intervention but commonly known as "Robodebt." This system cross-checked income reported to the tax office against welfare benefits received, using an automated data-matching algorithm to identify discrepancies. If the algorithm found a gap (for instance, inferring a welfare recipient had unreported income), it would automatically issue a debt notice to the individual, alleging overpayment of benefits. The problem was that the algorithm often misidentified debts by using flawed assumptions; notably, it averaged annual income over 26 fortnights, assuming steady work, which for many welfare recipients (who often have sporadic or casual employment) was not accurate. In essence, it frequently counted people as having earned more in a given period than they actually did, incorrectly flagging them as owing money (Royal Commission into the Robodebt Scheme, 2023).

The impact of Robodebt was massive: around 433,000 people were sent debt notices, and the government unlawfully collected AU$751 million from about 381,000 people based on these automated assessments. Many of those targeted were among society's most vulnerable unemployed or low-income individuals reliant on social security. The debts (often amounting to thousands of dollars) caused significant distress; there were reports of increased mental health strain and even suicides partly attributed to the pressure of false debt claims. Crucially, the affected individuals often struggled to challenge the algorithm's determination due to lack of transparency and a presumption of accuracy in the automated system, a stark accountability failure.

Legally, the scheme was eventually deemed unlawful. In 2019, after mounting public criticism and a class-action lawsuit, the Australian government suspended Robodebt and later agreed to a $1.2 billion settlement, admitting that debts had been wrongfully calculated. In 2023, a Royal Commission report called Robodebt "a shameful chapter" in public administration, excoriating the government for running a system that lacked both fairness and accountability. The Commission found that the algorithm's method was not just flawed but illegal, and it castigated officials for not exercising proper oversight or responding to clear evidence of problems.

Robodebt stands as a vivid example of how automated decision-making can perpetrate large-scale injustice if deployed without safeguards. The biases here were in the policy logic (treating fluctuating incomes as steady effectively a bias against those with irregular work, often lower-income individuals), which disproportionately harmed welfare recipients. It illustrated the need for human oversight for social decisions, transparency in algorithmic processes, and avenues for appeal. In the wake of this scandal, there have been calls worldwide to ensure algorithms used by governments undergo rigorous ethical review and that algorithmic impact assessments be conducted before implementation to avoid repeating such failures.

The UK A-Level Grading Algorithm Fiasco

In mid-2020, due to the COVID-19 pandemic, the UK cancelled the usual A-Level exams (national exams taken by 18-year-olds for university entry). To generate grades for students, the exam regulator Ofqual designed an algorithm intended to moderate teacher-predicted grades and prevent grade inflation. This algorithm took into account a school's historical performance data and the student's prior exam results (GCSEs). The outcome was a national uproar: when results were released, approximately 39% of students' grades were downgraded from teacher estimates, and it became clear that the standardization process had disproportionately lowered grades of high-achieving students from lower-ranked schools while often benefiting those in elite schools (Wikipedia contributors, 2024).

Effectively, the algorithm baked in historical outcomes so a talented student at a historically underperforming school was likely downgraded (because their school "normally" didn't produce many top grades), whereas a mediocre student at an elite school might be bumped up (because their school usually did well). This was perceived as fundamentally unfair: it penalized students for their school attended, which often correlates with socio-economic status. Students from disadvantaged backgrounds who had worked hard for top grades suddenly saw themselves marked down, in some cases losing university offers as a result. Meanwhile, some students from privileged backgrounds received grades they hadn't actually earned, solely due to their school's reputation.

The public outcry was immediate and intense. Within days, the government announced that it would scrap the algorithmic results and instead use the original teacher-assessed grades for both A-Levels and GCSEs. The incident underscored several points about algorithmic fairness and accountability:

- A one-size-fits-all statistical approach can overlook individual merit, thus violating individual fairness (students weren't treated on their own performance, but by group characteristics like school history).
- The lack of transparency about the algorithm (students and teachers didn't know exactly how grades would be determined until results came out) prevented scrutiny that might have identified the issues earlier.
- There was no effective appeals mechanism robust enough to handle the sheer volume of perceived errors, indicating poor accountability in design.
- Bias in this case was indirect but potent it mirrored and reinforced existing educational inequalities (schools in poorer areas generally have worse historical results, often due to fewer resources, and the algorithm enshrined that legacy in new outcomes).

This fiasco, beyond its impact on thousands of students' futures, had a silver lining: it greatly raised awareness among the public and policymakers that algorithms making consequential decisions must be vetted for fairness. In the aftermath, the UK and others became more cautious about using algorithms in education without oversight, and the term "algorithm" briefly became a household word in the UK, often with negative connotations regarding fairness. It demonstrated that even well-intentioned uses of AI (to ensure consistency and avoid grade inflation) can backfire if they do not account for biases and if they lack proper accountability and appeal structures (Wikipedia contributors, 2024).

Biometric Exclusion in India's Aadhaar System

India's Aadhaar program, the world's largest biometric identification initiative, was designed to streamline welfare and public service delivery through a centralized ID system based on fingerprints and iris scans. However, it has faced widespread criticism for contributing to the exclusion of marginalized communities, including homeless individuals, transgender persons, and those without formal documentation. In many cases, fingerprint mismatches or database errors resulted in people being denied food rations, pensions, or healthcare.

A report by CyberLeninka highlights that "vulnerable groups such as the homeless or transgender individuals face exclusion due to mismatched biometrics or lack of documentation, perpetuating social inequities." These failures illustrate how a one-size-fits-all system, when coupled with flawed biometric matching, can inadvertently deepen societal hierarchies. The Aadhaar case offers a postcolonial perspective on technological governance showing that fairness concerns in AI are not limited to algorithmic training data but also stem from infrastructural and implementation bias.

Additionally, constitutional challenges and data privacy lawsuits in India highlight a growing demand for accountability. Critics argue that citizens should have recourse when algorithms deny them essential services. This case demonstrates how digital exclusion and lack of redress mechanisms can turn a well-intentioned digital identity system into a source of algorithmic harm.

Kenya's Huduma Namba and Digital Statelessness

Kenya's Huduma Namba initiative aimed to provide a universal digital ID to all citizens. However, civil society groups and legal experts raised concerns that the system would exclude historically marginalized communities, such as the Nubian population, who often lack official documentation due to colonial-era policies. The risk, they argued, was the creation of "digital statelessness," where individuals unable to register would effectively be cut off from government services and legal recognition.

In 2020, Kenya's High Court halted the mandatory rollout of Huduma Namba, citing concerns about data protection and inclusion. The case demonstrates how seemingly neutral digital systems can exacerbate historical injustices unless accompanied by rigorous accountability, community engagement, and inclusive design. It also shows the role of courts and civil society in safeguarding fairness in AI-driven governance systems, especially in contexts with fragile or developing legal protections.

Racial Bias in Healthcare Risk Prediction

In 2019, a landmark study by Obermeyer and colleagues published in *Science* revealed racial bias in a widely used commercial healthcare algorithm that affected millions of patients in the United States. The algorithm in question was used by hospitals and insurers to identify patients who would benefit from "high-risk care management" programs and additional resources for patients with complex health needs (e.g., extra monitoring, nurse check-ins, care coordination). To decide who was high-risk, the algorithm did not directly use health metrics, but instead used a proxy: it predicted which patients would incur the highest healthcare costs in the future, assuming high cost = high need (Obermeyer et al., 2019).

This cost proxy introduced a bias because, due to unequal access to care, Black patients often incur lower healthcare costs than white patients with the same level of illness (if Black patients face barriers to receiving care, their spending stays artificially low). The study found that at a given risk score (cost prediction), Black patients were actually sicker (had more chronic conditions) than white patients. In other words, the algorithm was systematically underestimating the health needs of Black patients. The magnitude of the bias was substantial: the researchers estimated that this bias reduced the number of Black patients identified for extra care by more than half meaning that thousands of Black patients who should have been flagged for care management were not, solely because the algorithm equated lower spending with lower need.

The bias came from the choice of outcome to predict (future cost) a seemingly reasonable design decision that had a racially biased impact because of structural inequalities in healthcare utilization. Notably, the study showed that if the algorithm instead predicted a patient's number of chronic conditions or a direct health indicator, the racial bias dramatically reduced. This is an example where the designers likely did not intend any discrimination, but a lack of deeper analysis into how race intersects with the chosen proxy led to a significant fairness issue.

The company that developed the algorithm (unnamed in the study, but widely reported to be Optum) responded by working to change the algorithm to use direct health data instead of cost. This case had immediate practical impact given the algorithm's extensive use it was affecting care decisions for an estimated 70 million or more people. It has become a seminal example in academic and policy discussions about algorithmic bias. For one, it shows that neutral features (like cost) can mask biases; thus, bias audits need to include checking for differential outcomes by race (or other protected attributes) even if those attributes are not explicitly used. Secondly, it underscores that fairness and efficacy align here: by improving fairness (treating equally sick patients equally), one also improves the quality of care delivered to those who need it. The case prompted healthcare organizations to audit other clinical algorithms e.g., other risk scores, diagnostic tools, even medical devices (like pulse oximeters which were found to be less accurate on darker skin). It has contributed to an ongoing discussion in medicine about "algorithmic accountability" and the need for regulatory oversight for medical AI to ensure it does not widen health disparities (Obermeyer et al., 2019).

Wrongful Arrests Due to Facial Recognition

Perhaps the most visceral examples of AI bias have been the wrongful arrests of innocent people due to misidentifications by facial recognition systems. As of 2023, at least six known cases in the United States have come to light where an individual (all of whom were Black) was arrested based on a false face recognition match. One of the earliest and emblematic cases was that of Robert Williams, who in 2020 was arrested in front of his family in Detroit for a robbery he did not commit, after software matched his driver's license photo to grainy surveillance footage of the suspect. Williams spent over a day in jail. Only during interrogation, when police showed him the surveillance image, did they realize the computer had gotten it wrong. The suspect in the video looked noticeably different and the match was erroneous (ACLU, n.d.-a).

Subsequently, other Black men like Michael Oliver and Nijeer Parks also came forward with similar stories of being wrongfully arrested due to false face recognition matches. In August 2023, a notable case involved Porcha Woodruff, a Black woman who was eight months pregnant, falsely accused of carjacking based on face recognition making her the first Black woman publicly known to suffer this fate. These systems often were not thoroughly vetted for accuracy across demographics; studies (such as one by the U.S. NIST) have shown many facial recognition algorithms have higher false positive rates for Black and Asian faces compared to white faces, and for women compared to men. The technology therefore disproportionately risks misidentifying people of color (Grother et al., 2019).

From an accountability standpoint, these incidents revealed serious gaps. Police investigators often treated the face recognition match as actionable evidence without corroboration. In Detroit, for instance, officers sought warrants based largely on the algorithm's suggestion, and in at least three cases, innocent people were jailed. It raises the question: who is accountable when an AI error leads to someone's liberty being taken away? The opacity of vendor-provided algorithms and lack of mandatory disclosure in police reports meant defendants themselves sometimes only learned that face recognition was involved through dogged legal discovery and advocacy.

The fallout from these cases has been significant. Civil rights groups like the ACLU sued (Robert Williams, for example, sued Detroit Police and reached a settlement) and have called for bans or moratoria on police use of facial recognition. Some cities, such as San Francisco, Oakland, and Boston, responded by banning government use of facial recognition. At the very least, there are growing demands for stricter controls: if used, facial recognition matches should be considered an investigative lead only, not probable cause for arrest, and certainly not without human verification. There are also calls for bias testing and auditing of these algorithms and for public disclosure of their accuracy metrics broken down by race, gender, etc., as part of accountability. These cases have become a powerful narrative in the movement for "AI ethics in policing," underscoring how a lack of fairness (biased accuracy) and accountability (police uncritically trusting a flawed tool with no transparency) can directly harm individuals and erode public trust in criminal justice (ACLU, n.d.-a; ACLU, n.d.-b; Grother et al., 2019).

Having examined what can go wrong and the real harm caused when it does we turn in the next section to how society is responding. There is a growing consensus that we cannot rely on tech companies' voluntary efforts alone to ensure AI is fair and accountable; hence, governments and international bodies are developing regulatory frameworks to govern AI. The following section reviews the state-of-the-art in AI governance as of 2024, mapping out how different jurisdictions aim to tackle bias, fairness, and accountability through law and policy.

LEGAL AND REGULATORY FRAMEWORKS FOR ETHICAL AI

The period 2023–2025 has been pivotal for AI regulation globally. Recognizing both the transformative potential and the risks of AI, lawmakers have been racing to establish guidelines and rules to ensure AI systems are fair, transparent, and accountable. This section surveys major developments in AI governance, focusing on how these frameworks address bias, fairness, and accountability. We cover the European Union's AI Act, initiatives in the United States (which has taken a somewhat different approach), China's regulatory regime including its novel rules on generative AI, and moves in other jurisdictions including Brazil's proposed AI law, Canada's efforts, and the African Union's continental strategy.

European Union: The 2024 EU AI Act

The European Union has led the way in comprehensive AI regulation with its proposed Artificial Intelligence Act (AIA), which, after several years of debate, was formally approved by EU institutions in 2024 (IAPP, 2024). The EU AI Act is a landmark law: it introduces a broad risk-based regulatory approach to AI, meaning that the stringency of requirements for an AI system depends on the level of risk it poses to health, safety, or fundamental rights. Key aspects of the Act relevant to bias and fairness include:

- **Categorization of High-Risk AI:** The Act identifies certain uses of AI as "high-risk" (e.g., AI in employment decisions, education, credit scoring, law enforcement like facial recognition, etc.). These are areas where biased AI could seriously impact someone's life. Providers of high-risk AI systems must comply with strict requirements, one of which is to implement a risk management system and ensure data governance to minimize bias (PwC, 2024). Specifically, Article 10 of the Act requires that training data sets for high-risk AI are sufficiently relevant, representative, and free of errors and statistically unbiased (PwC, 2024).
- **Bias Mitigation Obligations:** The Act explicitly mandates efforts to reduce discriminatory impacts. For example, Article 10(2)(f) obligates providers of high-risk AI systems to identify, test, and mitigate biases that may result in discrimination or harm fundamental rights (PwC, 2024). In practice, this could mean testing an AI system for different outcomes by gender or ethnicity, and adjusting it if any unjustified disparities are found. The Act thus essentially encodes a requirement for algorithmic bias audits in high-risk cases.
- **Transparency and Explainability:** The EU law places emphasis on transparency. Users of a high-risk AI system (like a company deploying it) must be provided with information on how it works, its limitations, and how it was tested. High-risk AI systems also need to be "sufficiently transparent to enable users to interpret the system's output" meaning regulators or notified bodies should be able to understand and trace decisions for enforcement purposes. For certain AI, like remote biometric identification (facial recognition in public), even tighter transparency (such as logging requirements for each use) is mandated.
- **Accountability via Conformity Assessments:** Much like safety certification for products, high-risk AI systems under the Act will have to undergo conformity assessments before being placed on the EU market. This process forces providers to document their system's design, risk assessments, and mitigations including how they addressed bias risks. Some systems might require external third-party audits. This creates an accountability paper trail and potential liability if a system is later found non-compliant.
- **Prohibited Practices:** The Act bans a few AI uses outright, notably those that are deemed to cause unacceptable risk. Among banned practices are AI systems that deploy subliminal techniques to manipulate someone, or systems that allow "social scoring" of citizens by governments (a clear attempt to prevent something like a biased social credit system), and the use of real-

time remote biometric identification (facial recognition) in public for law enforcement except in special cases. These prohibitions underscore ethical red lines, many of which relate indirectly to fairness (social scoring could be seen as inherently unfair and oppressive).

- **Generative AI and Transparency:** In the final negotiations, the EU also added provisions addressing general-purpose AI and generative AI. Providers of large generative models (like GPT-type models) will have to disclose if content is AI-generated in certain contexts (especially for images or video addressing deepfakes). They also must publish summaries of the training data used and design information, to enhance transparency and accountability for potential biases or misuse.

The EU AI Act is expansive and has broad extraterritorial reach (AI systems outside the EU that affect people inside the EU can fall under it). Its approach to fairness is somewhat principles-based; it doesn't dictate exactly what fairness metric to use but demands that risks of bias are addressed in high-risk systems and that fundamental rights (which include non-discrimination as per EU treaties) are respected. The Act's adoption marks the first time a major regulator is imposing legally binding rules on AI fairness. It likely will set a precedent and influence AI governance globally, potentially becoming a template for other nations (similar to how the EU's GDPR influenced global privacy standards).

United States: Emerging Policy and Patchwork Regulation

In contrast to the EU's comprehensive statutory approach, the United States (as of 2025) has taken a more multi-pronged and less centralized approach to AI governance. No single federal law exclusively regulating AI systems has been enacted. Instead, the U.S. relies on a combination of existing laws (like anti-discrimination laws), executive branch guidance, agency enforcement, and state-level initiatives. Key developments include:

- **Blueprint for an AI Bill of Rights (2022):** The White House Office of Science and Technology Policy released a policy document outlining five core principles for AI systems: (1) Safe and Effective Systems, (2) Algorithmic Discrimination Protections, (3) Data Privacy, (4) Notice and Explanation, and (5) Human Alternatives, Consideration, and Fallback. Though not binding, this Blueprint serves as an ethical framework for federal agencies and industry, clearly stating that citizens should be free from abusive AI-driven practices and that AI decisions should be fair and not discriminatory. It calls for building equity assessments into AI development and emphasizes the importance of explanation and appeal mechanisms (The White House, 2022).

- **Executive Order 14091 (Feb 2023):** Titled "Further Advancing Racial Equity and Support for Underserved Communities Through the Federal Government," this order directed federal agencies to ensure their own use of automated systems advances equity. It required agencies to refrain from using AI that exacerbates discrimination and to implement best practices to mitigate algorithmic biases (The White House, 2023a).
- **Executive Order on Safe, Secure, and Trustworthy AI (Oct 2023):** In a sweeping EO, the Biden Administration addressed AI from national security to civil rights. It instructed agencies like the DOJ and Consumer Financial Protection Bureau to redouble efforts to identify and prevent algorithmic discrimination in areas like housing, credit, employment, and justice. It also pushed for development of best practices on the use of AI in policing and mandated federal agencies to consider fairness when deploying AI (The White House, 2023b).
- **Federal Agency Actions:** U.S. regulators are leveraging existing laws to tackle AI bias. The EEOC clarified in 2022 that AI hiring tools must comply with the Americans with Disabilities Act and Title VII, emphasizing liability for disparate impact (EEOC, 2022). The CFPB stated that lenders using AI must still provide adverse action notices under the Equal Credit Opportunity Act (CFPB, 2022). The FTC warned companies that claiming their AI is "neutral" while it causes harm could be an unfair or deceptive practice under Section 5 of the FTC Act (FTC, 2021).
- **NIST AI Risk Management Framework (2023):** Though not a regulation, the NIST AI RMF 1.0 is a voluntary framework widely referenced for responsible AI development. It identifies harmful bias as a core risk and recommends processes to mitigate it throughout the AI lifecycle. It emphasizes transparency, documentation, and governance structures to ensure accountability (NIST, 2023).
- **State and Local Laws:** Some U.S. jurisdictions have enacted fairness-focused AI laws. New York City's Local Law 144 requires that automated hiring tools be independently audited for bias and that the results be disclosed to applicants (NYC Council, 2021). Illinois passed a law regulating video interview analysis, requiring informed consent and demographic reporting. States like California and Colorado are also considering broader AI fairness legislation (Illinois General Assembly, 2020).
- **Proposed Federal Legislation:** Several federal bills remain under consideration. The Algorithmic Accountability Act (2022 reintroduction) would mandate bias impact assessments for automated decision systems (Congress.

gov, 2022). The ADPPA, though stalled, proposed requirements for transparency and fairness. Discussions around watermarking AI-generated content and regulating high-risk AI are ongoing.

One unique element in the U.S. has been the role of civil society and litigation. Groups like the ACLU have filed lawsuits and complaints that have directly led to policy changes (ACLU, n.d.-a; ACLU, n.d.-b). Additionally, procurement policies at city and federal levels increasingly require fairness benchmarks for AI, incentivizing ethical design through market pressure.

China: Regulating AI with an Emphasis on Control and Content

China has taken a strong regulatory stance on AI, albeit with different motivations and principles compared to Western approaches. The Chinese government's approach emphasizes social stability, content control, and alignment with state objectives, alongside some provisions about fairness and privacy. Throughout 2022–2023, China rolled out pioneering rules targeting recommendation algorithms, deepfakes, and generative AI:

- **Algorithmic Recommendation Rules (March 2022):** China was among the first to regulate recommendation engines. These rules, issued by the Cyberspace Administration of China (CAC), require recommendation algorithms to promote "positive energy," allow users to opt out of profiling, and ban algorithms that encourage addiction or spread harmful information. Providers must also file details of their algorithms with regulators for oversight (Cyberspace Administration of China, 2022).
- **Deepfake and Synthesis Media Regulation (January 2023):** To combat impersonation and misinformation, China implemented rules on "deep synthesis" technology. Providers must obtain consent to use someone's likeness in AI-generated content, and such content must be clearly labeled as AI-generated (Cyberspace Administration of China, 2023a).
- **Interim Measures for Generative AI Services (July 2023):** This regulation targets generative AI made available to the public. Models must not generate content that subverts state power or contains prohibited material. Importantly, it includes a non-discrimination clause AI must not generate biased content on the basis of race, gender, or ethnicity demonstrating regulatory concern with fairness (Cyberspace Administration of China, 2023b).
- **Fairness in Chinese AI Ethics Guidelines:** Chinese authorities have previously issued ethical guidelines like the Beijing AI Principles (2019) and Ministry of Science and Technology's 2022 AI ethics framework. These

stress high-level principles like fairness and justice, but the aforementioned regulations are enforceable and backed by penalties for non-compliance (Cyberspace Administration of China, 2023b).

China's regulatory model blends ethical goals with strong state control. For global firms operating in China, compliance requires adapting AI systems to these rules, often by incorporating additional filtering and censorship. While the non-discrimination clause aligns with global fairness norms, its enforcement may focus more on maintaining social harmony than empirical bias auditing.

Other Global Developments: Brazil, Canada, and the African Union

Beyond the big three (EU, US, China), many other jurisdictions have been actively developing AI governance initiatives. Here we highlight a few:

Brazil. In December 2024, the Brazilian Senate passed Bill No. 2338/2023, a comprehensive AI law emphasizing a human-rights and risk-based approach. The bill requires fairness, bias audits, and accountability mechanisms such as algorithmic impact assessments (Brazilian Senate, 2024). Brazil's data protection authority (ANPD) and public pressure, especially after biased facial recognition cases, have driven urgency around regulation.

Canada. Canada introduced the Artificial Intelligence and Data Act (AIDA) in 2022 as part of Bill C-27, aiming to govern high-impact AI and prevent bias. Although AIDA stalled in Parliament, the Canadian government already mandates Algorithmic Impact Assessments (AIAs) for federal use of AI, including transparency and bias evaluation (Innovation, Science and Economic Development Canada, 2022).

African Union. In 2024, the AU adopted the Continental AI Strategy promoting "ethical, responsible, and equitable AI." While non-binding, it encourages member states to build policies ensuring fairness, transparency, and human rights. It also emphasizes capacity building to help implement AI governance (African Union, 2024).

The evolving regulatory landscape may converge over time as countries learn from each other's experiences; for instance, companies operating globally might adopt the strictest standard across the board for simplicity (a "highest common denominator" approach). International cooperation is nascent in the U.S.-EU Trade and Technology Council has a working group on AI, and UNESCO's Recommendation on AI Ethics (endorsed by many countries) sets out values including non-discrimination and fairness.

EMERGING ISSUES: GENERATIVE AI AND ETHICAL CHALLENGES

The latest wave of AI notably large language models (LLMs) like GPT-4, and image generators like Stable Diffusion or DALL·E 2 has brought AI capabilities to new audiences and applications. These generative AI systems can produce text, images, audio, and video that often appear strikingly human-like or realistic. With their rise (especially in 2023), a new set of ethical concerns has become prominent, including AI "hallucinations" (confidently making up false information), the mass generation of misleading media (deepfakes and propaganda), and biases in generated content reflecting stereotypes present in training data. These issues have significant implications for fairness and accountability.

Hallucinations and Misinformation

LLMs like ChatGPT are prone to hallucinating meaning they may produce statements that sound plausible and authoritative but are factually incorrect or entirely fabricated (Ji et al., 2023). While an incorrect answer about a trivial question is one thing, hallucinations become an ethical issue when they involve real people or potentially harmful content. For example:

- **Defamation by AI:** In one instance, ChatGPT falsely claimed that a prominent law professor had been accused of sexual harassment, even citing a nonexistent Washington Post article as evidence (Romano, 2023). The professor, Jonathan Turley, had never been accused of such conduct. This hallucinated accusation could have seriously damaged his reputation if taken at face value. Turley described it as "chilling" and was at a loss on how to correct an allegation coming from an AI rather than a human source. Such cases raise accountability questions: if an AI libels someone, who is responsible? The user who prompted it? The developer of the AI? In 2023, at least two defamation lawsuits were filed in the U.S. against OpenAI for false statements ChatGPT generated about individuals (one case involved a radio host falsely implicated in a legal scandal by the AI). These are novel legal terrain. OpenAI and others include warnings about possible inaccuracies, but when an AI presents a lie with the specificity of a fake citation, it can be very convincing, as seen in that case.
- **Fake Citations and Legal Consequences:** A related incident saw attorneys use ChatGPT for legal research, and it fabricated court case citations that did not exist (Neumeister, 2023). The attorneys unknowingly submitted these in a brief, and the judge sanctioned them when the deception came to light (they

had not independently verified the cases, which was a violation of their duty). Here the AI's hallucination led to real-world penalties for users who trusted it too much. It underscores that until these systems are more reliable, users must exercise caution and verify AI outputs but not all users may be aware of this need.

- **Disinformation at Scale:** Hallucinations also generalize to concerns about misinformation. An AI could generate a very large volume of false but coherent narratives (e.g., a set of fake news articles, complete with fabricated statistics and quotes) that could mislead the public. Already in 2024, we saw some attempts at this: reports emerged of "news websites" with entire articles written by AI that contained inaccuracies or partisan spin (Kobie, 2024). If done maliciously, generative AI could be used by bad actors to flood social media with realistic-sounding misinformation, making it harder for people to discern truth.

To mitigate hallucinations, developers are trying several approaches: improving training data quality, fine-tuning models with a focus on factual accuracy, and integrating retrieval (where the model fetches real documents to base its answer on, rather than relying on its sometimes flawed memory) (Mialon et al., 2023). OpenAI's GPT-4, for instance, is generally more factual than GPT-3.5, but still not immune (OpenAI, 2023).

From an accountability standpoint, some have proposed that AI developers be held to certain standards for truthfulness if their models are used in contexts likely to produce actionable information (Weidinger et al., 2022). Alternatively, AI outputs might need to be flagged or verified by external fact-checkers in critical uses. Interestingly, OpenAI has a usage policy that forbids using their models to create disinformation campaigns (OpenAI, 2023). But detection is challenging: a user could ask for "an essay on why vaccine conspiracies are true" and the AI might produce it if not correctly constrained.

Companies are deploying filters to refuse obviously harmful requests, but they are imperfect. In terms of fairness, if certain communities are more exposed to or prone to trust AI-generated content, they might be disproportionately affected by hallucinations or disinformation. For example, non-native language speakers might trust translations or summaries from AI that could contain errors, affecting their access to accurate information.

Deepfakes and Synthetic Media

Generative AI can produce highly realistic images, videos, or audio that depict events or speech that never occurred, commonly known as *deepfakes*. The ethical concerns here revolve around deception, manipulation, and harassment:

- **Political Deepfakes:** In elections and politics, deepfakes could be used to create a fake video of a candidate saying or doing something scandalous. In early 2024, a political operative in the U.S. created a fake audio of President Biden telling voters to "save your vote for the general election" (Bond, 2024). This deepfake call was presented to voters in New Hampshire and led to criminal charges.
- **Non-consensual Pornography:** Deepfake technology has been used to create fake pornographic content of individuals without consent, often targeting women. The harm is significant victims suffer reputational and psychological damage.
- **Erosion of Trust (the "Liar's Dividend"):** As deepfakes proliferate, real content may be dismissed as fake. This has already happened in 2020, a real video of a foreign politician was dismissed as a deepfake (Chesney & Citron, 2019). This leads to an accountability crisis: proving authenticity becomes as important as detecting fakes.

To combat deepfakes, researchers are developing detection algorithms, but it's a cat-and-mouse game (Gu et al., 2023). Platforms like YouTube and Twitter have policies to remove or label harmful deepfakes. One technical solution is watermarking AI-generated content (Lyu, 2023). Another is cryptographic signing of real media at capture time initiatives like the Content Authenticity Initiative explore this (Adobe, 2023).

Recent developments: OpenAI, Google, and others prohibit malicious use in their policies (OpenAI, 2023). In 2023, the U.S. saw its first deepfake-related criminal conviction: a mother created fake images to harass her daughter's cheerleading rivals (BBC, 2023).

Yet, generative AI can also promote fairness when used responsibly for instance, to generate more inclusive synthetic data. Ultimately, generative AI presents both opportunities and ethical risks. Regulators like the EU, China, and U.S. lawmakers are actively responding with new frameworks.

SOLUTIONS AND RECOMMENDATIONS FOR AI FAIRNESS AND ACCOUNTABILITY

Ensuring fairness and accountability in AI is a multi-faceted endeavor, requiring technical, organizational, and regulatory interventions. In this section, we outline practical recommendations and tools that stakeholders from AI developers to deploying organizations to regulators can implement. These solutions aim to minimize bias in AI systems, increase transparency, and establish accountability for AI-driven outcomes. We divide the recommendations into several categories: algorithm and data interventions (to make the AI itself fairer and more explainable), process and governance measures (like audits and impact assessments), and regulatory and oversight mechanisms (policy approaches to enforce or incentivize ethical AI). Table 2 summarizes key strategies alongside their purpose.

Table 2. Strategies for Promoting AI Fairness and Accountability

Strategy/Tool	Description	Purpose	Examples/Use Cases
Bias Audits (Algorithmic Audits)	Independent or internal evaluations of AI systems to detect bias or disparate outcomes.	Identify and quantify unfair biases before/after deployment; provide evidence for accountability.	NYC law mandates bias audits for AI hiring tools; Facebook's 2020 civil rights audit examined bias in algorithms used on the platform (Facebook, 2020).
Explainability & Interpretability	Techniques to make AI decisions understandable (e.g., feature importance, interpretable models, or post-hoc explanations).	Enable humans to trust and verify AI decisions; facilitate appeals and debugging of AI outputs.	Credit scoring models providing reason codes to applicants; LIME/SHAP tools explaining ML predictions in healthcare diagnostics (Ribeiro et al., 2016; Lundberg & Lee, 2017).
Transparency Documentation	Detailed documentation of model development, data, and intended use (e.g., model cards for models, datasheets for datasets).	Improve accountability by informing users and regulators of AI's purpose, performance (including limitations), and design.	Google's Model Card for a vision model showing accuracy by demographic; IBM's FactSheets documenting an AI service's testing results (Arnold et al., 2019).
Fairness Constraints in Model Training	Adjusting model algorithms to satisfy fairness criteria (parity in outcomes or errors) or using re-weighting/pre-processing to reduce biases.	Technically reduce disparate impact while maintaining predictive performance as much as possible.	Training a hiring model with a constraint that selection rates for men and women are within 5%; pre-processing credit data to remove zip code influence (Feldman et al., 2015).

continued on following page

Table 2. Continued

Strategy/Tool	Description	Purpose	Examples/Use Cases
Diverse & Representative Data	Ensuring training and test datasets include diverse groups and are checked for representation issues; augmenting data if needed.	Prevent bias from imbalanced or unrepresentative data; allow AI to perform well across groups.	Adding more samples of darker skin tones in face recognition training (Buolamwini & Gebru, 2018); including schools from various socioeconomic backgrounds in an exam algorithm's dataset.
Human-in-the-Loop Decision Making	Keeping human oversight in critical AI decisions and establishing clear review processes for AI outputs.	Catch and correct AI errors or biases; provide a check against over-reliance on automation.	A hiring manager reviews all AI-rejected job candidates from underrepresented groups; a doctor double-checks an AI's diagnosis suggestion (Amann et al., 2020).
Algorithmic Impact Assessments (AIAs)	A process to evaluate and document the potential impacts, biases, and risks of an AI system before deployment (and periodically after).	Proactively identify ethical issues; engage stakeholders; serve as a record of due diligence for accountability.	Canada requires AIAs for federal AI systems (Treasury Board of Canada Secretariat, 2020); some cities require public impact assessments for surveillance tech.
Regulatory Sandboxes	Controlled environments for testing AI systems with regulatory oversight and safeguards before full deployment.	Allow innovation while monitoring for biases and unintended effects; adjust systems with feedback before broad release.	The EU AI Act encourages sandboxes for high-risk systems (PwC, 2021).
Governance and Ethics Boards	Organizational measures like internal AI ethics committees, bias "red teams," and training programs on AI ethics for staff.	Embed ethical consideration into AI development; assign responsibility and empower employees to flag issues.	A company's AI ethics board reviews high-risk AI projects; Twitter's 2021 "bias bounty" invited outsiders to find bias in its image cropping algorithm (Twitter, 2021).
Legal and Regulatory Compliance	Adhering to emerging laws (e.g., EU AI Act, U.S. anti-discrimination laws, China's AI rules) and industry standards (e.g., ISO, IEEE guidelines).	Provide external accountability and incentives (or mandates) for fairness; set baseline practices across industry.	Conducting required bias testing and documentation to comply with EU AI Act; following EEOC guidance for AI hiring (EEOC, 2023; Skadden, 2023).

Now, we discuss these in detail, with an emphasis on practical implementation and examples:

1. Bias Audits and Ongoing Monitoring

Just as financial firms undergo regular audits of their books, AI systems, especially those used in high-impact decisions should be subject to algorithmic bias audits. This involves evaluating the system's outputs across different demographic groups to check for disparate outcomes (Raji & Buolamwini, 2019). For example, before deploying a hiring algorithm, a company might test it on a diverse set of résumés to see if equally qualified female candidates are ranked lower than male candidates, or if candidates from certain colleges (perhaps correlating with socioeconomic status) are disadvantaged without business justification.

Such audits should be both pre-deployment (to catch biases in design) and post-deployment (to catch biases that may emerge in operation due to feedback loops or changing data) (Barocas et al., 2019). The methodology often includes statistical analysis (comparing error rates, selection rates, or output distributions between groups), and sometimes qualitative analysis (examining model features or even outputs for bias signs).

A leading example: New York City's Local Law 144 (2023) requires that any automated employment decision tool used in the city be audited annually by an independent auditor, and a summary of results (e.g., impact ratios by race and gender) must be publicly posted (New York City Council, 2021). This law has prompted vendors to commission formal bias audits and, in some cases, improve their models to pass muster. It also has an effect of informing candidates (via the posted summary) about the tool's fairness.

Another example: Facebook (Meta) underwent a civil rights audit in 2018–2020 where auditors examined its algorithms (like ad delivery algorithms) for bias. They found, for instance, that ads for jobs in the lumber industry were shown to a 90% male audience due to algorithmic targeting, raising concerns of disparate impact. Facebook agreed to change some practices as a result (Facebook, 2020). This shows how audits can highlight unintended bias and lead to concrete changes.

In ongoing monitoring, organizations can implement dashboards or periodic reports tracking key fairness metrics. For instance, a bank might monitor loan approval rates and loan performance by race and gender each quarter to ensure their AI model isn't drifting to be more favorable to one group unfairly (Holstein et al., 2019).

Bias audits also create accountability by producing documentation that can be shared with stakeholders (like regulators, customers, or affected groups). Publishing audit results (at least in summary) builds trust and holds organizations to account in the court of public opinion as well (Raji et al., 2020).

2. Improve Explainability and Transparency

Explainability and transparency are crucial for accountability. They ensure that AI decisions can be understood and, if necessary, contested or debugged.

For explainability: Many AI systems, especially those using complex models (deep learning, ensemble models, etc.), act as black boxes. Techniques to extract explanations include:

- Using interpretable models where possible. Some modern ML frameworks can produce a list of key features that influenced a specific decision (Doshi-Velez & Kim, 2017).
- Local explanation methods like LIME (Ribeiro et al., 2016) or SHAP (Lundberg & Lee, 2017) can show the contribution of each input feature to a particular output.
- Counterfactual explanations can tell a user what changes would have altered the outcome (Wachter et al., 2017).

Transparency goes beyond just explaining decisions. It's about the *process* and *limitations* of the AI system:

- Providing documentation like model cards (Mitchell et al., 2019) and datasheets for datasets (Gebru et al., 2021).
- User notifications: The EU's GDPR and U.S. laws like Illinois' AI Video Interview Act require informing users when AI is used (Goodman & Flaxman, 2017).
- Legal requirements: In credit, adverse action notices must include explanations, and under GDPR's right to explanation, companies often provide algorithmic decision logic summaries (Selbst & Powles, 2017).

Internally, explainability also helps developers debug models and flag potential bias (Slack et al., 2020). Contestability mechanisms, such as human review options, enhance fairness and accountability by allowing decisions to be challenged (Binns, 2018).

3. Fairness Through Model Design and Data

At the technical level, a rich area of research and practice is designing models and datasets to be as fair as possible:

- **Pre-processing (Data):** This involves modifying the training data to reduce biases. Techniques include:
  - Re-sampling: e.g., if a dataset has 80% male and 20% female instances, up-sample the female instances (or down-sample males) to balance it, so the model doesn't learn a strong gender prior.
  - Re-labeling: if labels themselves reflect bias (like arrests vs. actual crime rates), sometimes one can relabel or weight samples to correct known distortions (though this is complex).

- Removing sensitive attributes or proxies: simply excluding features like race or gender is a naive approach, because the model can still infer from proxies. However, one can transform data into a representation where the sensitive attribute is not easily inferable (using techniques from adversarial learning to remove sensitive info from features). An example is learning "fair representations" that capture needed info for prediction but are decorrelated from sensitive attributes.
- Synthetic data augmentation: e.g., generating additional data for under-represented groups using simulation or data augmentation (perhaps using generative AI itself) to balance the dataset.

- **In-processing (Algorithmic):** Alter the model training procedure to incorporate fairness goals. This might mean adding a term to the loss function that penalizes differences in outcomes between groups, or using constraint optimization (maximize accuracy subject to a fairness constraint). For example, one could require equal false negative rates across races in a disease prediction model by adding a Lagrange multiplier for the difference in false negative rates. There is academic work like "reducing bias using adversarial training" where a subnetwork tries to predict the protected attribute from the model's outputs and the main model is trained to make that difficult, thus encouraging removal of bias.

One concrete application: A credit scoring company might discover their model has a 5% higher denial rate for equally qualified minority applicants. They could introduce a constraint in training that forces the model to achieve parity in positive prediction rates for a given credit score threshold across groups (essentially calibrating it better). Some open-source libraries like IBM's AI Fairness 360 provide implementations of such in-processing methods that practitioners can use.

- **Post-processing (Adjusting outputs):** After a model is trained, one can adjust its decisions to improve fairness. For instance, one could set different thresholds for different groups to equalize acceptance rates or error rates. The classic example is "threshold adjustment" in classification: if group A has a higher base rate and thus more positive predictions, one might use a slightly higher cutoff for them or a lower cutoff for group B to balance outcomes. This is controversial because it explicitly uses group information at decision time, which in some contexts is legally or politically sensitive (the specter of "quotas" or affirmative action arises). However, in some domains it's considered (e.g., some university admissions algorithms after the UK fiasco considered adjusting by school type to not disadvantage those from historically low-performing schools).

Another post-processing approach is *output randomization*: e.g., if two groups have disparate selection rates, randomly select some additional candidates from the disadvantaged group (among those near the cutoff) to balance. This was proposed in some hiring contexts to introduce a measure of fairness without strictly enforced quotas.

Important: In regulated domains, any such intervention must comply with law. For example, in the U.S. context, explicit use of protected class in decision could violate the Civil Rights Act unless it's a narrow remedial measure. Thus, many companies prefer to adjust the model inputs or training process so that the final model *naturally* doesn't discriminate, rather than explicitly tweaking outputs by group.

LIMITATIONS AND FUTURE DIRECTIONS

While this chapter provides a broad examination of bias, fairness, and accountability in AI, it is essential to acknowledge the limitations that remain both in current research and in practical implementation. These challenges must be explicitly addressed to guide the next phase of responsible AI development.

Limitations of Fairness Definitions

Despite extensive scholarly effort, there is no universally accepted definition of fairness in AI. The literature includes over twenty formal definitions from statistical parity and equalized odds to individual fairness and calibration each with different ethical assumptions and trade-offs. It is mathematically impossible to satisfy all fairness criteria simultaneously when base rates differ between groups, meaning that any selected fairness metric implicitly reflects value judgments (Friedler et al., 2021). This limitation underscores the normative nature of fairness: what is deemed fair in one context (e.g., equal opportunity) may be inappropriate in another (e.g., compensatory fairness). As a result, claims of an AI system being "fair" must be qualified by which definition is used, under what assumptions, and for whose benefit.

Difficulty of Measuring Bias at Scale

Detecting and quantifying bias in large-scale AI systems remains an ongoing challenge. First, measurement requires access to reliable and representative demographic data, something often constrained by privacy laws, data sparsity, or organizational silos. Second, modern AI models operate on high-dimensional inputs with millions or billions of parameters, making it difficult to isolate biased behaviors or systematically audit outcomes across all subgroups. Bias may manifest only under rare conditions or in intersectional groups (e.g., Black women over 60), requiring substantial computational and domain expertise to detect. Furthermore, bias is not static: as AI systems are updated, retrained, or deployed in new environments, their fairness profiles can shift a phenomenon known as **fairness drift**. This highlights the insufficiency of one-time audits and the need for continuous monitoring throughout the AI lifecycle.

Challenges with Black-Box Models and Generative AI

Many AI systems are "black boxes" highly complex models whose internal logic is opaque even to their creators. In such systems, tracing the source of a biased output or explaining a decision becomes extraordinarily difficult, hindering accountability. For instance, if a neural network with hundreds of layers generates discriminatory outputs, current explainability tools may only offer partial or misleading insights. These challenges are magnified in generative AI, where models like large language models (LLMs) or image generators produce open-ended outputs. Evaluating fairness in such systems is context-dependent and lacks standardized benchmarks. Studies have shown that generative models may produce biased or offensive outputs depending on prompt phrasing, reinforcing the view that fairness in generative systems cannot be easily formalized or guaranteed (Bender et al., 2021). Post-hoc interpretability methods often fall short and can themselves introduce bias, raising concerns about the reliability of current accountability mechanisms.

Future Research and Innovation Avenues

Despite these limitations, several promising directions are emerging to advance fairness and accountability in AI:

- **Fairness in Multimodal and Complex AI Systems**: As AI increasingly operates on combinations of text, image, video, and audio (e.g., in healthcare diagnostics or surveillance), future work must explore how biases manifest across modalities and design new fairness criteria accordingly.
- **Global Standards and Auditing Protocols**: There is growing consensus on the need for international standards to guide algorithmic audits. Collaboration between researchers, governments, and standardization bodies (e.g., IEEE, ISO/IEC) can lead to third-party certification systems and fairness ratings, similar to existing models in finance or cybersecurity.
- **Interpretable and Explainable AI**: A vital frontier lies in improving model transparency. New generations of explainable AI tools ("XAI 2.0") could be designed to test bias explicitly for example, by simulating counterfactual sce-

narios (e.g., how would the model behave if the input's race or gender were different?) to highlight discriminatory behavior.

- **Contextual and Participatory Approaches**: Fairness should not be treated as an abstract technical metric alone. Instead, participatory design approaches involving impacted communities in development, auditing, and oversight can ensure context-appropriate definitions of fairness and bolster legitimacy.
- **Continuous Monitoring and Lifelong Learning**: As AI systems adapt and evolve, bias monitoring must evolve with them. Real-time fairness dashboards, alert systems, and feedback loops can ensure that AI models remain accountable throughout their operational lifespan, not just at launch.

CONCLUSION

Bias, fairness, and accountability in AI decision-making have moved from peripheral concerns to central imperatives in the design, deployment, and governance of intelligent systems. This chapter examined how biases in AI can produce discriminatory outcomes from hiring tools that disfavor women to welfare algorithms that penalize the poor and how accountability mechanisms often fall short when opaque systems go unchallenged. These failures are not technological inevitabilities but often reflect correctable flaws in data, model design, policy context, or institutional oversight.

Encouragingly, the past few years have seen significant progress. Revelations of algorithmic harm have sparked reforms, such as the redesign of healthcare risk tools and the rollback of facial recognition use in policing. Governments around the world from the European Union and the United States to Brazil, Kenya, and India are developing governance frameworks that embed principles of fairness, explainability, and human rights. These regulatory efforts signal a paradigm shift: from unregulated "black box" systems to an expectation that AI must be auditable, contestable, and justifiable.

Yet as AI evolves, so too do the challenges. Generative AI, multimodal systems, and globally deployed models introduce new frontiers of risk from hallucinated content to exclusionary biometric ID systems. Addressing these challenges requires a commitment to fairness not as a fixed formula but as a context-sensitive and value-laden goal one that varies across legal systems, cultural settings, and affected communities. It also demands interdisciplinary insight, drawing from feminist ethics, critical race theory, and postcolonial studies to surface the deeper structural inequities that AI systems may reflect or reinforce.

Ultimately, ensuring fairness and accountability in AI is not a one-time fix but an ongoing responsibility. It means asking hard questions at each stage: Who is represented and who is left out? What values guide the model's optimization? Who bears the consequences of its errors? And who has the power to intervene? Embedding processes like bias audits, participatory design, impact assessments, and continuous monitoring can help institutionalize these reflections and build systems that uphold democratic values.

Looking forward, the goal is not merely to make AI less harmful but to make it affirmatively just: to ensure it enhances equity, expands opportunity, and amplifies marginalized voices. That will require ongoing research, cross-sector collaboration, global coordination, and above all, ethical resolve. As AI becomes increasingly enmeshed in the infrastructure of daily life, the legitimacy and sustainability of algorithmic decision-making will hinge on our ability to constrain power with accountability and to match innovation with integrity.

REFERENCES

ACLU. (2020). Williams v. City of Detroit | Face Recognition False Arrest. American Civil Liberties Union. https://www.aclu.org/cases/williams-v-city-of-detroit-face-recognition-false-arrest

Adobe. (2023). Content Authenticity Initiative. https://contentauthenticity.org

African Union. (2024). Continental artificial intelligence strategy. https://au.int/en/documents/20240809/continental-artificial-intelligence-strategy

Amann, J., Blasimme, A., Vayena, E., Frey, D., & Madai, V. I. (2020). Explainability for artificial intelligence in healthcare: A multidisciplinary perspective. *BMC Medical Informatics and Decision Making*, *20*(1), 310. DOI: 10.1186/s12911-020-01332-6 PMID: 33256715

Angwin, J., Larson, J., Mattu, S., & Kirchner, L. (2016). Machine bias: There's software used across the country to predict future criminals. And it's biased against blacks. ProPublica. https://www.propublica.org/article/machine-bias-risk-assessments-in-criminal-sentencing

Arnold, M., Bellamy, R. K. E., Hind, M., Houde, S., Mehta, S., Mojsilović, A., & Varshney, K. R. (2019). FactSheets: Increasing trust in AI services through supplier's declarations of conformity. IBM Research. https://arxiv.org/abs/1808.07261

Barocas, S., Hardt, M., & Narayanan, A. (2019). Fairness and machine learning. https://fairmlbook.org/

Bartlett, R., Morse, A., Stanton, R., & Wallace, N. (2022). Escaping the Subprime Trap in Algorithmic Lending. arXiv. https://arxiv.org/abs/2502.17816

BBC. (2023, March 3). Mother sentenced in deepfake cheerleader case. https://www.bbc.com/news/technology-56404038

Bender, E. M., Gebru, T., McMillan-Major, A., & Shmitchell, S. (2021). On the dangers of stochastic parrots: Can language models be too big? In *Proceedings of the 2021 ACM Conference on Fairness, Accountability, and Transparency* (pp. 610–623). Association for Computing Machinery. https://doi.org/DOI: 10.1145/3442188.3445922

Benjamin, R. (2019). Race After Technology: Abolitionist Tools for the New Jim Code. Polity. https://www.researchgate.net/publication/351607959_Benjamin_Ruha_2019_Race_After_Technology_Abolitionist_Tools_for_the_New_Jim_Code_Medford_Polity_Press_172_pages_eISBN_9781509526437

Binns, R. (2018). Fairness in machine learning: Lessons from political philosophy. *Proceedings of the 2018 Conference on Fairness, Accountability and Transparency*, 149–159. DOI: 10.1145/3287560.3287598

Bond, S. (2024, May 23). A political consultant faces charges and fines for Biden deepfake robocalls. NPR. https://www.npr.org/2024/05/23/nx-s1-4977582/fcc-ai-deepfake-robocall-biden-new-hampshire-political-operative

Brazilian Senate. (2024). Brazil AI Act: Bill No. 2338/2023. https://artificialintelligenceact.com/brazil-ai-act/

Buolamwini, J., & Gebru, T. (2018). Gender shades: Intersectional accuracy disparities in commercial gender classification. Proceedings of the 1st Conference on Fairness, Accountability and Transparency, 77–91. https://proceedings.mlr.press/v81/buolamwini18a.html

Business Insider. (2023, August). In every reported false arrest based on facial recognition, that person has been Black. https://www.businessinsider.com/in-every-reported-false-arrests-based-on-facial-recognition-that-person-has-been-black-2023-8

Chesney, R., & Citron, D. K. (2019). Deep fakes: A looming challenge for privacy, democracy, and national security. *California Law Review*, *107*(6), 1753–1820. https://www.californialawreview.org/print/deep-fakes-a-looming-challenge-for-privacy-democracy-and-national-security

China Briefing. (2023, April 11). China releases new draft regulations on generative AI. https://www.china-briefing.com/news/china-releases-new-draft-regulations-on-generative-ai/

Cyberspace Administration of China. (2022). Internet Information Service Algorithmic Recommendation Management Provisions. https://digichina.stanford.edu/work/translation-internet-information-service-algorithmic-recommendation-management-provisions-effective-march-1-2022/

Cyberspace Administration of China. (2023a). Provisions on the Administration of Deep Synthesis Internet Information Services. https://www.chinalawtranslate.com/en/deep-synthesis/

Cyberspace Administration of China. (2023b). Interim Measures for the Management of Generative Artificial Intelligence Services. https://www.chinalawtranslate.com/en/generative-ai-interim/

Dastin, J. (2018, October 10). Amazon scraps secret AI recruiting tool that showed bias against women. Reuters. https://www.reuters.com/article/us-amazon-com-jobs-automation-insight-idUSKCN1MK08G

Doshi-Velez, F., & Kim, B. (2017). Towards a rigorous science of interpretable machine learning. arXiv preprint arXiv:1702.08608. https://arxiv.org/abs/1702.08608

Equal Employment Opportunity Commission (EEOC). (2023). Technical assistance on AI and employment discrimination. https://www.eeoc.gov/meetings/meeting-january-31-2023-navigating-employment-discrimination-ai-and-automated-systems-new

European Commission. (2024, August 1). AI Act enters into force. https://commission.europa.eu/news/ai-act-enters-force-2024-08-01_en

Facebook. (2020). Facebook's civil rights audit – Final report. https://about.fb.com/wp-content/uploads/2020/07/Civil-Rights-Audit-Final-Report.pdf

Feldman, M., Friedler, S. A., Moeller, J., Scheidegger, C., & Venkatasubramanian, S. (2015). Certifying and removing disparate impact. *Proceedings of the 21th ACM SIGKDD International Conference on Knowledge Discovery and Data Mining*, 259–268. DOI: 10.1145/2783258.2783311

Friedler, S. A., Scheidegger, C., & Venkatasubramanian, S. (2021). The (Im)possibility of fairness. *Communications of the ACM*, *64*(4), 136–143. DOI: 10.1145/3433949

Government of Canada. (2025, March). AI strategy for the federal public service 2025–2027: Overview. https://www.canada.ca/en/government/system/digital-government/digital-government-innovations/responsible-use-ai/gc-ai-strategy-overview.html

Grother, P., Ngan, M., & Hanaoka, K. (2019). *Face recognition vendor test (FRVT) Part 3: Demographic effects (NIST Interagency/Internal Report 8280).* National Institute of Standards and Technology., DOI: 10.6028/NIST.IR.8280

Gu, Z., Chen, X., Wang, T., & Zhao, Y. (2023). Deepfake detection algorithm based on dual-branch data augmentation and modified attention mechanism. *Applied Sciences (Basel, Switzerland)*, *13*(14), 8313. DOI: 10.3390/app13148313

Illinois General Assembly. (2020). Artificial Intelligence Video Interview Act, 820 ILCS 42. https://www.ilga.gov/legislation/ilcs/ilcs3.asp?ActID=4015&ChapterID=68

Innovation, Science and Economic Development Canada. (2022). Artificial Intelligence and Data Act (AIDA). https://ised-isde.canada.ca/site/innovation-better-canada/en/artificial-intelligence-and-data-act

Ji, Z., Lee, N., Frieske, R., Yu, T., Su, D., Xu, Y., & Yang, Y. (2023). Survey of hallucination in natural language generation. *ACM Computing Surveys*, *55*(12), 1–38. Advance online publication. DOI: 10.1145/3571730

Kobie, N. (2024, February 6). Fake AI-generated news sites are spreading online. Wired. https://www.wired.com/story/ai-generated-fake-news-is-coming-to-an-election-near-you/

Lundberg, S. M., & Lee, S.-I. (2017). A unified approach to interpreting model predictions. Advances in Neural Information Processing Systems, 30. https://proceedings.neurips.cc/paper_files/paper/2017/file/8a20a8621978632d76c43dfd28b67767-Paper.pdf

Mialon, G., ... (2023). Augmented language models: A survey. arXiv preprint. https://arxiv.org/abs/2302.07842

Mitchell, M., Wu, S., Zaldivar, A., Barnes, P., Vasserman, L., Hutchinson, B., & Gebru, T. (2019). Model cards for model reporting. *Proceedings of the Conference on Fairness, Accountability, and Transparency*, 220–229. DOI: 10.1145/3287560.3287596

National Institute of Standards and Technology. (2022). Towards a Standard for Identifying and Managing Bias in Artificial Intelligence (NIST Special Publication 1270). https://nvlpubs.nist.gov/nistpubs/SpecialPublications/NIST.SP.1270.pdf

National Institute of Standards and Technology (NIST). (2023). Artificial Intelligence Risk Management Framework (AI RMF 1.0). https://nvlpubs.nist.gov/nistpubs/ai/nist.ai.100-1.pdf

Neumeister, L. (2023, June 22). Lawyers submitted bogus case law created by ChatGPT. A judge fined them $5,000. AP News. https://apnews.com/article/artificial-intelligence-chatgpt-fake-case-lawyers-d6ae9fa79d0542db9e1455397aef381c

New York City Council. (2021). Int. No. 1894-A: Automated employment decision tools. https://legistar.council.nyc.gov/LegislationDetail.aspx?ID=4344524&GUID=B051915D-A9AC-451E-81F8-6596032FA3F9

New York City Council. (2021). Local Law Int. No. 1894-A. https://legistar.council.nyc.gov/LegislationDetail.aspx?GUID=B051915D-A9AC-451E-81F8-6596032FA3F9&ID=4344524

New York State Department of Financial Services. (2021, March). Report on Apple Card Investigation. https://www.dfs.ny.gov/reports_and_publications/202103_report_apple_card_investigation

Obermeyer, Z., Powers, B., Vogeli, C., & Mullainathan, S. (2019). Dissecting racial bias in an algorithm used to manage the health of populations. *Science*, *366*(6464), 447–453. https://www.science.org/doi/10.1126/science.aax2342. DOI: 10.1126/science.aax2342 PMID: 31649194

Open, A. I. (2023). Usage policies. https://openai.com/policies/usage-policies

PricewaterhouseCoopers (PwC). (2024). EU's AI Act: What regulators should know. https://www.pwc.com/us/en/services/consulting/cybersecurity-risk-regulatory/library/tech-regulatory-policy-developments/eu-ai-act.html

Ribeiro, M. T., Singh, S., & Guestrin, C. (2016). "Why should I trust you?": Explaining the predictions of any classifier. Proceedings of the 22nd ACM SIGKDD International Conference on Knowledge Discovery and Data Mining, 1135–1144. DOI: 10.1145/2939672.2939778

Romano, A. (2023, April 5). ChatGPT lied about a law professor, and he's not the only one. The Washington Post. https://www.washingtonpost.com/technology/2023/04/05/chatgpt-lies/

Royal Commission into the Robodebt Scheme. (2023). Report of the Royal Commission into the Robodebt Scheme. https://robodebt.royalcommission.gov.au/publications/report

Selbst, A. D., & Powles, J. (2017). Meaningful information and the right to explanation. *International Data Privacy Law*, *7*(4), 233–242. DOI: 10.1093/idpl/ipx022

Skadden. (2023). EU Artificial Intelligence Act – Overview and impact. https://www.skadden.com/insights/publications/2023/06/quarterly-insights/ai-and-the-workplace

Slack, D., Hilgard, S., Jia, E., Singh, S., & Lakkaraju, H. (2020). Fooling LIME and SHAP: Adversarial attacks on post hoc explanation methods. In Proceedings of the AAAI/ACM Conference on AI, Ethics, and Society (pp. 180–186). Association for Computing Machinery. DOI: 10.1145/3375627.3375830

The White House. (2022, October 4). Blueprint for an AI Bill of Rights. Biden White House Archives. https://bidenwhitehouse.archives.gov/ostp/ai-bill-of-rights/

The White House. (2023a, February 16). Executive order on further advancing racial equity and support for underserved communities through the federal government. Biden White House Archives. https://bidenwhitehouse.archives.gov/briefing-room/presidential-actions/2023/02/16/executive-order-on-further-advancing-racial-equity-and-support-for-underserved-communities-through-the-federal-government/

The White House. (2023b, October 30). Executive order on the safe, secure, and trustworthy development and use of artificial intelligence. Biden White House Archives. https://bidenwhitehouse.archives.gov/briefing-room/statements-releases/2023/10/30/fact-sheet-president-biden-issues-executive-order-on-safe-secure-and-trustworthy-artificial-intelligence/

Treasury Board of Canada Secretariat. (2020). Directive on Automated Decision-Making. https://www.canada.ca/en/government/system/digital-government/digital-government-innovations/responsible-use-ai/algorithmic-impact-assessment.html

Twitter. (2021). Twitter's first algorithmic bias bounty challenge. https://blog.twitter.com/engineering/en_us/topics/insights/2021/algorithmic-bias-bounty-challenge

Wachter, S., Mittelstadt, B., & Russell, C. (2017). Counterfactual explanations without opening the black box. *SSRN*, *31*(2), 841–887. DOI: 10.2139/ssrn.3063289

Walsh, B. (2020, August 19). How an AI grading system ignited a national controversy in the U.K. Axios. https://www.axios.com/2020/08/19/england-exams-algorithm-grading

Weidinger, L., Uesato, J., Rauh, M., Griffin, C., Huang, P.-S., Mellor, J., Glaese, A., Cheng, M., Balle, B., Kasirzadeh, A., Biles, C., Brown, S., Kenton, Z., Hawkins, W., Stepleton, T., Birhane, A., Hendricks, L. A., Rimell, L., Isaac, W., & Gabriel, I. (2022). Ethical and social risks of harm from language models. *ACM FAccT*, *2022*, 214–229. Advance online publication. DOI: 10.1145/3531146.3533088

Wikipedia contributors. (2020, December 4). Robodebt scheme. Wikipedia. https://en.wikipedia.org/wiki/Robodebt_scheme

Wikipedia contributors. (2024, February 14). Ofqual exam results algorithm. Wikipedia. https://en.wikipedia.org/wiki/Ofqual_exam_results_algorithm

KEY TERMS AND DEFINITIONS

Accountability: In the context of AI, accountability refers to the obligation of organizations and individuals to answer for the outcomes of automated decision systems. It involves establishing clear responsibility for how AI systems operate and ensuring mechanisms (like audits and oversight) are in place to address errors, biases, or harms caused by these systems. Organizations implementing AI should have accountable parties who can explain decisions and take corrective action if the AI behaves inappropriately.

Algorithmic Bias: A systematic and unfair tendency of an AI system to favor certain outcomes or groups over others based on extraneous characteristics. Algorithmic bias often stems from biased training data or flawed model design and results in discriminatory or inequitable decisions (e.g., an AI that recommends higher-paying jobs to men more than women due to learned historical bias).

Automated Decision-Making: The process by which decisions are made by an algorithm or AI system with little or no human intervention. This can range from simple rule-based systems to complex machine learning models. Examples include software that decides whether to approve a loan, filter job applicants, or set insurance premiums without a human directly making each decision.

Bias Audit: An evaluation process where an independent party examines an AI system for potential biases and disparate impacts on protected or sensitive groups. A bias audit typically involves statistical testing of the system's outcomes (e.g., checking if a hiring algorithm's selection rates differ by gender or race) and results in a report of findings and recommendations to mitigate any bias discovered.

Explainability (Explainable AI): The quality of an AI system that enables its workings or decisions to be understood by humans. Explainability methods provide understandable reasons for an AI's output, often by highlighting important factors or providing simplified approximations of the model. High explainability helps users trust the system and is crucial for debugging and accountability.

Fairness (Algorithmic Fairness): The principle that an AI system's decisions are just, equitable, and impartial with respect to different individuals or groups. In practice, fairness can be defined in many ways (equal opportunity, demographic parity, etc.), but generally it means the AI does not systematically disadvantage people based on protected attributes like race, gender, or age. Achieving fairness may involve removing biases or ensuring balanced outcomes.

Governance (AI Governance): The framework of policies, processes, and oversight that directs how AI systems are developed and used within an organization or society. AI governance covers setting ethical guidelines, compliance with regulations, risk management, and the roles of various stakeholders (e.g., an AI ethics board) to ensure AI deployment aligns with legal and ethical norms. Good governance helps maintain control over AI's impact and prevent misuse or harm.

Model Card: A document that provides essential information about a machine learning model, including its intended use, performance across various datasets or subgroups, limitations, and ethical considerations. Model cards are a transparency tool recommended for responsible AI practice, allowing stakeholders to understand appropriate and inappropriate uses of the model and any potential biases in it.

Protected Characteristic: An attribute of a person that is legally or ethically safeguarded from discrimination. Common protected characteristics include race, gender, ethnicity, religion, age, disability, and sexual orientation. In the context of AI, fairness efforts focus on ensuring that models do not produce worse outcomes for individuals primarily because of these protected traits.

Transparency: The openness and clarity with which an AI system's workings, data, and decision criteria are communicated to stakeholders. Transparency can involve disclosing that a decision was made by AI, providing information about the algorithm's design or data sources, and, when feasible, offering insights into how the algorithm arrived at a specific decision. Transparency is a precursor to accountability and informed consent, helping affected individuals know when and how AI is influencing them.

Chapter 4
The Role of AI in Policy and Regulation: Navigating the Legal Landscape

Vidisha Shekhawat
http://orcid.org/0009-0009-7816-3591
Manipal University Jaipur, India

Pranjal Khare
http://orcid.org/0000-0002-9937-9588
Jindal Global Law School, O.P. Jindal Global University, India

Kiet Le Hoang
http://orcid.org/0009-0002-9968-1952
Vietnam National University, Hanoi, Vietnam

ABSTRACT

The chapter reports a developing picture of the role of Artificial Intelligence (AI) in policy and regulation, and the legal limitations and obstacles presented by AI technology. It shows the landscape of AI as a regulatory technology, establishing the need for legal regulation and policy, and dealing with a multiplicity of factors indicative of the emergent nature of digital technology. It discusses the foundational elements of AI, its implications for law and governance, and the conceptualizations that researchers are working with, particularly in relation to algorithmic accountability. A brief overview of different regulatory regimes worldwide, like the EU AI Act and segmental approaches taken in the US, leads to the potential roles of AI as a policymaking tool. The chapter also identified key regulatory challenges and gaps in law, including ineligibility and liability, and examined a number of sector

DOI: 10.4018/979-8-3373-1509-6.ch004

specific regulatory frameworks. The chapter concludes by suggesting a number of future pathways for regulation of AI that are developing new ways to address its emergent nature

INTRODUCTION

The theoretical aspects of science fiction about artificial intelligence (AI) have transformed into a fundamental power that is currently restructuring institutions throughout the world and their economic systems. Transformational changes have taken AI beyond experimental phases as it now operates through core decision systems across private and public organizations through automated legal research and predictive policing as well as algorithm-driven healthcare and financial decisions (Boden, 1996). The new era of technology has opened broad possibilities for enhanced efficiencies; however, it has caused numerous unexpected difficulties in legal aspects and social justice rights with requirements for greater ethical accountability. AI technology advances toward greater autonomy within infrastructure systems creating an intense demand for legal supervision and government regulation. Current attempts to monitor AI system development and deployment fail due to its fast-paced nature in the regulatory domain (Ertel, 2024). The traditional regulatory frameworks are often rigid and reactive and are not efficient enough to address the fluid and adaptive transnational nature of AI systems. However, the legislators have the responsibility to strike a balance between encouraging innovation with AI and ensuring that public trust, fairness, and human rights are taken care of. The challenge of dealing with the rapid advancement of AI has multiple stakeholders involved including the government, corporations, technologists, and civil societies that are operating within different legal frameworks (Perc et al., 2019). So, in this context, the emergence of AI is not just an evolution of technology but also a governance resolution requiring a paradigm shift in how societies regulate and adjudicate the digital era.

There are critical questions raised in the context of the legal implications of AI across every sector; those include the liability that arises in case of automated vehicles causing harm, the ways to ensure transparency in algorithmic decision-making if AI systems should be granted legal personality and many more. They are not merely theoretical questions but are becoming fundamental to international and national legal discussions. AI has challenged the foundational assumptions of legal systems being a socio-technical phenomenon, calling for new forms of reasoning and regulatory design. Adaptive policymaking with legal oversight can play a crucial role in the responsible deployment of AI as they rely more on agility, foresight, and interdisciplinary collaboration rather than conventional regulatory approaches that rely on fixed rules and slow-moving legislative cycles. There is an involvement of critical debates among legal scholars, ethicists, and policymakers over the development of binding international norms and ethical principles that can contribute to evolving AI in a way that is consistent with democratic values and human dignity. This chapter delves into the ongoing conversations by examining the role of AI in regulation and policy critically with a specific focus on how legal systems are responding or how should they respond to the threats and challenges posed by AI technologies. It aims to identify key regulatory conundrums, map the existing legal landscape, explore the sector-specific governance models, and propose future courses for responsive and inclusive AI regulation. Multiple studies are focusing on the technical or ethical dimensions of AI, however, this chapter Accenture the policy and legal aspects, emphasizing the need for a compatible and principled regulatory framework that is context-sensitive and forward-looking (McDermid et al., 2021).

To accomplish this objective, the structure of the chapter begins with laying the conceptual foundations of AI governance, and typologies, offering definitions and theoretical frameworks that inform the regulation of AI in the legal sense. The chapter then turns to a comparative analysis of global approaches to AI regulation by examining different strategies adopted by Countries such as the European Union, the United States, China, and India. The next section examines the use of AI within public policy as a tool for policy formulation and implementation resulting in questions about accountability in democratic systems, algorithmic bias, and public participation. This is followed by an elaborative discussion concerning legal and regulatory challenges posed by AI that include issues of liability, cross-border governance and explainability, and human rights. Further, the chapter underscores the legal framework of specific sectors like healthcare, finance, and criminal justice, identifying innovative policy experiments. The chapter then presents surfacing models of AI governance including co-regulation, human-in-the-loop mechanisms, and international norm setting to evaluate their efficiency. Finally, the chapter outlines a framework for responsive AI regulation that offers policy recommendations and proposes directions for international cooperation and further research (Upadhyay & Romashkin, 2019). The growth of AI and its integration into governance structures and policy domains has been inconsistent and rapid. The European Union (EU) has drafted an Artificial Intelligence Act to demonstrate regulatory leadership classifying AI mechanisms as per their level of risk while imposing corresponding legal obligations. The approach followed by the EU is a rights-based model striving to align the General Data Protection Regulation (GDPR) and the Charter of Fundamental Rights with AI governance. However, the U.S. has adopted a more disintegrated approach that relies on existing laws and agency-specific guidelines, nevertheless, it has also made efforts to shift the approach towards a more comprehensive legal framework through the "Blueprint for an AI Bill of Rights" (Bal & Gill, 2020).

China on the other hand preferred a model of techno-regulation which is more centralised providing the state more control in shaping the direction of AI through national strategies, standards, and governance based on surveillance. Meanwhile, India aims to follow an approach that balances innovation with the need for ethical safeguards, especially in the finance, health, and public welfare sectors. These eclectic approaches underline the global regulatory bifurcations that characterize the current state of AI governance. The demand for a transparent, accountable, and fair policy is growing rapidly, however, the implementation of the existing principles has significant variations across jurisdictions. The formulation of a regulatory framework has become the most complex challenge of this century compelling the policymakers to rethink the core legal concepts and embrace new tools for effective implementation. This chapter aims to provide a consecrated roadmap for steering the ever-evolving landscape that is grounded in theory, comparative analysis, and policy innovation. The imperative is clear as we stand at the intersection of innovation and constitution i.e., the future of AI should be governed by legal responsibility and not only technological possibility (Dixon, 2022).

FOUNDATIONS OF AI AND LEGAL GOVERNANCE

The 21st century has witnessed the most transformative technology i.e., artificial intelligence, yet it has not been defined as it remains fluid and dependent on context. It is pertinent to understand the foundational characteristics from a legal and policy standpoint for effective governance. If we understand its core, it can be referred to as a system or machine that can mimic human intelligence to perform tasks such as learning, reasoning, perception, and problem-solving. However, the definition is not exhaustive as it varies across institutions, organizations, and disciplines. For instance, the European Commission defines AI as a system that takes actions to achieve specific objectives, with some level of autonomy, by demonstrating intelligent behaviour through analysing the context and environment. Similarly, the Organisation for Economic Co-operation and Development (OECD) has defined AI as a machine-based tool that functions for a set of human-defined goals, making predictions, recommendations, and decisions by influencing real or virtual environments. These interpretations evaluate AI by its autonomous functionality and adaptive capabilities together with its programming-constrained independent operation which creates regulatory complexity.

Typology AI stands as the leading framework for understanding matters related to legal and governance during legal or governance analysis. Policy experts together with scholars identify three primary categories of AI which include narrow AI and general AI alongside Superintelligent AI. Narrow AI operates under the category of weak AI for performing single functions while existing mainly as a prolific technology today. The deployed systems boost well-known applications including digital assistants as well as recommendation engines and fraud detection algorithms along with recognition platforms (Paschen et al., 2020).

These limited AI systems are extensively present throughout healthcare and financial institutions as well as criminal justice departments and public administration sectors. Dissimilarly, Artificial General Intelligence (AGI) machines can understand, learn, and apply knowledge across a wide range of tasks at such a level that is comparable to that of a human. Research and debates about AGI continue with intensity because scientists have not yet developed it fully, especially regarding ethical concerns control measures, and social impact predictions. Artificial Superintelligence (ASI) seeks to advance conceptual limits with technology that would attain superior mental abilities to the human mind in every aspect. Theoretically, discussions about ASI raise significant legal questions about oversight, accountability, and even the nature of personhood. AI functions as an advanced network of sub-domains that provide technical prowess for modern programs while surpassing class organization systems. The dominant technology within these capabilities is Machine Learning (ML) because it enables computers to gain knowledge from data without explicit programming. These types of models identify patterns within large datasets to create decisions and predictions from their discovered insights. To date, supervised learning represents many AI implementations in government projects for 2024 with a portfolio value that exceeds 70% and major applications appearing in predictive law enforcement and medicine diagnostics along with tax fraud discoveries. Natural Language Processing (NLP) stands as one of the important subfields that enable machines to understand natural human language and provide them the ability to interpret and create human language. Multiple tools utilize NLP including virtual assistants alongside feeling analysis engines automated content moderation systems and chatbots (Fahad et al., 2024).

The extensive use of artificial intelligence has led to major public worries about false information circulation as well as breaches of personal data privacy and distorted public sentiment particularly during political election seasons. These definitions and operational characteristics need thorough comprehension by academics because they form a necessary base to help develop efficient policy frameworks. AI systems need quick standardized terminology along with classification methods due to their expanding complexity within private as well as government decision-making processes. Lawmakers face dual dangers when they target the wrong parts of the system or miss important high-risk areas because they lack a complete understanding of Artificial Intelligence. Using current AI development patterns to establish AI standards functions as an essential policy implementation tool. AI and governance characteristics have become a contested legal discussion point that remains aggressive in modern legal discussions. AI technologies continue their penetration into government agencies and commercial operations which triggers changes to decision-making aspects and responsibility management systems. The analysis of massive data sets and pattern detection abilities combined with probability prediction functions of AI systems have established them as tools that boost policy operations and legal accuracy as well as administrative operational capacity. This transformative capability brings forth critical problems which involve algorithmic bias along with opaqueness and discrimination as well as infringements on human rights and procedural rights. The implementation of computer systems and artificial intelligence within law practice benefits courts by reducing the expenses and time needed to deliver legal solutions through AI-assisted analytical tools for anticipating litigation results and automated legal document screening (Karmaza et al., 2021).

Dependency on AI systems creates doubts, mainly because it leads to diminishing human involvement in legal reasoning decisions. Some U.S courts employ COMPAS to predict criminal reoffending even though the widespread use of this and other AI systems continues to grow in judicial assistance applications. Professionals and specialists criticize AI applications in law enforcement because they perpetuate social discrimination even though they intend to make judicial procedures standard and uniform which questions fundamental ideals of legal justice and equal rights under the law. Policymaking has experienced a significant impact from the introduction of AI because governments now use data-centric governance for creating targeted policy solutions. The AI systems display the capacity to predict economic patterns besides controlling complex systems including healthcare services and traffic control systems with optimized resource distribution. Predictive models served during COVID-19 times through supply chain support of essential medical items while offering guidance about public health strategies. The HM Revenue and Customs division of the United Kingdom implemented AI technology to detect tax-related anomalies and frauds, thus, boosting policy performance measures. This development generates significant issues about both its validity and its ability to include all parties. The decision processes of AI systems remain challenging to understand and challenge by citizens particularly when the algorithms incorporate proprietary machine learning methods with unclear operation. Political authorities need to revise their regulatory systems to adopt AI technology. Traditional post hoc enforcement mechanisms with transparent legal rules face challenges in adapting to fast-evolving AI systems because the systems exceed their regulatory speed. Thus, a regulatory time-lapse persists. Large technology companies have acquired excessive power in AI advancement while causing a need for governmental frameworks to coordinate responsibility with technical progress. The required governance models need to integrate ethical guidelines voluntary standards and soft law instruments with formal AI Act legislation passed by the European Union (Brunette et al., 2009).

The AI Act applies risk-based regulations because it divides AI systems into threat classifications corresponding to their effects on fundamental rights citizen safety and public trust before assigning relevant obligations. The worldwide governance framework receives shape through two primary initiatives: the Organisation for Economic Co-operation and Development's AI principles and the Global Partnership on AI (GPAI). These organizations focus on maintaining human-centric development along with transparency democratic supervision and accountability systems. Although non-binding these principles have a significant influence on national and global artificial intelligence strategy development. Different jurisdictions use separate regulatory methods since the United States creates conditions encouraging innovation whereas the EU conducts stricter oversight programs and China supports unified governmental control. The multiple regulatory approaches created problems in building unified international structures because the diverse approaches stem from deeply conflicting ideological positions. Public officials need to maintain technological advancement without letting either democratic values legal principles or human dignity disappear. The future sustainability of governance depends on organizations learning to establish ethical and right-oriented norms that should be embedded into AI system architecture.

Artificial intelligence continues to penetrate social, economic, and legal systems through which appropriate frameworks need to be developed to regulate and comprehend its effects. The most essential aspects for understanding AI technology include technology governance alongside algorithmic accountability together with the concept of AI legal personhood. These guidelines support public servants alongside academics and legal professionals to comprehend the connection between technological developments that outpace established legal and policy norms (Marinova, 2024).

Technology governance functions as the established managerial systems and spontaneous administrative processes used to direct computer innovation while ensuring its alignment with social principles and priorities. AI systems need technology governance which includes processes together with institutions and principles that determine the development path and operational monitoring of AI systems. The hallmark of efficient technology governance consists of transparent processes that integrate a wide range of stakeholders alongside the capacity to adjust. The system requires active participation by stakeholders from the industry sector and civil organizations together with end-users with governments and their regulatory bodies representing only part of the framework. The essential regulatory problem involving AI development occurs from its global nature because it exceeds national regulatory abilities thereby emphasizing multi-faceted governance systems between national AI strategic frameworks and cross-border principles supporting diverse legal systems while accounting for social requirements. Algorithmic accountability represents the mandatory duty regarding fair operation alongside transparent and responsible implementation of AI technologies by their creators as well as users and deployers. The growing awareness of AI systems causing and deepening existing biases in society led to the development of algorithmic accountability standards. This issue specifically affects legal justice and financial credit systems and job placement decisions because AI-based decisions have important implications for individual rights.

Explainability, auditability, and redress mechanisms are the essential and core elements of algorithmic accountability. Jurisdictions around the globe have started experimenting with different regulatory approaches necessitating impact assessments, requirement of documents, and algorithmic transparency to enforce accountability. Acquiring legal personhood status for AI represents the most intricate theoretical and legal concept in regulating AI technologies. This approach seeks to offer legal status to AI systems by following the principles used for corporations as legal persons which enables assigning rights and responsibilities to them. This theoretical framework remains both extremely controversial and speculative for cases that need to function independently of human supervision as well as situations with complex liability assignments. The responsibility for liability needs definition in cases where AI generates intellectual property or when vehicles operate automatically. The critics maintain that extending personhood to AI systems would make developers less accountable because a personhood status could protect them. Defenders maintain that AI systems performing independently in specific cases need legal recognition for specific applications which allows for addressing outstanding legal issues and delivering justice (Busuioc, 2021).

The confusion accompanying AI liability and eligibility is evident in new legal disputes. For example, in the case of autonomous vehicles, the responsibility of potential liability for accidents moves from a human driver to a potentially multi-layers of liability between the AI developer, software developer, automaker, or even the owner, creating serious difficulties for traditional tort law. Similarly, there are issues of copyright eligibility and copyright infringement as they relate to AI-generated content. If an AI system creates a piece of art that is original, who holds the copyright - the AI (if granted personhood), the programmer, or the user who initiated the prompt? All of these challenges are illustrative of the limits of existing legal precedents and the need to develop new frameworks that will ensure responsibility and equitable rights in an increasingly AI driven world.

CURRENT GLOBAL APPROACHES TO AI REGULATION

As the world is seeing quick progress and development of artificial intelligence systems, there also comes countervailing responsibility to regulate it. Countries worldwide have urged implementing actions on governing the artificial and judicious use of AI so that its full potential can be explored safely and within ethical standards. There is an evolving tapestry of laws around the world along with increased collaboration amongst international organizations for mitigating the risks associated with the use of AI.

European Union (EU) - At the EU level, they have adopted the Artificial Intelligence Act [Regulation (EU) 2024/1689]. The scope of this framework encompasses extraterritorial jurisdiction with its application in the European Union and other countries. Notably, this was the first comprehensive legal regulation which dealt with risks associated with AI and setting standards with respect to accountability, transparency, safety and fairness in artificial intelligence applications. Such measures would allow high-risk AI systems to receive training with clearly identifiable information. The AI platform needs to showcase its decision-making process comprehensively and identify human intervention procedures as well.

In the Act, there is categorization of risks according to their respective regulatory mechanisms. The four levels of risk defined in the Act are unacceptable, high, limited, and minimal (Turk, 2024). Subsequently, the Act has provisions for AI system providers with high-risk whereby they need to establish stringent risk control measures and data management protocols. Additionally, they also need to incorporate open system functionality and human monitoring requirements in their AI applications. Similarly, there is a countervailing responsibility of deployers to demonstrate sufficient AI knowledge and to execute fundamental rights evaluation tests and follow precise compliance systems and monitoring protocols (Khare & Raghuwanshi, 2025). Thus, the implementation of these requirements would subsequently establish responsibility between developers and enhance new solutions.

With respect to compliance, there are provisions for follow-up requirements, including extensive technical record keeping and monitoring systems. For overseeing enforcement and imposing penalties, a European AI Office is established which is a regulatory body. This authority is empowered to levy fines for noncompliance in order to ensure that organizations have robust governance systems. Other bodies like the AI Board, the Scientific Panel, and the Advisory Forum are responsible for steering and advising on the governance of the Act.

Another framework which co-exists with the AI Act is the European Union's General Data Protection Regulation (GDPR). A peculiar observation to be noted is that AI does not appear directly in the GDPR, yet AI systems are directly impacted by multiple GDPR provisions. The two legal frameworks in the EU share fundamental concepts that provide protection for individual rights during AI-led personal data processing operations (Tarafder & Vadlamani, 2024). Each of these regulations showcases similar characteristics which establish a basic foundation for accountability. Additionally, both features fairness and transparency along with accurate data management. However, it is pertinent to observe that the GDPR has acted as guiding principles through which provisions of the AI Act have evolved.

United States - The United States of America at present lacks comprehensive federal laws or regulations specifically governing AI development or its restriction. In January 2025, President Trump issued an Executive Order for removing barriers to American leadership in AI ("Removing Barriers EO"). The purpose of Removing Barriers EO is to direct federal departments and agencies to revise or eliminate any Biden administration policies, directives, regulations, and other actions which are deemed 'inconsistent' with enhancing America's global AI dominance. While numerous policies from the Biden EO were in place, the extent of future changes remains uncertain. Although the U.S. does not have unified AI regulation, the federal government has implemented various guidelines and frameworks for the AI sector. This includes Executive Orders like the one focused on maintaining American leadership in AI and the Promoting the Use of Trustworthy AI in the Federal Government Act.

The US Congress has considered various AI-related bills addressing a broad range of issues. However, it is yet to be determined whether the Republican controlled Congress will prioritize AI legislation, or it will be focusing on other matters. On the other hand, many proposed bills emphasize the creation of voluntary guidelines and best practices for AI systems. These indicate a cautious regulatory approach aimed at promoting innovation without imposing strict mandatory rules. This approach is driven by concerns about hindering technological advancement and maintaining competitiveness, especially in comparison to countries like China, which reportedly produces significantly more STEM graduates than the US.

Furthermore, various scholars (Litwin & Racabi, 2024) have observed that The United States is implementing AI governance by addressing specific cases as they arise, rather than adopting a broad, overly cautious regulatory stance. The benefit of this approach is that it allows for flexibility and avoids potentially stifling innovation in the rapidly evolving field of artificial intelligence. Therefore, in the absence of specific legislation for artificial intelligence, the monitoring and control of its implementation are handled by the governmental bodies of the US. These authorities tasked with overseeing AI guidelines majorly consist of the Federal Trade Commission, Department of Justice, Consumer Financial Protection Bureau, Equal Employment Opportunity Commission, etc.

While regulatory frameworks are being developed for AI across the world, the realities of jurisdictions' implementation and enforcement differ greatly within and between jurisdictions. For example, the AI Act in the European Union is large in scope and encompasses multiple sectors, but two unique challenges arise in their jurisdictions. First, the complexity of having to harmonize how agencies will implement and enforce the act within their member states. Second, the need for new regulatory entities like the AI Office and mechanisms for market surveillance to demonstrate compliance and recognition of non-compliance with an expansive framework also becomes an important development priority. In the US, this results in enforcement relying on the mandates of the agency that has jurisdiction over the non-legal compliance, since there is rarely a statute or specialized guidelines that are truly applicable to the legal use of AI tools. As such, companies are left to comply with current consumer protection laws or adaptation of antitrust laws based on their traditional and non-legal industry's historical experiences, leading to a patchwork of compliance that often differs with sector or AI tool. Such compliance structures also require organizations to engage in individual thoughtful proactive risk assessments and evolving agency guidelines with each time period, whereas in other jurisdictions this focused approach may be easier to navigate with a simple compliance structure.

China - A country which is excelling amongst some of the world's first and most spectacular artificial intelligence regulations is China. Its regulations cover areas like recommendation algorithms which is the most widespread AI online, and synthetic media including ChatGPT-style chatbots. This evolving AI governance framework in China will redefine how the technology is created and used domestically and internationally. This will definitely have a positive effect on both its tech exports and global AI research networks. China is among the leading nations actively establishing regulations for artificial intelligence (Liu et al., 2024). Even as they develop an overarching AI framework, existing rules already govern various specific AI applications, like the management of algorithmic recommendations. They also include ethical standards for new-generation AI and views on strengthening the ethical oversight of science and technology.

However, in the West, China's regulations are often dismissed as insignificant or solely seen as a geopolitical race to set AI rules. These regulations instead merit careful analysis to understand their influence on China's AI path and the potential lessons for policymakers worldwide. Despite varied and contrasting views on specific regulatory content, nations can still learn from each other's approaches.

India - Just like the US, India also does not have a dedicated comprehensive legal framework that exclusively governs artificial intelligence. Instead, the regulatory landscape in the country is evolving and is characterized by a pro-innovation approach. The primary focus is on ethical guidelines, sector-specific regulations, and the application of existing digital laws to address concerns related to AI. It is pertinent to note that India has implemented various policy measures in response to the rise of artificial intelligence, particularly as its government aims to establish the nation as a leading global AI developer.

The policies in India span from investing in public infrastructure to foster market-driven AI creation, to nationalizing data collections. The basic idea behind this is to facilitate AI-powered big data analysis. Scholars (Joshi, 2024) contends that India's AI policy and governance framework establishes and validates a globally prevalent model of informational capitalism. This model relies on framing data as a key resource for an information-centric economy and promotes industry self-regulation of negative consequences. This is in light of the background that the state seeks to exert significant control, via legislation and infrastructure, over the market structure and objectives.

Despite the diverse approaches to AI regulation across different jurisdictions, certain fundamental principles are emerging as a common theme. These principles aim to ensure that AI systems are developed and executed in a manner that is ethical, responsible, and beneficial to society as a whole. AI system functions along with data usage and decision-making processes that must be transparent according to principles so that people can both trust these systems and perform adequate inspection. AI regulations from EU, US and China incorporate this principle. The principle of non-discrimination together with fairness works to stop AI from upholding prejudices. The intervention of bias detection requires precise handling of data alongside bias attenuation measures according to the EU AI Act, the US AI Bill of Rights and Indian draft legislation. The core purpose of safety principle depends on creating dependable AI operation along with risk reduction below acceptable thresholds. Robustness and cybersecurity form a part of the EU risk-based approach. The principle manifests through US system safety requirements and Chinese content safety regulations. The principle of accountability maintains AI responsibility through provider-deployer oversight roles which demonstrates responsibility mechanisms in the EU framework. Soft law instruments serve as essential tools for developing the accountable deployment of AI according to ethical standards (Buczynski et al., 2024). The instruments for soft law enforcement encompass guidelines in addition to ethical charters alongside technical standards as well as industry codes of conduct. The non-binding nature of these soft law instruments does not prohibit their usefulness because they create ethical standards and support best practices in addition to developing new norms for AI governance.

AI IN PUBLIC POLICY FORMULATION AND IMPLEMENTATION

Public policy is the foundational structure upon which societies and their futures are based upon and governed accordingly. The governance structure of public policy continues to become more influenced together with being transformed through the execution of artificial intelligence systems. AI serves as an effective policy development tool because it enables prediction of social evolutions and maximizes resource distribution alongside warning about policy inadequacies before execution. The complete potential of AI as a policymaking instrument needs examination of its complex role in public policy development (Saheb & Saheb, 2024). Studies must evaluate significant cases from multiple sectors to analyze complex regulations along with the potential biases as well as algorithmic injustices which emerge in the face of this transformative change.

The core function of AI as a policymaking tool is to perform predictive analytics. Through the combination of statistical algorithms and machine learning techniques, the system evaluates extensive datasets to search for patterns which enable future trend prediction. The capability stands as a revolutionary construct within public policy operations according to Hossin et al. (2023). The modern government has access to AI systems which enable them to predict social disruptions. Basic information analysis of social media platforms combined with economic measurements and population distribution data exists as an accessible process for AI models. The identified hotspots of social unrest enable organizations to allocate necessary resources for preemptive interventions that minimize risks effectively. Further, AI can also be used in predicting disease outbreaks. AI models have specific features which help them in analyzing health records, travel patterns, and environmental factors. This has the capability to enable early detection and prediction of disease outbreaks. If this is flagged timely, the government can then help facilitate timely public health responses, resource mobilization, and targeted interventions to contain epidemics. In the infrastructure department, AI tools are majorly used to predict future demands on transportation networks, energy grids, and water resources, enabling planning in advance and investment in necessary infrastructure upgrades. This is done with AI analyzing population growth, urbanization trends, and infrastructure usage data. Even for estimating the impact of policy interventions, AI proves effective. These models can simulate potential effects of policy on different segments of society, even before the new policy is implemented. This helps policymakers to constructively assess its likely success, identify potential unintended consequences, and further refine the policy design for optimal outcomes.

The other aspect of efficacious policy making is a deep understanding of human behavior. AI-powered behavioral modeling comes to the rescue in this case as they offer unprecedented insights into how individuals and groups respond to different incentives and regulations. AI analyzes large-scale behavioral data from various sources like online platforms, mobile phone usage, and transactional records. The objective consists of discovering behavioral patterns among humans concerning their energy use and healthcare selections together with regulatory compliance.

The modeling process for computational AI relies on sampling techniques to model many autonomous agents in action. The environment contains agents which act as representatives of both people and organizations. Legislators in a country's government can understand detailed impacts of various policies on communal outcomes through easy assessment. It is necessary to demonstrate both variations in policy parameters and changes in agent behavior. In this instance, we can use this approach to develop tailored policy interventions for cohorts or individuals by understanding their behavioral profiles. Digital platforms can provide personalized nudges; this is an example of the type of approach that could be used to nudge into healthier lifestyle choices. Political decision-makers must test the likelihood of policy successes before expending significant amounts of money on investment that might affect large groups of people.

As AI simulations provide an effective method of developing policy scenarios, they will allow decision makers to build virtual testing environments. Specifically, an AI system will build authentic digital twins that replicate almost every aspect of what exists in reality. Using simulation modeling, policymakers can evaluate the likely ramifications of functionally implemented policy programs with respect to performance measures such as economic growth, environmental sustainability, and social justice. For accurate virtual modeling, simulation trials will expose latent needs or undesirable consequences that may occur in implementing a specific policy. The simulation processes allow the organization to enact strategic alterations and protective mechanisms that reduce impact on reality (Zuiderwijk et al., 2021).

The potential of AI in public policy is not just in theory but is rapidly being ushered into real world applications across various sectors like healthcare, welfare distribution, urban governance, etc. AI is bringing a revolutionary change in healthcare systems worldwide when the entire world is grappled with challenges like rising costs, aging populations, and the need for personalized care. The foremost contribution of AI in healthcare has been diagnosis and prediction of disease. AI algorithms are skilled in analyzing medical images like X-rays and MRIs, genomic data, and patient records. After analyzing these data, it helps in the early and accurate diagnosis of diseases like cancer, diabetes, cardiovascular conditions, etc. Some AI algorithms can even identify individuals at high risk of developing certain diseases, enabling proactive interventions and preventative care. There have been major cases worldwide where patients could not have their ailments diagnosed for years but simply identified with the help of an AI. Also, it is seen in several countries that AI-powered systems are used to analyze mammograms. As a matter of fact, these AI systems can analyze with higher accuracy than human radiologists, in detecting early signs of breast cancer, potentially saving lives through earlier diagnosis (Khare & Raghuwanshi, 2025). The role of AI does not just end at diagnosis but also covers drug discovery and development. AI has the potential to accelerate the process of identifying potential drug candidates, predicting their efficacy and toxicity.

It is observed that social welfare programs aim to provide a safety net for vulnerable populations. However, there are many shortcomings in these programs as well. In this context, AI can enhance the effectiveness of these programs on several fronts. AI algorithms utilize training techniques to analyze large datasets to identify the people that need the most help. This system ensures the allocation of resources goes only to eligible recipients who require help, therefore limiting fraudulent events. The system will conduct an assessment method in determining their needs and provide customized individualized welfare plans that offer the most appropriate assistance. These AI systems have superior abilities for identifying households, or families that are suffering from the effects of poverty or are at risk of more social exclusion. The ability to employ interventions at various times with precautionary intentions allows the ability to tackle problems. The blend of AI gives politicians the ability to quantify the impacts of welfare policy, while earmarking resources to create maximum possible impact to combat poverty and provide positive social outcomes. Just about all jurisdictions globally take advantage of AI systems to assess social benefit claims. Processes lead to rapid assessment, decreased administrative contingencies which leads to accurate and objective identification of appropriate candidates. AI technology is being employed to design sustainable smarter cities in association with urban governance initiatives.

Further, the impact of AI can also be seen in urban governance where it is used for building smarter and more sustainable cities. Cities are grappling with complex challenges, and urbanization is only expected to increase during this time. The world needs to build better sustainable and livable urban environments. AI has an expansive role as it considers real-time traffic information to more efficiently manage traffic patterns while expanding public transit capabilities (Yu at al., 2025). Its analysis of energy consumption patterns, its ability to efficiently integrate renewable energy sources, and its ability to project future energy demands allow for both parked distribution networks and supply systems. Overall, this means that better long-term planning can advance more sustainable, resilient energy systems. The impact of AI systems upon solid waste management optimizations is also significant. When it comes to reliable public safety systems, crime forecasting systems are supported through AI capabilities. AI systems analyze police statistics, social media trends, and environmental conditions to develop forecasts regarding crime locations and types of crime patterns that better inform law enforcement operations (Khare, 2025). AI-powered surveillance camera systems have become common, as they can monitor video feeds and record unexplained incidents. The system enables faster emergency service responses and may assist in identifying possible criminal incidents (Cugurullo & Xu, 2025).

SECTOR-SPECIFIC REGULATORY FRAMEWORKS AND POLICY INNOVATIONS

The increased availability of Artificial Intelligence (AI) in various industries is creating a new wave of change that promises to increase efficiency, provide new solutions, and change the way we live. Tedious tasks will be easier, new tools will increase our creative capacities, and we will have opportunities to reinvent sectors and industries. However, as in any rapid transition, AI is creating a challenge in a matrix of ethical, legal, and social dilemmas that will require a solid and often flexible regulatory structures. Eventually, as AI systems spread into the most critical sectors such as healthcare, finance, criminal justice, and labour and employment, it will require not only regulations specific to the sector but new and original policy approaches. It is important to explore the regulatory challenges ahead and the policy innovations in these areas.

Healthcare - The healthcare sector stands on the verge of a massive transformation powered by AI. AI in healthcare has the potential to revolutionize diagnostic services, treatment pathways, drug discovery, and personalized medicine. Algorithmic diagnostic systems that analyze medical images capture our imagination because they can complete a more detailed analysis long before it would ever occur to a human medical professional, possibly accelerating the identification of critical conditions like cancer. AI-powered tools may also produce treatment plans based upon data from a range of patients that are specific to an individual's data which could advance new treatment options (Bajwa et al., 2021).

Despite the benefits AI can provide to healthcare, applying AI in medical care is not without significant regulatory issues. Patient health records are one of the most sensitive, private types of information. There are regulations in place in the United States, such as the Health Insurance Portability and Accountability Act (HIPAA), and in Europe, there is the General Data Protection Regulation (GDPR), to safeguard patient records. However, AI has specific characteristics that make it necessary to create new interpretations and modifications to that specific regulation, such as the re-identification of anonymized data, or the sharing of data in order to develop an algorithm.

In addition, the safety and effectiveness of medical AI devices and algorithms should be regulated. Existing regulation for medical devices may be inadequate for medical AI systems that learn, change, and adapt while in clinical use, such as apps and wearables. Toward those ends, FDA continues to develop a framework for evaluating and authorizing medical AI, as well as considering ongoing transparency, explainability, and provisions for marketing surveillance and updates. Algorithmic fairness and bias issues are another concern to address. In healthcare, biases related to age, race, and socio-economic class can often negatively impact diagnostic accuracy and appropriateness of treatment across different demographic groups. Such bias in medical AI will further perpetuate the existing health consequences arising from health inequities. Regulatory structures must require strong bias testing and provide or promote the advancement of just and fair AI algorithms. Some recent policy innovations include the creation of explicit policies governing the use of AI in clinical contexts, the establishment of data governance frameworks to facilitate the responsible sharing of data for AI development while also protecting patient privacy, and the establishment of policies and pathways to facilitate the ethical and transparent deployment of AI-based tools for diagnosis and treatment. Regulatory sandboxes can also facilitate the responsible development of AI and may yield valuable lessons for future regulation by allowing AI innovations to be tested in controlled settings.

Finance - The financial service industry has already adopted AI to drive more efficient and accurate processes, enhance the ability to identify and mitigate fraud, and create more tailored and customized consumer services. In terms of credit scoring, AI algorithms can be trained on large datasets and sophisticated statistical analysis, using more inputs than traditional assessment methods, to provide a determination of an individual consumer's credit worthiness. For fraud detection, AI can automatically detect anomalous patterns and transactions beyond what standard change detection and other methods can identify. In the financial service industry, regulatory sandboxes, a policy innovation designed to support policymakers to experiment with new technologies, are becoming more popular, as these allow fintech companies to develop AI-driven financial products and services in a regulated space, under regulatory supervision (Roy et al., 2025).

Contemplating the immense opportunities that AI brings to the finance sector, there are considerable regulatory implications that emerge when discussing and applying AI controls to critical financial use cases. Areas like credit scoring and loan approvals with regard to product fairness, transparency, and explainability of AI decisions are paramount to the consumer outcome and confidence in the use of AI technologies in finance. However, it is also noted that some AI algorithms maintain a degree of transparency but, nonetheless, complexity make the explanatory processes more difficult for not only the consumer but, for the financial participant considering liability as a mitigating factor. These situations raise a number of regulatory topics that may have further regulatory implications. Thus, as regulatory guidance evolves, it must also contemplate the responsible development of "explainable" AI. However, it is also imperative for financial institutions to assess how to articulate a reasonable explanation for the AI decision impacting a consumer so that intermediaries and other stakeholders are informed in the process.

The subject of data security and privacy is also particularly important in the finance sector because financial data can be uniquely sensitive. Legislation such as the Gramm-Leach-Bliley Act (GLBA) in the United States and the General Data Protection Regulation (GDPR) in Europe throws some light on legal protection, but without additional guidance tailored to the intricacies of AI-oriented data analytics, we face both the challenge of harm from AI and the opportunity for enhanced protections against financial services cyberattacks aimed at AI systems. Additionally, the use of AI for high-frequency trading and algorithmic trading creates concerns about systemic risk and market instability. It will be necessary for regulators to address and reduce these risks and instability and ensure that AI-oriented systems are robust, resilient and have no likelihood of creating instability or market manipulation opportunities.

Public policy in the finance sector involves innovations like guidelines for explaining AI determinations in credit scoring and lending practice, requirements for cybersecurity standards for AI-oriented financial services, and proposals for monitoring and lowering systemic risk in markets with AI capabilities. With priority given to establishing regulations for licensed institutions that operate with AI-based systems in the finance sector, regulators can still consider using existing regulatory sandboxes to promote innovation with responsible and caution while educating themselves about emerging technologies. Regulators may retain the option to exercise their discretion to emphasize and educate themselves regarding the risks and rewards of new AI capabilities as they continue to develop in the finance sector.

Criminal Justice - The use of artificial intelligence in criminal justice, through predictive policing, facial recognition, and surveillance laws, raises some of the most ethically and legally complex dilemmas. Predictive policing algorithms examine historical crime data in order to predict likely future crime locations and allocate resources accordingly. Facial recognition technology identifies and tracks an individual using that person's biometric data, while surveillance laws cover the collection, retention, and use of data utilized for law enforcement. While these technologies can potentially improve public safety and efficiency for law enforcement, they can also create significant risks connected to algorithmic bias, violation of people's privacy, and possibly discriminatory outcomes. Predictive policing algorithms rely on historical data, whether biased or not, to identify criminal patterns, and can exacerbate and perpetuate current patterns of discriminatory policing activities to pursue further surveillance and intervention to marginalized communities. Facial recognition technology has also been found to have various levels of accuracy based on demographic groups, resulting in potential false identification and accusation of innocent people. The advancement of surveillance technology overall can violate people's rights to privacy and rights to associate freely (Barrington & Farid, 2023).

Regulatory mechanism in this space must emphasize on protecting fundamental freedoms, civil rights, and fairness and accountability in any use of AI. This means creating restrictions on the extent to keep and reuse sensitive forms of biometric data, requiring transparency and explainability on the development and use of predictive policing algorithms, and developing oversight structures and mechanisms to address bias and prohibited discrimination. Policy innovations include developing explicit legal standards of admissibility of AI-produced evidence and information in court, creating independent bodies to protect and monitor AI use in law enforcement and builds in protections for data protection of AI-enabled data collection and information, which presents unique risks. And further, we must participate in public consultations and engagement as we embark on the implementation of AI in the criminal justice system to ensure where technologies are aligned with societal values and ensure core rights are protected.

Labour and Employment - AI is rapidly changing the nature of labour and employment, including hiring, algorithmic management, and raising concerns about jobs lost. Employers are using AI systems to screen applications, conduct initial interviews, and even hire employees. Algorithmic management is free of hiring decisions; it monitors and measures performance in real time. The ever-increasing automation of tasks through AI raises concerns of significant job loss across a number of sectors. The integration of AI into labour and employment will require a regulatory approach that establishes rules or codes of conduct regarding algorithmic bias, data privacy of workers, and the future of work. Algorithms used in hiring or recruitment are biased if they embed the existing biases found in the training data. Algorithmic management raises concerns regarding the surveillance of workers, loss of autonomy of workers, and the risk of punitive performance evaluations. The potential for widespread job loss requires meaningful policy measures to assist workers to transition to a new worker role and ensure fair distribution of the benefits from automated employment sectors.

Regulatory frameworks in this area should be focused on transparency and fairness in AI-assisted hiring, on establishing some rules for the ethical application of algorithmic management systems, and on establishing social safety nets and retraining to help take the edge off the displacement of wages and jobs. This might include requiring employers to audit their AI hiring tools for bias, granting workers' rights to the data being used by algorithmic management systems, and providing funding for education and training programs to give workers the skills needed in the future economy. Further, policy innovations may include creating legal regimes to counteract algorithmic discrimination in hiring, establishing rights of workers in algorithmic management contexts, and exploring universal basic income and other social welfare alternatives that would offset the scope of possible employment displacement. These kinds of policies will require governments, employers, and labour unions to convene in tripartite discussions in response to instability in the world of work to mitigate the possibility that these transitions are not just and fair (Wang et al., 2024).

The advent of AI across multiple industries presents both great opportunities and considerable challenges. In order to seize the opportunities associated with AI and mitigate the risks, the development of sector-specific regulatory frameworks, and models of original policy, are necessary. In the healthcare sector, regulating AI and machine learning (ML) raises the questions of balancing the benefits of innovation with ensuring patient safety and privacy of patient data. In the financial sector, a regulator is faced with interpreting moral dilemmas related to efficiency against risk. Within the criminal justice sector, regulators face a judicious balancing act of addressing bias in the AI system while ensuring protected rights. Last, in the labour and employment sector, regulators must ensure fairness and competence in hiring and firing decisions, and an awareness of the ever-present challenges of automation. As we move toward an incredible range of new legal doctrines and novel adaptable regulatory frameworks, being equipped to move into the algorithmic age and ensuring AI is imbued with justice and morality is a must. Ongoing communication, cooperation, and commitment to ethical principles and policies would all contribute to the advancement of AI regulation and policy.

REGULATORY CHALLENGES AND LEGAL GAPS

AI development speed has produced new regulatory difficulties and legal vacuum because of its distinctive attributes in modern society. There are many models of AI systems, one of them is called "black boxes" which is based on deep learning and poses a pressing concern of opacity and explainability as such models provide outcomes without justifying the reasons for their decisions. Data obscuring led to immediate violations of the legal rights protected by due process, accountability, and the right to a fair trial. Invisible AI-driven decisions throughout healthcare and financial systems as well as criminal justice facilities result in serious consequences for people due to an insufficient understanding of decision origin and process. The current legal bases recognize decision makers should produce grounds for their decisions, so AI systems remain exempt from this basic principle. The current gap requires addressing two issues: the ongoing discussion about giving AI legal personage and unresolved matters regarding liability attributions for AI system defaults. One essential gap exists in determining liability responsibility while scientists continue to disagree on whether to grant legal personhood to AI systems. The determination of legal responsibility becomes difficult during incidents of autonomous system-caused damage between faulty autonomous vehicles and flawed medical diagnostic AI systems. The causality in AI-driven cases spreads among the entities responsible for development, deployment, and data provision because traditional human-oriented tort laws depend on either intention or negligence. Complications arise when attempting to file litigation and seek compensation because there is no defined entity responsible. The creation of a special legal position for AI remains controversial among scholars because it poses threats to both moral hazard risks and diminished human accountability (Kingston, 2016).

The worldwide structure of AI technologies creates difficulties for regulatory bodies to enforce their rules because data management needs proper jurisdiction determination. AI programs draw their training data from worldwide sources which gets stored on multiple server locations across several countries while performing operations in jurisdictions having divergent legal frameworks. Cross-border data flows create problems because they require determining the applicable laws regarding data protection and consumer rights as well as algorithmic duty. The fragmented securities regulatory environment enables business entities to position their operations within jurisdictions that maintain less oversight so they can continue offering services to users based in regions with rigorous regulatory frameworks (Popescu & Yu, 2024). Article 28(3) of GDPR has faced multiple hurdles when trying to enforce EU standards on global AI operators because the transnational digital space remains bounded by territorial law enforcement methods. The insufficient alignment between AI laws along with their incompatible nature increases the overall complexity. States create regulatory frameworks through frameworks that vary across nations because of different economic interests' institutional capabilities and varying core values. The proposed EU AI Act prioritizes human rights with risk management features, yet the US and China follow different paths with innovation freedom for the US and state-driven strategic development for China. Multiple global standards create confusion within the legal structures because they complicate compliance for multinational developers while making it difficult to create shared standards for safety ethics and accountability (Pesapane et al., 2018).

FUTURE PATHWAYS: A BLUEPRINT FOR RESPONSIVE AI REGULATION

Regulating artificial intelligence consists of more than managing technology since it defines how society will cooperate with progressively independent systems. The need for dynamic regulation emerges as the foremost priority when viewing the situation. AI development moves quickly yet traditionally stiff laws develop slowly which makes such instruments inadequate for AI changes. Regulatory sandboxes that provide regulators with monitored testing spaces for AI applications represent an effective solution for promoting innovation while performing oversight functions. Regulatory sandboxes give regulators as well as innovators space to work together so they can gain real-time learning and adapt laws accordingly. Sunset clauses set automatic deadline systems on specific regulations thereby assuring regulatory laws will be continually evaluated after technological changes occur. These mechanisms demonstrate the broader movement toward anticipatory governance since they show that regulation needs flexibility with an iterative approach that evolves parallel to the targets of regulation (Truby et al., 2022).

Forward-thinking regulatory systems require the combination of ethics-by-design principles and legal requirements during the complete artificial intelligence development process. System architecture needs to incorporate legal and ethical factors at the first stage before continuing these components until systems reach deployment and subsequent use. All ethical rules beginning with fairness and stretching to human-centricity accountability and transparency must exist in systems before and after development rather than only in audits or reviews. Ethical principles should function as guiding elements throughout design protocols while they also determine the selection of data inputs and model training as well as system updates. Through internalised governance mechanisms harm can be anticipated while public trust is developed. The implementation of ethical principles necessitates mandatory standards backed by surveillance systems and organizational support for academic relationships between ethicists technologists and legal scholars. Institutionalization stands as a necessary condition to avoid ethics becoming a superficial protective measure instead of a genuine protective mechanism (Scherer, 2015).

A worldwide standardized AI regulatory system is now recognized as being essential to achieve. AI development and deployment throughout different geographical zones require more than national regulations on their own. National fragmentation of legal systems creates both regulatory burdens for companies and allows agencies to exploit different regulations to neglect fundamental rights protection. The need for a worldwide AI regulatory treaty has become a pressing demand because it resembles established international environmental and trade agreements. Such a treaty would determine fundamental principles together with unacceptable hazard thresholds as well as establish processes for international disputes and inter-jurisdictional enforcement. The project faces obstacles from geopolitical differences and regulatory inconsistencies, but recent international organization activities provide essential direction from both UNESCO through its Recommendation on Artificial Intelligence ethics as well as the OECD and its AI Principles. A globally accepted legal structure stands as the most effective approach to secure universal human values combined with regulatory uniformity and prevention of unethical technological competition (Ahuerma, 2023).

To create effective governance structures, it is indispensable for both inclusion and equity to be prioritized. The populations who mostly encounter AI-related harms retain limited participation in regulatory discussions and standard-development activities. Inclusive governance entails not only geographic and demographic diversity but also epistemic pluralism, the recognition and integration of multiple ways of knowing, including Indigenous knowledge systems, feminist perspectives, and global south insights. AI technology gets enriched through various ways of seeing things which extend knowledge about damage criteria alongside fairness practices and responsible action guidelines. The concept of equitable governance requires mechanisms to spread redistributive financial power which tackles the economic concentration of tech giant power and secures universal benefits from AI (Raghuwanshi et al., 2025). The approach requires investments towards public-interest technologies together with backing open-source AI projects and promoting community-based creative innovation. The multiple complex relationships of AI governance require scientists from law and technology to join forces with public policy experts for meaningful collaboration. The establishment of effective AI governance requires law experts to work with tech scientists for technical parameters comprehension and government officials to combine ethical principles with sociological understanding to make regulatory decisions. A multi-disciplinary integration approach is essential for governance because it functions as a fundamental requirement. A regulatory framework that combines interdisciplinary collaboration will create structures that are both legitimate and responsible enough to sustain AI development for human community welfare (Ahuja et al., 2020).

CONCLUSION

Artificial intelligence's regulatory needs rank among the foremost important challenges that modern civilization faces during this twenty-first century. AI systems actively moderate critical choices within health circles and further critical choices in education and law enforcement along with finance and governance domains which raises legal stakes to enormous levels. The discussion in this chapter examined AI regulatory concepts alongside actual practical issues that emerge from using modern AI technology applications. AI represents both technology and social-technical elements that demonstrate and modify power structures together with knowledge frameworks and institutional systems. Regulation of AI exceeds practical boundaries that would justify sectoral or technical-based solutions. A holistic approach to AI governance comes from treating it as an initiative devoted to normative management which establishes democratic responsibility while serving human rights and enforcing social equity.

Artificial intelligence’s regulatory needs rank among the foremost important challenges that modern civilization faces during this twenty-first century. AI systems actively moderate critical choices within health circles and further critical choices in education and law enforcement along with finance and governance domains which raises legal stakes to enormous levels. This chapter has established an exploration of AI regulation which starts with conceptual foundation and practical frameworks of different AI applications and ends with discussions about existing legal shortcomings and ethical disputes. AI represents both technology and social-technical elements that demonstrate and modify power structures together with knowledge frameworks and institutional systems. Regulation of AI exceeds practical boundaries that would justify sectoral or technical-based solutions.

This governance project should originate from democratic foundations and human rights principles while serving social justice needs. Multiple policy recommendations should be applied considering the current situation. Governments need to invest resources toward developing institutions that monitor algorithms through independent auditors, technical regulators, and public interest technology offices. The implementation of ethical governance should take place through legally enforceable standards which also include participatory processes and punishments for violations. International nations should fully cooperate through a global AI framework agreement for setting fundamental standards alongside allowing different enforcement at regional levels (Yu et al., 2024). The study requires additional research about AI harm intersectionality as it impacts vulnerable populations and how to develop new liability frameworks that handle distributed responsibility and causation of AI systems. The pressing question is not about AI regulation but how we should implement this approach which will support specific interests or groups. The time requires legislative preparedness in AI because it serves as both an imperative for lawmaking authorities and an opportunity to restructure digital lawmaking practices that uphold justice and human rights. Modern technology stands at a turning point between AI usage as an enabling instrument and a restraining force that determines exclusion and control. The duty to accomplish this task rests as a joint responsibility between scholars, practitioners, policymakers, and communities.

REFERENCES

Ahuerma, F. M. (2023). Recommendation of the OECD Council on Artificial Intelligence: Inequality and Inclusion. *Research Gate*. https://www.researchgate.net/publication/374508986_Recommendation_of_the_OECD_council_on_artificial_intelligence_inequality_and_inclusion

Ahuja, A. S., Reddy, V. P., & Marques, O. (2020). Artificial Intelligence and COVID-19: A Multidisciplinary Approach. *Integrative Medicine Research, 9*(3), 100434. DOI: 10.1016/j.imr.2020.100434 PMID: 32632356

Bajwa, J., Munir, U., Nori, A., & Williams, B. (2021). Artificial Intelligence in Healthcare: Transforming the Practice of Medicine. *Future Healthcare Journal, 8*(2), e188–e194. DOI: 10.7861/fhj.2021-0095 PMID: 34286183

Bal, R., & Gill, I. S. (2020). Policy Approaches to Artificial Intelligence Based Technologies in China, European Union and the United States. *SSRN*. https://dx.doi.org/DOI: 10.2139/ssrn.3699640

Barrington, S., & Farid, H. (2023). A Comparative Analysis of Human and AI Performance in Forensic Estimation of Physical Attributes. *Scientific Reports, 13*(1), 4784. DOI: 10.1038/s41598-023-31821-3 PMID: 36959267

Boden, M. A. (Ed.). (1996). Artificial Intelligence. *Elsevier,* 161-182. https://core.ac.uk/download/pdf/82671696.pdf

Brunette, E. S., Flemmer, R. C., & Flemmer, C. L. (2009). A Review of Artificial Intelligence. *ResearchGate*. 385-392. https://dx.doi.org/DOI: 10.1109/ICARA.2000.4804025

Buczynski, W., Steffek, F., Cuzzolin, F., Jamnik, M., & Sahakian, B. J. (2024). Hard Law and Soft Law Regulations of Artificial Intelligence in Investment Management. *Cambridge Yearbook of European Legal Studies*, 262-293. https://papers.ssrn.com/sol3/papers.cfm?abstract_id=4786682

Busuioc, M. (2021). Accountable Artificial Intelligence: Holding Algorithms to Account. *Public Administration Review, 81*(5), 825–836. DOI: 10.1111/puar.13293 PMID: 34690372

Cugurullo, F., & Xu, Y. (2025). When AIs Become Oracles: Generative Artificial Intelligence, Anticipatory Urban Governance, and the Future of Cities. *Policy and Society, 44*(1), 98–115. DOI: 10.1093/polsoc/puae025

Dixon, R. B. L. (2022). Artificial Intelligence Governance: A Comparative Analysis of China, the European Union, and the United States. *University of Minnesota*. https://hdl.handle.net/11299/229505

Ertel, W. (2024). Introduction to Artificial Intelligence. *Springer Nature*. https://unidel.edu.ng/focelibrary/books/Introduction%20to%20Artificial%20Intelligence%20(%20PDFDrive%20).pdf

Fahad, M., Basri, T., Hamza, M. A., Faisal, S., Akbar, A., Haider, U., & Hajjami, S. E. (2024). The Benefits and Risks of Artificial General Intelligence (AGI). *ResearchGate*, 27-52. https://doi.org/DOI: 10.1007/978-981-97-3222-7_2

Hossin, M. A., Du, J., Mu, L., & Asante, I. O. (2023). Big Data-Driven Public Policy Decisions: Transformation Toward Smart Governance. *SAGE Open*, *13*(4), 21582440231215123. Advance online publication. DOI: 10.1177/21582440231215123

Joshi, D. (2024). AI governance in India – Law, Policy and Political Economy. *Communication Research and Practice*, *10*(3), 328–339. DOI: 10.1080/22041451.2024.2346428

Karmaza, O. O., Koroied, S. O., Makhinchuk, V. M., Strilko, V. Y., & Iosypenko, S. T. (2021). Artificial Intelligence in Justice. *Linguistics and Culture Review*, *5*(S4), 1413–1425. DOI: 10.21744/lingcure.v5nS4.1764

Khare, P. (2025). Investigating the Influence of AI in Social Media Governance. *IGI Global Scientific Publishing*. 365-388. https://doi.org/DOI: 10.4018/979-8-3693-9904-0.ch015

Khare, P. & Raghuwanshi, V. (2025). Analysis of Privacy Laws Impacting Healthcare Data Across Different Regions. *IGI Global Scientific Publishing*. 393-410. https://doi.org/DOI: 10.4018/979-8-3693-7630-0.ch016

Khare, P. & Raghuwanshi, V. (2025). Legal Frameworks Surrounding the Use of AI in Online Content Moderation. *IGI Global Scientific Publishing*, 235-260. https://doi.org/DOI: 10.4018/979-8-3693-9904-0.ch011

Kingston, J. K. (2016). Artificial Intelligence and Legal Liability. *Springer International Publishing*. 269-279. https://doi.org/DOI: 10.1007/978-3-319-47175-4_20

Litwin, A. S., & Racabi, G. (2024). Varieties of AI Regulations: The United States Perspective. *Industrial & Labor Relations Review*, *77*(5), 799–812. DOI: 10.1177/00197939241278956a

Liu, M., Zhang, H., & Sui, Y. (2024). Workplace Artificial Intelligence Regulation in China: Between Efficiency and Social Stability. *Industrial & Labor Relations Review, 77*(5), 813–824. DOI: 10.1177/00197939241278956b

Marinova, N. (2024). Effects of the Artificial Intelligence Ethical Regulations on Society. *IDEAS*, (1), 99-106. https://ideas.repec.org/a/nwe/iitfed/y2024i1p99-106.html

McDermid, J. A., Jia, Y., Porter, Z., & Habli, I. (2021). Artificial Intelligence Explainability: The Technical and Ethical Dimensions. *Philosophical Transactions. Series A, Mathematical, Physical, and Engineering Sciences, 379*(2207), 20200363. DOI: 10.1098/rsta.2020.0363 PMID: 34398656

Paschen, U., Pitt, C., & Kietzmann, J. (2020). Artificial intelligence: Building Blocks and an Innovation Typology. *Business Horizons, 63*(2), 147–155. DOI: 10.1016/j.bushor.2019.10.004

Perc, M., Ozer, M., & Hojnik, J. (2019). Social and Juristic Challenges of Artificial Intelligence. *Palgrave Communications, 5*(1), 61. DOI: 10.1057/s41599-019-0278-x

Pesapane, F., Volonte, C., Codari, M., & Sardanelli, F. (2018). Artificial Intelligence as a Medical Device in Radiology: Ethical and Regulatory Issues in Europe and the United States. *Insights Into Imaging, 9*(5), 745–753. DOI: 10.1007/s13244-018-0645-y PMID: 30112675

Popescu, C. R., & Yu, P. (Eds.). (2024). *Intersecting Environmental Social Governance and AI for Business Sustainability*. IGI Global., DOI: 10.4018/979-8-3693-1151-6

Raghuwanshi, V., Khare, P., & Sharma, P. (2025). Global AI Governance and Collaboration: The Role of International Bodies in AI Regulation. *IGI Global Scientific Publishing*, 225-250. https://doi.org/DOI: 10.4018/979-8-3373-1210-1.ch009

Roy, P., Ghose, B., Singh, P. K., Tyagi, P. K., & Vasudevan, A. (2025). Artificial Intelligence and Finance: A bibliometric review on the Trends, Influences, and Research Directions. *F1000Research, 14*, 122. DOI: 10.12688/f1000research.160959.1 PMID: 39911727

Saheb, T., & Saheb, T. (2024). Mapping Ethical Artificial Intelligence Policy Landscape: A Mixed Method Analysis. *Science and Engineering Ethics, 30*(2), 9. https://10.0.3.239/s11948-024-00472-6. DOI: 10.1007/s11948-024-00472-6 PMID: 38451328

Scherer, M. U. (2015). Regulating Artificial Intelligence Systems: Risks, Challenges, Competencies, and Strategies. *SSRN, 29*(2), 353–400. DOI: 10.2139/ssrn.2609777

Tarafder, A., & Vadlamani, A. (2024). Will the EU AI Regulations Give Rise to Another 'Brussels Effect'? Lessons from the GDPR. *Journal of Development Policy and Practice*, *10*(1), 45–60. DOI: 10.1177/24551333241247670

Truby, J., Brown, R. D., Ibrahim, I. A., & Parellada, O. C. (2022). A Sandbox Approach to Regulating High-risk Artificial Intelligence Applications. *European Journal of Risk Regulation*, *13*(2), 270–294. DOI: 10.1017/err.2021.52

Turk, Z. (2024). Regulating Artificial Intelligence: A Technology-independent Approach. *European View*, *23*(1), 87–93. DOI: 10.1177/17816858241242890

Upadhyay, N.K., & Romashkin, S. (2019). Using Artificial Intelligence to Address Criminal Justice Needs. *Legal Analytics*, 21-34. https://doi.org/DOI: 10.1201/9781003215998-3

Wang, X., Chen, M., & Chen, N. (2024). How Artificial Intelligence Affects the Labour Force Employment Structure from the Perspective of Industrial Structure Optimisation. *Heliyon*, *10*(5), e26686. DOI: 10.1016/j.heliyon.2024.e26686 PMID: 38434398

Yu, P., Mulli, J., Syed, Z., & Umme, L. (Eds.). (2024). *Facilitating Global Collaboration and Knowledge Sharing in Higher Education with Generative AI*. IGI Global., DOI: 10.4018/979-8-3693-0487-7

Yu, P., Wong, S., & Prabhakar, A. (Eds.). (2025). *AI Strategies for Social Entrepreneurship and Sustainable Economic Development*. IGI Global., DOI: 10.4018/979-8-3693-6392-8

Zuiderwijk, A., Chen, Y. C., & Salem, F. (2021). Implications of the Use of Artificial Intelligence in Public Governance: A Systematic Literature Review and a Research Agenda. *Government Information Quarterly*, *38*(3), 101577. Advance online publication. DOI: 10.1016/j.giq.2021.101577

Chapter 5

From "Do No Evil" to "Can't Do Evil": AI-Blockchain Technology as a Transformative Paradigm for Kenya - How To Build Public Trust, Accountability, and Institutional Integrity With AI-Blockchain Technology

James Mulli
http://orcid.org/0009-0001-0423-9225
European Business Institute of Luxembourg, Luxembourg

ABSTRACT

This chapter examines the transformative potential of integrating blockchain and artificial intelligence (AI) to address systemic governance, financial, and electoral challenges in Kenya. Grounded in theories of algorithmic trust, machine ethics, and programmable constraints, it explores how blockchain's immutability and AI's adaptive intelligence can enhance transparency, optimize public services, and reduce corruption. Using case study analysis and comparative review of international models, the chapter proposes practical applications in Kenya's e-procurement, land registry, tax compliance, and healthcare systems. A layered voting model—featuring decentralized digital identity, blockchain-backed registration, and AI-assisted audit trails—is presented to strengthen electoral integrity. While acknowledging regulatory, technical, and social adoption hurdles, it argues that AI can mitigate these

DOI: 10.4018/979-8-3373-1509-6.ch005

through simulation, compliance automation, and capacity-building. This synergy offers Kenya a leapfrogging opportunity to pioneer inclusive, accountable, and innovation-led governance reform.

INTRODUCTION

Global governance systems are under increasing pressure to rebuild public trust, enhance transparency, and combat endemic corruption. In Kenya, where institutional inefficiencies and systemic corruption hinder development, blockchain technology presents a transformative opportunity. By enabling immutable, decentralized record-keeping, blockchain can enhance public accountability, reduce bureaucratic opacity, and provide tamper-proof systems that deter manipulation (Brookings Institution, 2024).

When integrated with Artificial Intelligence (AI), the potential impact of blockchain expands significantly. AI introduces intelligent automation, real-time decision-making, and data analytics capabilities—allowing for dynamic responsiveness to governance challenges. Together, these technologies create what scholars increasingly refer to as algorithmic trust: a system of trust derived not from interpersonal or institutional relationships, but from code, cryptographic consensus, and transparency-by-design (Kshetri, 2023); (Zwitter & Boisse-Despiaux, 2022).

This paper investigates how AI-enhanced blockchain systems, grounded in theories of machine ethics, algorithmic governance, and programmable constraints, can be applied in Kenya to improve transparency, service delivery, and electoral integrity. The objective is to conceptualize a "trustless trust protocol"—a governance model where ethical behavior is encoded and enforced by technology itself, rather than reliant on fallible institutions.

BLOCKCHAIN AND AI IN GOVERNANCE

We need to come to terms with the understanding that artificial intelligence can work synergistically with blockchain to overcome several barriers and unlock new functionalities. While blockchain provides a secure and transparent record-keeping system, AI adds the capability to analyze, interpret, and make decisions based on the vast amounts of data stored on blockchains (Finjan, 2024). For instance, AI algorithms can detect patterns of fraud or corruption by analyzing transaction histories in real time, enabling governments to act preemptively.

The convergence of blockchain and AI offers a novel socio-technical architecture for reimagining governance. Blockchain ensures data immutability, consensus-based validation, and traceability. AI, on the other hand, processes these secure datasets to deliver insights, detect fraud, and optimize public services in real-time. AI-powered blockchain systems are increasingly being deployed in e-procurement, land registry, healthcare logistics, and tax compliance. In Chile, for example, blockchain-based e-procurement systems reduced corruption and saved billions in public funds (World Economic Forum, 2024). In Ghana and Estonia, land and health records are now stored on distributed ledgers, reducing fraud and increasing citizen access (e-Estonia, 2023); (Agaba & Aboagye, 2022).

In Kenya, where corruption remains a significant impediment to progress, AI-powered analytics combined with blockchain can enhance systems like e-procurement, ensuring that public funds are allocated effectively. A compelling example of this integration is evident in the Kenya ICT Authority's 2025 pilot project on blockchain-enabled land titling in Nairobi and Mombasa. The pilot resulted in a 35% reduction in fraudulent transactions and a 40% faster processing time for land title registrations (Kenya ICT Authority, 2025). This initiative not only validates the practical viability of blockchain in Kenya's governance systems but also addresses the critical gap in region-specific empirical evidence. It underscores the potential of blockchain to provide secure, transparent, and tamper-proof records of property ownership—long a source of legal disputes and public mistrust.

Additionally, blockchain can significantly improve tax collection systems by reducing revenue leakages and ensuring compliance through real-time monitoring and audits (Capital FM, 2018). In the healthcare sector, it can enhance drug supply chain management, preventing counterfeit medications from entering the market and ensuring equitable distribution of medical resources. Furthermore, blockchain can be applied in electoral processes to support secure, transparent, and tamper-proof voting systems as an additional complementary layer, restoring public confidence in election outcomes (Korea Tech Today, 2024). These applications demonstrate the transformative potential of blockchain in addressing systemic inefficiencies and enhancing public trust across critical sectors in Kenya. As Finjan (2024) notes, "the application of supervised learning algorithms to blockchain data sets yields predictive governance tools that operate autonomously and impartially."

Essentially, this synergy enables the creation of programmable constraints—smart contracts or protocols that enforce ethical and legal norms by default. For instance, public procurement systems could automatically reject bids that violate legal thresholds, while tax systems could auto-reconcile inconsistencies in declared earnings based on cross-platform data. Such mechanisms ensure that governance is not only transparent but self-enforcing, minimizing reliance on human discretion.

These developments are underpinned by the emerging field of machine ethics, which explores how AI systems can be trained to recognize and act upon ethical principles (Anderson & Anderson, 2023). Drawing from both deontological and utilitarian ethics, these systems can be designed to prioritize fairness, equity, and public welfare—especially critical in countries where institutional trust is often fragile (Zwitter & Gstrein, 2022).

On the basis of the above, it should not be overlooked that blockchain's impact on governance and public integrity worldwide is irrevocable. For example, through the implementation of an e-procurement system, the Chilean government increased savings from $180 billion to $280 billion between 2010 and 2012 (World Economic Forum, 2024).

ADDRESSING CHALLENGES WITH AI-ENHANCED SOLUTIONS

There are, nevertheless, challenges in integrating blockchain technology with government services, including political and social resistance to disruptive change, compatibility with existing government systems, regulatory hurdles, and lack of adequate personnel training (Brookings Institution, 2024). Before proceeding with policy decisions related to blockchain technology, governments should be amenable to the idea and equipped with the knowledge of how exactly the technology can be applied—or misapplied—and ensure that the populations who are meant to benefit from these technologies are also fluent in their use and have access to the necessary digital public infrastructure (The Guardian, 2024). As witnessed in the passage of the U.S.A GENIUS (Guiding and Establishing National Innovation for U.S. Stablecoins) Act., leadership must have buy-in and be knowledgeable enough to understand these complex tools being applied. Acceptance and exposure will therefore allow lawmakers to create a broader system and approach in dealing with blockchain—also called Distributed Ledger Technology (DLT)—so that its benefits can be instrumentalized in service of democratic governance.

AI can mitigate these challenges in integrating blockchain technology by offering adaptive solutions:

- Compatibility: AI algorithms can facilitate seamless integration of blockchain with legacy systems by mapping data structures and automating interoperability (Finjan, 2024).
- Regulatory Compliance: AI-powered compliance tools can ensure that blockchain applications adhere to evolving regulations, reducing the risk of non-compliance (Capital FM, 2018).
- Training and Adoption: AI-driven educational platforms can provide interactive training programs tailored to different stakeholders, from government officials to citizens (Exonum, 2024).

Coupled with blockchain, integrating AI into public services could revolutionize governance in Kenya. AI can simplify complex governmental workflows, ensure compatibility with existing systems, and address the knowledge gap among personnel. Smart contracts, powered by AI, could enforce regulations autonomously, eliminating the risk of human error or bias. Additionally, AI can tailor digital interfaces for citizens, making blockchain-based systems more accessible and user-friendly, thereby bridging the digital divide.

AI can also assist policymakers in crafting strategies for blockchain adoption. By simulating scenarios, AI can predict the social, economic, and political outcomes of blockchain implementation. This enables lawmakers to design policies that maximize benefits and minimize risks, ensuring that populations not only benefit from these technologies but are also prepared to use them effectively. As Zwitter and Gstrein (2022) emphasize, a decentralized digital identity infrastructure is critical to enabling inclusive governance transformations.

BLOCKCHAIN AND AI FUNDAMENTALS

Blockchain technology, while revolutionary, is not an easy concept to grasp due to its inherent complexity—yet this complexity is precisely what makes it unique (Finjan, 2024). At its core, blockchain operates as a decentralized consensus mechanism that provides a system that is secure, transparent, and immutable. Its potential to fundamentally alter political, social, and economic systems cannot be understated.

When combined with artificial intelligence, the capabilities of blockchain are significantly amplified. AI brings adaptability, intelligence, and data-driven decision-making to blockchain systems. While blockchain ensures secure and transparent record-keeping, AI enables the interpretation, analysis, and proactive use of the vast amounts of data stored on blockchains (Kshetri, 2023). Together, these technologies present a transformative tool for combating corruption, streamlining processes, and fostering trust in governance.

Understanding the social and economic implications of AI-enhanced blockchain technology on corruption—and examining proof-of-concept use cases currently in action—is critical for public and institutional buy-in. Blockchain and AI, when leveraged effectively, hold the promise of reshaping how transactions such as voting, finance, and record-keeping are conducted, laying the groundwork for a more equitable and efficient future.

Beyond the speculative "gold rush" of cryptocurrency, the true promise lies in the underlying blockchain infrastructure and its integration with AI. This combination has the potential to redefine trust, accountability, and efficiency in governance, finance, and societal systems (Zwitter & Boisse-Despiaux, 2022). Unfortunately, the cryptocurrency craze and its volatility have significantly distracted public discourse from the real revolution—decentralized data security and algorithmic trust.

RUDIMENTARY UNDERSTANDING

How is it that we have come to accept voting results, financial transactions, bills of lading, letters of credit, and other documents scripted on paper as valid? How is it that we trust the validity of elections, financial statements such as balance sheets or income and cash flow statements presented by auditors? When we review bank statements and audit reports, we place faith in these records because they reflect the value of what we hold in trust—even when that trust is distributed among multiple intermediaries.

The failure of these intermediaries is a fundamental issue in both developing and developed countries. Where checks, balances, and enforcement mechanisms are weak or absent, blockchain technology has emerged as a new trust protocol. As discussed by Kshetri (2023), blockchain provides decentralized, tamper-resistant systems that reduce the risk of corruption and enable traceability. Although complex, this technology must be understood—not dismissed—due to its critical importance in contemporary governance and finance.

WHAT IS ARTIFICIAL INTELLIGENCE?

Artificial intelligence refers to the simulation of human intelligence by machines, particularly computer systems designed to perform tasks that typically require cognitive functions such as reasoning, learning, perception, and decision-making. AI systems operate through algorithms trained on large datasets, enabling them to recognize patterns, predict outcomes, and make autonomous decisions (Anderson & Anderson, 2023).

Machine learning, a subset of AI, improves system performance by analyzing vast amounts of data. For example, AI systems can analyze voter behavior to identify potential anomalies or detect fraud in government procurement. As Binns, Veale, and Edwards (2023) highlight, the deployment of AI in such sensitive areas requires meaningful human oversight to avoid discriminatory outcomes.

AI can be categorized into two main types: narrow AI—designed for specific tasks like facial recognition—and general AI, which aims to replicate a full range of human cognitive abilities, though it remains largely theoretical. When combined with blockchain technology, AI enhances transparency, accountability, and efficiency. AI can interpret data stored on blockchains to detect anomalies, automate workflows, and optimize service delivery.

Together, blockchain and AI create systems that are both trustworthy and intelligent—capable of transforming public administration, enhancing service delivery, and combating fraud in ways previously unattainable through traditional systems.

WHAT IS THE BLOCKCHAIN?

A simple explanation of blockchain—a complex computing concept—is that it is an electronic ledger. In financial accounting, a ledger is a document that records transactions through a system of credits and debits, which are audited to verify and reconcile all inflows and outflows. Blockchain transactions follow a similar process; however, each transaction is unique and identifiable through a hashed encryption code with 64 characters. A hash is a function that maps data of any size into a fixed size (Zwitter & Gstrein, 2022).

Consider that each time a document is transmitted over the internet, what is sent is typically just a copy. Blockchain disrupts this paradigm by enabling the transmission of only one "original" version through the assignment of a hashed encryption code for unique identification. This encryption code is called a Secure Hash Algorithm (SHA), akin to a digital signature, but significantly more complex. The SHA-256 hash is one of several cryptographic hash functions used to secure blockchain transactions. It generates an almost-unique, fixed-size 256-bit (32-byte) output. Due to the mathematical complexity involved, the security of SHA-256 is extremely robust, boasting 36^64 possible values or approximately 330 bits of key strength (Kshetri, 2023).

Now take this a step further: imagine a document being continuously monitored and audited in a digital ledger. This ledger belongs to a larger file system called the blockchain. The auditing is not conducted by a single centralized authority, but by a globally distributed network of independent nodes using sophisticated encryption technology. This decentralized and distributed architecture makes the system virtually impossible to compromise. Its security and immutability are central to its appeal and are what make it the foundational technology for cryptocurrencies (Zwitter & Boisse-Despiaux, 2022).

While blockchain offers a foundational layer of transparency and accountability, the integration of artificial intelligence provides the intelligence and adaptability necessary to address modern governance challenges. Together, these technologies can reshape public services, fight corruption, and reestablish trust between citizens and institutions (Anderson & Anderson, 2023); (Kshetri, 2023).

For Kenya, adopting the synergy of AI and blockchain could signal a pivotal transformation in governance—enhancing societal resilience, promoting economic development, and setting a global precedent for innovation-driven reform (Zwitter & Gstrein, 2022).

Several real-world examples already demonstrate the power of this combination. In supply chain management, AI-enhanced blockchain systems verify product authenticity and track goods through every stage of production. In electoral systems, AI can flag irregular voting patterns, while blockchain ensures each vote remains secure and immutable (Binns, Veale, & Edwards, 2023).

These cases underscore the transformative capacity of AI-integrated blockchain infrastructure to create transparent, secure, and highly accountable systems

As of the writing of this chapter, the application of Blockchain is multifold, as illustrated in some of the following real-world cases

REAL-WORLD USE CASES

Georgia – Blockchain Land Registry

Georgia has implemented blockchain technology to register land titles and validate property transactions. This initiative has improved transparency in land ownership, reducing fraud and corruption (U4 Anti-Corruption Resource Centre, 2019).

United Kingdom – Blockchain-as-a-Service, Welfare Payments

The UK government offers blockchain-as-a-service through its Digital Marketplace. In 2016, the Department for Work and Pensions trialed a mobile app for benefit claimants, enabling them to receive and spend payments securely. Transactions are recorded on a distributed ledger, enhancing financial management and reducing benefit fraud. The UK's chief scientific adviser has highlighted the potential of blockchain in efficiently safeguarding critical operational infrastructure and asset registration on a distributed ledger (UK Government, 2016).

Estonia – Blockchain Identity Management, e-Voting, Electronic Health Records

Estonia is a pioneer in blockchain adoption. Its citizens receive a secure digital ID card, granting access to public services. The integrity of government records is maintained on the blockchain, allowing citizens to control access to their information. Recently, Nasdaq tested a blockchain voting system in Estonia, and the government is using blockchain to secure healthcare records, preventing fraud and providing real-time alerts to security breaches to secure records, thereby ensuring integrity (e-Estonia, 2023).

Singapore – Blockchain Interbank Payments

The Monetary Authority of Singapore (MAS) successfully completed a proof-of-concept pilot for using blockchain in interbank payments, aiming to streamline transactions and enhance security in the financial sector (MAS, 2024).

Dubai – Global Blockchain Council

Dubai established the Global Blockchain Council, consisting of public and private sector members, to explore blockchain applications. They have launched multiple proofs-of-concept covering health records, diamond trade, title transfers, and more. The Crown Prince announced a strategic plan to secure all government documents on blockchain, potentially saving 25.1 million hours of economic productivity annually (UAE Government, 2024).

Delaware, USA – Smart Blockchain Contracts, Public Archives

In 2016, Delaware became the first U.S. state to adopt distributed ledger technology, allowing for the use of smart contracts and improving the management of public archives, enhancing transparency and efficiency in governmental processes (State of Delaware, 2016).

Brazil – Blockchain for Public Accountability

Brazil is exploring blockchain to enhance transparency in public spending. The government is testing a blockchain-based system to track the allocation and use of public funds, aiming to combat corruption and improve accountability (Agaba & Aboagye, 2022).

South Korea – Blockchain for Voting and Public Services

South Korea has launched initiatives to use blockchain for secure voting systems and to enhance public service delivery. The government aims to reduce fraud and improve citizen trust through transparent record-keeping (KoreaTechToday, 2024).

India – Blockchain for Land Titles and Agricultural Supply Chains

India is piloting blockchain initiatives to secure land titles and enhance agricultural supply chains. These projects aim to prevent fraud in land ownership and ensure that farmers receive fair prices for their produce (IJIIRD, 2024).

Nigeria – Blockchain for Identity Verification

Nigeria is implementing blockchain technology for identity verification in its electoral process. This move aims to reduce electoral fraud and enhance the integrity of the democratic process (NIMC, 2024).

Canada – Blockchain for Land Registries and Identity Verification

Canada is exploring blockchain for land registry systems in several provinces. These initiatives aim to streamline the property transfer process and reduce fraudulent claims. Additionally, some provinces are considering blockchain for secure identity verification in public services, enhancing trust and efficiency (DIACC, 2024).

Australia – Blockchain for Supply Chain Transparency

Australia is utilizing blockchain technology to improve transparency in supply chains, particularly in agriculture and food production. This initiative helps ensure that products are sourced ethically and sustainably, thereby reducing corruption and enhancing consumer trust (University of Derby, 2024).

Finland – Blockchain for Public Records

Finland is experimenting with blockchain technology for public records management. The government aims to create a transparent and immutable record of official documents, improving access and reducing the risk of document fraud (ReliefWeb, 2024).

Chile – Blockchain for Public Procurement

Chile has initiated a blockchain project aimed at enhancing transparency in public procurement processes. By recording all transactions on a blockchain, the government seeks to reduce corruption and ensure that public funds are used appropriately. For example, through the implementation of an e-procurement system, the Chilean government increased savings from $180 billion to $280 billion from 2010 to 2012 (World Economic Forum, 2023).

Mexico – Blockchain for Healthcare Records

Mexico is piloting blockchain initiatives to secure healthcare records, allowing patients to control access to their medical information while enhancing data integrity and security. This effort aims to combat medical fraud and improve healthcare outcomes (IEEE, 2024).

Russia – Blockchain for Government Services

Russia is exploring the use of blockchain technology to improve government service delivery. Initiatives include the development of a blockchain-based platform for public services, aimed at increasing transparency and reducing bureaucratic inefficiencies (The Moscow Times, 2017).

Ukraine – Blockchain for Voting and Land Registries

Ukraine has been implementing blockchain solutions for secure voting and land registries. These efforts are focused on improving electoral integrity and streamlining land ownership processes to reduce corruption (CSIS, 2023).

Philippines – Blockchain for Remittances and Public Services

The Philippines is looking into blockchain technology to facilitate remittances and improve public service delivery. This initiative aims to lower transaction costs for overseas workers sending money home while increasing transparency in public transactions (GovInsider, 2024).

Germany – Blockchain for Energy Trading

Germany is testing blockchain technology for decentralized energy trading, allowing consumers to buy and sell energy directly. This initiative promotes transparency in energy transactions and encourages the use of renewable energy sources (Industry Europe, 2024).

Thailand – Blockchain for Agricultural Supply Chains

Thailand is utilizing blockchain to enhance traceability in its agricultural supply chains. This initiative aims to ensure fair pricing for farmers and reduce corruption by making the entire supply chain more transparent (OpenGov Asia, 2024).

South Africa – Blockchain for Anti-Corruption and Diamond Traceability

In South Africa, blockchain is being proposed as a tool to fight corruption. Mining giant De Beers also plans to use the platform to provide a foolproof record of the source of diamonds to ensure they are not from conflict zones (Anglo American, 2024).

WHAT IS THE TRUST PROTOCOL?

In all of the aforementioned cases, Blockchain has been implemented due to failings of efficiency in the role of multiple intermediaries. Therefore, checks, balances and enforcement were weak or non-existent and Blockchain presented new applications for trust. Trust in society is the bedrock of all transactions. Without trust, intermediaries, who act as custodians, would not exist. Let us begin by continuing our understanding with a reliance on historical perspective.

The Historical Origins of Centralized Trust

If we were to wind back the clock and gaze through the historical lens into what we claim to be early systems of trust between individuals and their transactions, we can find what led to trust acceptance by society. Among others, it was a trust-based accounting system in Hammurabi's Babylon, in the form of accounting ledgers, that established this trust as the bedrock of civilization (Aiken & Lu, 1993). Included is the double-entry system of bookkeeping established by Luca Bartolomeo de Pacioli in the late 1400s in Europe, which went on to improve new ways of structuring the economy. The value of these systems has proved itself today in the corporate financial world (Mattessich, 2000).

Let us not fail to mention early trust systems established in other ancient kingdoms like China in the Western Zhou Dynasty that predate the aforementioned and reached a peak of sophistication as early as 1100–771 B.C., the Egyptian Civilization during 1000–3000 B.C., and early Islamic accounting systems dating back to 652 A.D., which also established similar foundations for commerce (Al-Awa, 1995).

Modern Centralized Trust Failings

In the face of history, another new form of bookkeeping might seem like a dull achievement. However, when we consider monumental thefts involving tampered records, such as Enron, WorldCom, or Wells Fargo with fake accounts, or consider Lehman Brothers and repo 105 being endorsed by its auditor Ernst & Young, the need for another trust protocol becomes clear (Brookings Institution, 2024). With Lehman Brothers, faith in a centralized system triggered the biggest financial crisis in 2008, with banks in the U.S. and Europe misreporting their positions. While the trust order and centralization of transactions are being redefined into our modern norms, much of the focus remains on the fanfare associated with cryptocurrencies.

NEW DEVELOPMENT PARADIGMS

As stated before, the real promise of blockchain technology lies not in the speculative markets but in creating what Thomas Friedman describes as a "flat world." If blockchain technology becomes universally accessible, lagging regions can catch up. For example, over two billion people are locked out of essential services such as banking due to the inability to ascertain identities and assets. Blockchain decentralizes trust, offering new opportunities (The Guardian, 2024). These bottom billion in the developing countries of the world have new hopes when trust becomes decentralized.

Development Paradigms, Infrastructure, Brief Colonial History, Neo Colonialism

Let us apply our understanding of technology to the early underpinning of Walter W. Rostow and his 5 stages of economic development:

1. traditional society,
2. preconditions to take-off,
3. take-off,
4. drive to maturity, and
5. age of high mass consumption Rostow, as cited in Kshetri (2023).

Essentially, all technology enables a leap-frog or at the very least an acceleration effect within these stages. Traditional society therefore may indeed bypass the preconditions and witness a takeoff. Do nation states need landline telephones or do mobile telephones suffice? Do small businesses need a bank in order to conduct financial transactions or does mobile banking suffice? This discussion can be extended to other social, economic, and political forums.

The Harrod-Domar theory of saving and productivity of capital as well as discussions of warranted growth, actual growth, and natural rate of growth all need to be revisited to allow for a discussion on labor-intensive and capital-intensive productivity. How does Artificial Intelligence enhanced Blockchain fit in the Ricardian Growth Model? Is comparative advantage a reality in a borderless system of trade where location is not the rule? In this very respect Eurocentricity is a non-item and perhaps as antiquated as Malthusian theory in the context of population, land and the debunking role of technology (Pasquale, 2023).

At the heart of this discussion on the penetration of technology, it is without a doubt that infrastructure challenges are self-evident not in the lack of seabed fiber optic cables and satellite dishes, but inward penetration and proliferation. Although mobile technology penetration has emerged like a beacon of light, penetration lags behind. These failings have perhaps ironically also been the root cause for the aforementioned emergence of a well-established mobile banking culture in East Africa. Without landlines mobile technology leapt forward and reframed our understanding of how mobile telecommunication, that is more pervasive than in more developed countries, can propel banking and thereby economic growth.

So, what does Artificial Intelligence enhance the Blockchain promise, in terms of adoption for early adaptors? If mobile technology was a leapfrog into mobile banking and financial empowerment, can Artificial Intelligence enhanced Blockchain prove to be the spring-board effect for a redefinition of political, social and economic development paradigms?

FINANCIAL REAL-WORLD USE CASES AND APPLICATION IN AFRICA

If technology lives up to its speed, by the time this paper is read the data and facts on Africa will have moved forward very fast.

Cross-Border Blockchain Transactions in Kenya

In December 2017 Kenya's cross-border blockchain transactions via LocalBitcoins.com exceeded KES 80 million per week, with annual transactions averaging KES 1.728 billion. However, LocalBitcoins ceased operations in February, 2023. Since then, other peer-to-peer (P2P) platforms like Paxful and Binance P2P have gained popularity in Kenya for cryptocurrency trading (Journal of International DBA Studies, 2023). While specific figures for Kenya's current cross-border blockchain transaction volumes are scarce, Kenya remains one of the top cryptocurrency markets in Africa. Approximately 6.1 million Kenyans, or about 10.7% of the population, own cryptocurrencies. Cryptocurrencies are increasingly used for remittances, e-commerce, and as a store of value, indicating a growing role in cross-border transactions (Africa Tech Summit, 2024); (LinkedIn Insights, 2024).

Mobile Money Transactions in Kenya

Kenya's mobile money sector continues to thrive and evolve. In 2024 mobile money agents facilitated transactions totaling KES 8.7 trillion, reflecting a 9.4% increase from the previous year's KES 7.95 trillion. This figure represents approximately 53% of Kenya's GDP, underlining the sector's centrality to the national economy (BitcoinKE, 2024); (TechAfrica News, 2024). By December 2024 the number of active mobile money agents had reached 381,116, up from 322,404 in 2023—a significant 18.2% increase. This sustained expansion has cemented mobile money platforms, particularly M-Pesa, as the most widely used financial tools in Kenya, driving unprecedented levels of financial inclusion (GSMA, 2024).

The integration of Artificial Intelligence and blockchain into Kenya's mobile money ecosystem marks the next frontier of innovation. In 2024. Safaricom conducted a cross-border payments pilot using AI-enhanced blockchain technology for M-Pesa transactions across Kenya, Uganda, and Tanzania. The pilot achieved a 22% reduction in transaction costs and maintained 99.8% uptime, setting a new standard for speed, security, and cost-efficiency in digital remittances (Safaricom, 2024). These metrics demonstrate how AI-powered blockchain systems optimize routing, fraud detection, and compliance checks in real time—vastly improving the reliability and scalability of digital financial infrastructure.

This transformation is particularly impactful in the realm of international remittances, where high transaction fees and delayed transfers disproportionately affect low-income households. With blockchain's decentralized ledger ensuring tamper-proof transactions, and AI dynamically managing transaction paths and risk scoring, cross-border mobile money has become more accessible, traceable, and trustworthy. For Kenya—where remittances make up a critical share of household income—these innovations are not just technical upgrades but social and economic enablers.

In a broader context, AI-enhanced blockchain payment systems also offer a promising solution to one of the most persistent challenges in the Global South: financial corruption. Though no technology can fully eradicate corruption, the immutable nature of blockchain significantly raises the cost of manipulation, while AI ensures early anomaly detection and real-time auditability. The Governance Lab at New York University has emphasized the social value of blockchain's traceability in safeguarding public resources, noting that transparent transaction histories disincentivize misuse and empower public scrutiny (Zwitter & Boisse-Despiaux, 2022).

These technologies are already influencing governance frameworks elsewhere. In Ukraine, the Ministry of Finance launched a blockchain-based auction system—Auction 3.0—for the sale of government property, aiming to eliminate bribery and underpricing by ensuring open, traceable transactions. Ukrainian Finance Minister Alexander Danilyuk emphasized that blockchain's transparency "leaves no place for corruption" in the management of public assets (CSIS, 2023). Such examples offer valuable lessons for Kenya as it seeks to modernize its public finance systems.

While corruption is often associated with developing economies, it persists globally and thrives in environments with poor monitoring and weak accounting information systems (Pasquale, 2023). In contrast, blockchain-based mobile money eliminates multiple intermediaries and enables the use of digital identities authenticated via cryptographic key pairs. This design not only lowers costs but also creates an indelible audit trail, allowing authorities to trace every transaction back to its origin without compromising user privacy under properly designed systems.

In sum, the convergence of AI and blockchain in Kenya's mobile money ecosystem represents a transformative leap—not just in transaction efficiency, but in public accountability, cross-border financial integration, and anti-corruption enforcement. These developments position Kenya as a digital finance pioneer in Sub-Saharan Africa and a model for other emerging economies.

LEGAL, ETHICAL, AND SECTORAL IMPLICATIONS OF AI-ENHANCED BLOCKCHAIN IN KENYA

1. Legal and Regulatory Landscape for Blockchain and AI in Kenya

As blockchain and AI technologies grow in relevance, Kenya must grapple with their implications within existing legal frameworks. While Kenya has made commendable strides in digital regulation, significant gaps persist in laws specifically addressing the complexities of blockchain and AI applications.

The current regulatory landscape includes Kenya's Data Protection Act (2019), modeled on the European Union's General Data Protection Regulation (GDPR), which provides a comprehensive framework for safeguarding personal data. However, blockchain's core feature—immutability—poses a legal challenge to this regime. The "right to be forgotten," a foundational aspect of data protection, becomes nearly impossible to enforce when data recorded on a distributed ledger cannot be altered or deleted (Binns, Veale, & Edwards, 2023).

Additionally, the Central Bank of Kenya (CBK) has maintained a cautious stance on blockchain-related financial instruments. Although it has not banned cryptocurrencies outright, the CBK has repeatedly warned against their risks, including potential avenues for money laundering, consumer exploitation, and regulatory arbitrage. Despite this, peer-to-peer trading platforms such as Binance and Paxful remain widely used across Kenya. While the CBK's 2022 discussion paper on Central Bank Digital Currencies (CBDCs) showed initial enthusiasm for blockchain-enabled monetary policy, especially in promoting financial inclusion, the initiative was not pursued beyond consultation stages (CBK, 2022).

To manage innovation without sacrificing oversight, Kenya needs to scale up its use of regulatory sandboxes. Though the Capital Markets Authority (CMA) introduced such a sandbox in 2019 for financial innovation, its application to blockchain and AI remains limited. Broadening the sandbox's scope to include decentralized finance (DeFi), AI-driven public service platforms, and smart contracts could foster experimentation within safe legal boundaries, encouraging compliance and innovation in parallel.

More crucially, the 2025 East African Community (EAC) Guidelines for AI and Blockchain Governance provide Kenya with a timely regional framework to harmonize its domestic regulatory strategy. These guidelines recommend interoperable standards for cross-border blockchain systems—particularly in areas like remittances, digital identity, and e-governance—while emphasizing AI ethics, algorithmic transparency, and inclusive design (EAC, 2025). This regional push for coherent governance structures encourages member states like Kenya to enact tailored legislation that aligns with continental norms while addressing local challenges, such as data sovereignty and electoral integrity.

Kenya can build on this regional momentum to formulate a national blockchain and AI policy, encompassing not only financial regulation but also applications in land titling, tax compliance, identity verification, and public procurement. Drawing inspiration from Estonia's X-Road infrastructure, which facilitates secure inter-agency data sharing using blockchain principles (e-Estonia, 2023), Kenya could create a standards-based architecture that ensures interoperability among state registries, healthcare databases, and biometric ID systems.

In summary, Kenya's legal and regulatory posture must evolve beyond general-purpose data protection and financial caution to embrace sector-specific, use-case-driven, and regionally harmonized policies. This shift is essential to unlock the transformative potential of AI and blockchain while safeguarding public interest, institutional accountability, and cross-border digital cooperation.

2. AI Use Cases in Governance and Finance

AI holds immense potential to modernize governance, streamline financial systems, and especially improve much-needed service delivery in Kenya. From fraud detection to predictive analytics in resource allocation, AI's transformative capabilities are manifold and very much needed.

In the current rapid adoption of e-Government services and public finance, AI can be deployed in procurement processes to detect anomalies in bidding patterns or identify collusion among contractors. In Brazil, the Supreme Audit Institution uses AI-powered software to screen for signs of corruption in public contracts (Organisation for Economic Co-operation and Development [OECD], 2020). Kenya's Public Procurement Oversight Authority could adopt similar systems to enforce compliance with procurement laws.

As the Government in Kenya struggles to collect taxes, machine learning models can identify underreported income by cross-referencing mobile money records, bank data, and utility bills. AI has also been employed in Rwanda's tax authority to improve compliance and revenue forecasting (African Development Bank [AfDB], 2021).

AI can also enhance efficiency in public health by predicting outbreaks using epidemiological data and optimizing supply chain logistics for pharmaceuticals. For example, Zipline uses AI and drones to deliver blood and vaccines in Rwanda and Ghana. Kenya could integrate such technologies within the National Hospital Insurance Fund (NHIF), now SHA (Social Health Authority), and the Ministry of Health to improve service delivery.

In education, AI can identify students at risk of dropping out, enabling proactive interventions. UNESCO's "AI in Education" initiative offers blueprints for data-driven planning in low-resource settings, which Kenya can adapt for its newly implemented Competency-Based Curriculum (CBC) (UNESCO, 2023).

3. The Trust Protocol Revisited: Ethics and Philosophical Foundations

Blockchain and AI present a paradigm shift in how societies perceive trust. Traditionally, trust has been centralized in institutions—banks, governments, or religious bodies. However, repeated scandals and failures have eroded public confidence, creating a demand for decentralized, self-verifying systems.

Trustless Trust: A Paradigm Shift

Blockchain represents a system of "trustless trust," where verifiability supplants reliance. This model aligns with David Chaum's vision of cryptographic protocols as an alternative to institutional oversight. AI further enhances this by offering intelligent surveillance and predictive capabilities, albeit with ethical implications (Zwitter & Boisse-Despiaux, 2022); (Kshetri, 2023).

AI systems can reflect and amplify societal biases if trained on skewed data. For example, facial recognition algorithms often misidentify people of color, raising concerns about discriminatory outcomes in voter ID verification or public service delivery. These risks necessitate rigorous ethical audits and inclusive data practices (Pasquale, 2023); (Anderson & Anderson, 2023).

Privacy is another critical concern. While blockchain ensures transparency, it can also expose sensitive data if not designed properly. Kenya must adopt privacy-preserving protocols such as zero-knowledge proofs or homomorphic encryption to balance openness with confidentiality (Binns et al., 2023).

For postcolonial states like Kenya, the introduction of immutable and decentralized systems offers an opportunity to bypass corrupt intermediaries and establish technological sovereignty. Blockchain and AI can function as tools of decolonization by redistributing power from centralized authorities to local communities, fostering democratic accountability (Zwitter & Gstrein, 2022).

4. Blockchain in Supply Chains and Anti-Counterfeit Measures

Blockchain can radically transform supply chain integrity and transparency across various Kenyan sectors.

Kenya's agricultural exports, such as coffee, tea, and fresh produce, face traceability challenges that affect market access in the EU and North America. Blockchain can log every step of the supply chain—from farm to port—ensuring that origin, processing, and handling standards are met. IBM's Food Trust and AgUnity have implemented such systems globally (IBM Food Trust, 2020).

Counterfeit medicines account for up to 30% of pharmaceuticals sold in Africa. Blockchain can provide tamper-proof authentication of drug batches using QR codes linked to distributed ledgers. This would enhance the integrity of the Ministry of Health's supply chains and save thousands of lives (World Health Organization [WHO], 2021).

Manufacturing and Trade

Smart contracts can be used to automate payments in export contracts once goods are verified through blockchain-logged shipping milestones. This enhances trust among trade partners and reduces fraud.

5. Blockchain and AI for Sustainable Development Goals (SDGs)

AI and blockchain have the potential to advance several of the United Nations' SDGs.

- SDG 1 (No Poverty): Through decentralized finance (DeFi), unbanked populations gain access to savings, loans, and insurance without traditional intermediaries.
- SDG 3 (Good Health and Wellbeing): AI-enhanced blockchain systems in healthcare ensure reliable, timely access to essential drugs and patient records.
- SDG 16 (Peace, Justice and Strong Institutions): Transparent governance platforms reduce opportunities for corruption and improve civic engagement.

Integrating these technologies into Kenya's development planning could fast-track achievement of Vision 2030 goals. Kenya's Vision, 2030 outlines a blueprint for economic transformation through investment in ICT infrastructure and governance reforms. AI-enhanced blockchain systems align closely with this vision by providing tools for transparency, efficiency, and inclusivity (Zwitter & Boisse-Despiaux, 2022); (Kshetri, 2023).

VOTER FRAUD AND THE BLOCKCHAIN

Voter apathy, electoral fraud, and institutional failures are the bane of any democratic form of government. While the voting process in Kenya has taken a technological leap forward, it remains vulnerable to manipulation. This paper not only proposes a drastic reduction in voter apathy by leveraging Kenya's over 81% mobile penetration rate (Capital FM, 2018), but also urges the implementation of a blockchain-based trust protocol for voting. It advocates for at least 90% of Kenya's registered voters to participate in the 2027 general election.

A ballot cryptographically represented within the blockchain can dramatically shift the cost-benefit calculus of electoral fraud. This technology not only reduces election administration costs but also improves accessibility and turnout, as voting can be conducted securely via mobile phones. Blockchain voting enables end-to-end verifiability, empowering voters to track and confirm their vote without compromising anonymity. Because blockchain is a distributed ledger, each transaction—or in this case, each vote—is redundantly stored across many independent nodes. This distribution of authentication renders the process tamper-proof: once a vote is hashed and recorded, modifying it would require control over a majority of the network's computational power—an unfeasible prospect under decentralized conditions (Zwitter & Boisse-Despiaux, 2022); (Kshetri, 2023).

Moreover, recent developments in AI have strengthened blockchain's capabilities in safeguarding electoral integrity. A 2024 case study by Ochieng and Njoroge analyzing Kenya's 2022 by-elections demonstrated that AI-powered anomaly detection systems flagged 17% of voting patterns as suspicious, allowing electoral officials to initiate investigations in real-time (Ochieng & Njoroge, 2024)). These findings underscore how AI can work in tandem with blockchain: while blockchain ensures data integrity and traceability, AI algorithms can dynamically monitor and detect irregularities, such as unusual voting spikes, ghost voters, or vote pattern manipulation. This layered defense system—cryptographic immutability paired with real-time fraud detection—offers a powerful framework for electoral transparency and public trust.

In addition, the cornerstone of any secure voting system is robust identity verification. Traditional credentials such as national ID cards, passports, or driver's licenses remain susceptible to forgery, duplication, or unauthorized use. A modern and more secure alternative lies in Decentralized Identity (DID) systems, which enable individuals to own and control their digital identities. Using cryptographic credentials such as AnonCreds (Anonymous Credentials), these systems can confirm voter eligibility—like age, citizenship, or registration status—without revealing extraneous personal information. Such privacy-preserving mechanisms are critical for enhancing trust, minimizing identity theft, and reducing coercion or vote-buying risks (Zwitter & Gstrein, 2022).

Together, blockchain and AI technologies present an unprecedented opportunity to transform Kenya's electoral system into one that is secure, transparent, inclusive, and resilient against both human error and malicious intent.

Secure Voter Registration

Even when a voter's identity is securely established, fraudulent voter registration remains a persistent threat. Issues such as fake identities, multiple registrations, and manipulated voter rolls have undermined elections worldwide.

The proposed solution integrates digital identity verification with blockchain-based voter registration. Blockchain's immutability ensures that once a voter is registered, the record cannot be altered without detection. Each new entry is timestamped, cryptographically signed, and transparently accessible to authorized entities for verification. This drastically reduces the risk of duplicate or unauthorized registrations (Anderson & Anderson, 2023).

Unforgeable Ballots and Secure Voting Process

Traditional paper ballots, while tangible and familiar, suffer from issues like chain-of-custody breaches, vote-buying, and ballot box tampering. Conversely, purely digital systems can be compromised by hacking and malware. Therefore, a hybrid model is the solution proposed.

In this approach, each voter receives a paper ballot that is simultaneously logged into a blockchain ledger. Additional layers of security include:

- Biometric verification (e.g., fingerprint, iris scan) to ensure that each vote is cast by a verified individual.
- Trusted Execution Environments (TEE) embedded in mobile devices to secure voting apps from tampering.

This model preserves the advantages of both formats while minimizing their vulnerabilities (Pasquale, 2023); (Binns, Veale, & Edwards, 2023).

Secure Voting and Verification Mechanisms

Finally, the voting process should allow voters to cast ballots using a secure application or at physical voting booths. The paper ballots serve as tangible evidence, while blockchain provides a transparent, tamper-proof digital record. Consider the blockchain therefore as a complementary feature and not an outright replacement.

Verification mechanisms include:

- Receipts enabling voters to confirm their vote was recorded correctly.
- Cross-verification during audits using both the paper and blockchain records.
- Triple redundancy, where data is stored across paper records, blockchain entries, and local TEEs.

This ensures that no single point of failure can compromise the integrity of the election (Anderson & Anderson, 2023); (Binns et al., 2023).

Auditability and Transparency

Transparency is a core democratic value, and blockchain's architecture inherently supports it. Each vote is individually verifiable, traceable (without revealing identities), and cryptographically secured.

Key auditing benefits include:

- End-to-end verifiability, enabling stakeholders to validate every phase of the election.
- Real-time access to voting data for authorized monitors.
- Fraud detection algorithms, which can identify anomalies in voting patterns or registration logs (Zwitter & Gstrein, 2022); (Kshetri, 2023).

There are always challenges in integrating blockchain technology with government services, including political and social resistance to disruptive change, compatibility with existing government systems, regulatory hurdles, and lack of adequate personnel training. Before proceeding with policy decisions related to blockchain technology, as previously stated, governments should be amenable to the idea and should be equipped with the knowledge of how exactly the technology can be applied (or misapplied) and ensure that the populations who are meant to benefit from these technologies are also fluent in their use and have access to the necessary digital public infrastructure (The Guardian, 2024); (Brookings Institution, 2024).

National ID System as a Prerequisite

A reliable national identification infrastructure is a prerequisite for any blockchain-based voting system. In many countries, legacy ID systems are fragmented or non-digital, undermining the feasibility of voter authentication (Zwitter & Gstrein, 2022).

Governments must therefore:

- Develop secure, universally recognized digital ID systems.
- Address public concerns about surveillance or misuse of digital IDs (Pasquale, 2023).
- Ensure equitable access to digital tools across urban and rural populations ((The Guardian, 2024).

Without a strong identity backbone, even the most advanced voting technologies cannot succeed. Decentralized identifiers (DIDs), as emphasized by Zwitter and Gstrein (2022), enable individuals to own and control their digital identities without relying on centralized registries. These identifiers, anchored to blockchain, enhance privacy and security while ensuring eligibility verification.

Furthermore, building a reliable national identity system requires public trust, legal safeguards, and interoperability across platforms—a goal Kenya must prioritize as it prepares for future electoral reforms (Anderson & Anderson, 2023).

Implementation Challenges

Despite its promise, blockchain alone is not a panacea. Several obstacles must be addressed:

- Technological complexity may hinder adoption, especially in low-tech environments (Zwitter & Boisse-Despiaux, 2022).
- Governance frameworks must be established to manage blockchain infrastructure, set validation rules, and ensure electoral neutrality (Anderson & Anderson, 2023).
- Integration of technologies like biometrics, Trusted Execution Environments (TEEs), and cryptographic interfaces is non-trivial and requires inter-agency coordination (Binns, Veale, & Edwards, 2023).
- Public awareness and trust are essential; without citizen buy-in, even secure systems may face resistance or underuse (Pasquale, 2023).

A successful implementation depends on a phased approach, stakeholder education, and sustained political will. Kenya must also develop digital public infrastructure and establish ethical standards to prevent misuse or exclusion (Zwitter & Gstrein, 2022).

Historically, when a new technology is introduced to the general population, there is not only anxiety, but also a disruption to the norm. In addition, the increased costs of adoption are generally borne by early adopters seeking to leverage the change. These fears were evident with the concerns raised by the media when Internet banking was first introduced more than two decades ago. Internet banking is now a gold standard in Kenya with mobile transactions accounting for close to 44% of GDP of $63.4Billion (Capital FM, 2018). The cost efficiency of internet banking and Blockchain voting is the common denominator. Blockchain technology and the use of the protocol can therefore be leveraged to bring an effective closure to electoral fraud and protect voter credentials through the creation of a secure end-to-end voting platform that not only offers convenience of use through, for example mobile applications, but also essentially ensures unparalleled transparent auditing through the aforementioned networked ledgers.

Voters would eliminate the casting of votes in centralized locations; thereby reducing the 3.7 trillion Shillings costs associated with these centers for the 2017 elections used for the purchase of equipment and mobilization of personnel. In addition, the convenience of this virtual vote mechanism when applied to a mobile platform would effectively raise participation rates with the unique quality of real-time viewing.nbsp;Participation would be further augmented when the underlying Blockchains auditability features underpin the faith and trust in the results. Present day elections are amenable to influence where voters can possibly be intimidated to vote against their will. In many instances, the trustworthiness of the election process is itself uncertain. In such a situation, we need an election process that is fair, convenient, transparent, and inexpensive. Blockchain technology provides a possibility to, as a compliment, attain a highly dependable and certifiable election process.

CONCLUSION

Any system has its shortcomings and the Blockchain is no exception. However, the cost-benefit analysis dispels this uncertainty.

While blockchain technology is not without its challenges, a cost-benefit analysis highlights its immense potential to address systemic inefficiencies and enhance transparency—a transformation from "do no evil" to "can't do evil" (Zwitter & Boisse-Despiaux, 2022). By integrating blockchain on the back-end for secure and transparent processes while maintaining user-friendly interfaces, its adoption becomes both practical and impactful. Moreover, the inclusion of Artificial Intelligence (AI) further amplifies blockchain's capabilities, offering data-driven, adaptive solutions to governance challenges (Kshetri, 2023). AI can analyze blockchain data in real-time, detecting fraud and optimizing resource allocation (Anderson & Anderson, 2023). In Kenya, where corruption and inefficiencies hinder progress, this synergy holds transformative promise.

The future of secure and inclusive democratic processes depends on layered security, not singular solutions. By combining decentralized digital identities, blockchain-based registration and audit mechanisms, hybrid paper-digital voting interfaces, and transparent governance policies, nations can radically strengthen electoral integrity. While no system is foolproof, such an approach drastically reduces avenues for fraud, increases transparency, and enhances public confidence (Pasquale, 2023); (Binns et al., 2023).

The transformation of voting from an opaque, trust-based process to a cryptographically verified, transparent, and participatory system is within reach—provided we invest in the infrastructure and education needed to implement it effectively (Zwitter & Gstrein, 2022).

By leveraging the synergy of AI and blockchain, Kenya can address its governance challenges, from corruption to electoral fraud, and emerge as a global leader in innovation-driven reform. The integration of these technologies promises not only to enhance transparency and accountability but also to foster economic growth and societal resilience, paving the way for a more equitable and efficient future (Kshetri, 2023).

From secure land registries and efficient tax systems to tamper-proof voting and improved healthcare supply chains, blockchain and AI together can build trust, foster economic growth, and create resilient, transparent systems—paving the way for a more equitable and innovative future (Zwitter & Boisse-Despiaux, 2022); (Agaba & Aboagye, 2022).

REFERENCES

U4 Anti-Corruption Resource Centre. (2019). *Anti-corruption reforms successful in Georgia—Blockchain stealing limelight.* Chr. Michelsen Institute. https://www.u4.no/blog/anti-corruption-reforms-successful-in-georgia-blockchain-stealing-limelight

African Development Bank (AfDB). (2021). *Taxation and technology in East Africa.*

Agaba, E., & Aboagye, E. (2022). Blockchain for land administration in Africa: Lessons from Ghana. *African Journal of Science, Technology, Innovation and Development, 14*(3), 558–570.

Aiken, M., & Lu, W. (1993). The role of accounting in early civilizations. *The Accounting Historians Journal, 20*(2), 163–186. DOI: 10.2308/0148-4184.20.2.163

Al-Awa, M. (1995). The role of Islamic accounting in commerce and civilization. *Arab Accounting Review, 15*(3), 45–67.

Anderson, M., & Anderson, S. L. (2023). *Machine ethics: The design and governance of ethical AI systems.* MIT Press.

Binns, R., Veale, M., & Edwards, L. (2023). Automated decision-making and meaningful human control: A review of recent EU law and scholarship. *Computer Law & Security Review, 49*, 105738.

Brookings Institution. (2024). *Corruption and the global financial crisis.* https://www.brookings.edu/opinions/corruption-and-the-global-financial-crisis/

Capital, F. M. (2018). *Kenya's mobile commerce transactions value hits Sh1.1 trillion.* https://www.capitalfm.co.ke/business/2018/04/kenyas-mobile-commerce-transactions-value-hit-sh1-1trillion/

Central Bank of Kenya (CBK). (2022). *Discussion paper on Central Bank Digital Currency.*

e-Estonia. (2023). *X-Road: Interoperability services.* https://e-estonia.com

Exonum. (2024). *Blockchain land registry in Georgia.* https://exonum.com/story-georgia

Food Trust, I. B. M. (2020). *Improving supply chain transparency.* https://www.ibm.com/blockchain/solutions/food-trust

International Journal of Innovative Research and Development (IJIIRD). (2024). *India's blockchain initiatives in land titles and agriculture.* https://ijiird.com/wp-content/uploads/030222.pdf

Korea Tech Today. (2024). *South Korea to invest $14.5M in blockchain innovation.* https://www.koreatechtoday.com/south-korea-to-invest-14-5m-in-blockchain-innovation-across-public-and-private-sectors/

Kshetri, N. (2023). 1.5 decades of blockchain research in business and management: A review. *Journal of Business Research, 161*, 113839.

Mattessich, R. (2000). *The beginnings of accounting and accounting thought.* Taylor & Francis.

Monetary Authority of Singapore (MAS). (2024). *Project Ubin: Blockchain for interbank payments.* https://www.mas.gov.sg/schemes-and-initiatives/project-ubin

. OECD. (2020). *AI in public integrity systems: Case studies.*

Pasquale, F. (2023). A rule of persons, not machines: The limits of legal automation. *Yale Journal of Law & Technology, 25*(2), 110–152.

ReliefWeb. (2024). *Finland's blockchain experiment for public records.* https://reliefweb.int/report/finland/how-finland-using-blockchain-revolutionise-financial-services-refugees

The Guardian. (2024). *Blockchain's role in reducing corruption.* https://guardian.ng/features/executive-briefs/records-and-information-management-play-a-critical-role-in-fighting-corruption/

The Moscow Times. (2017). *Russia's blockchain platform for public services.* https://www.themoscowtimes.com/2017/10/19/russian-government-announces-first-blockchain-project-for-moscows-property-registries-a59321

UNESCO. (2023). *Artificial intelligence in education: Challenges and opportunities.*

World Economic Forum. (2024). *Chile initiates blockchain for public procurement.* https://www3.weforum.org/docs/WEF_Blockchain_Government_Transparency_Report_Supplementary%20Research.pdf

World Health Organization (WHO). (2021). *Counterfeit drugs in Africa.*

Zwitter, A., & Boisse-Despiaux, M. (2022). Blockchain for humanitarian action and development: Ethical considerations. *Nature Communications, 13*(1), 355. DOI: 10.1038/s41467-022-27985-9 PMID: 35039495

Zwitter, A., & Gstrein, O. J. (2022). Decentralized digital identity: Essential infrastructure for the future. *Big Data & Society, 9*(1). Advance online publication.

KEY TERMS AND DEFINITIONS

Algorithmic Trust: A form of technologically mediated trust based on cryptographic verification, distributed consensus, and algorithmic logic rather than social or institutional relationships. This concept reconfigures trust as a computational output rather than a relational input (Kshetri, 2023; Zwitter & Gstrein, 2022).

Decentralized Identity (DID): A form of digital identity that is owned and controlled by the individual rather than a central authority. DIDs use blockchain to ensure identity verification while preserving privacy (Zwitter & Gstrein, 2022).

Digital Public Infrastructure (DPI): Foundational systems such as digital ID, payment systems, and public data registries, which serve as the backbone for the delivery of public and private digital services.

Distributed Ledger Technology (DLT): The underlying structure of blockchain that distributes data across multiple nodes in a network, eliminating central points of failure and enabling consensus-based validation of transactions.

Immutable Ledger: A digital record that, once created and confirmed on a blockchain, cannot be altered or deleted without consensus—ensuring data permanence and auditability.

Layered Security Model: A multi-faceted approach to security in digital systems that integrates decentralized identity, biometrics, cryptographic encryption, and redundant records (e.g., blockchain + paper) to ensure integrity and resilience.

Leapfrogging: The phenomenon where developing countries bypass intermediary stages of technological development (e.g., going straight from no banking to mobile banking), often driven by innovations like mobile money or blockchain.

Machine Ethics: The study of how autonomous systems (e.g., AI) can be programmed to make decisions aligned with moral principles, including fairness, accountability, and harm reduction (Anderson & Anderson, 2023).

Programmable Constraints: Pre-defined rules embedded into blockchain smart contracts or AI protocols that automatically enforce compliance with legal or ethical standards, thus limiting the potential for discretionary abuse or corruption (Binns et al., 2023).

Programmable Constraints: Rules encoded into smart contracts or AI systems that automatically enforce legal, ethical, or policy-based conditions—reducing discretion and potential for abuse (Binns, Veale, & Edwards, 2023).

Regulatory Sandbox: A controlled environment established by regulators to allow innovators to test new technologies (e.g., blockchain, AI) with temporary relaxations on standard regulations—intended to foster innovation while managing risks.

Secure Hash Algorithm (SHA256): A cryptographic hash function that generates a fixed-size (256-bit) output from data of any size. It is widely used in blockchain to ensure data integrity and secure identity verification.

Smart Contracts: Self-executing contracts with the terms of the agreement directly written into code. They run on blockchain platforms and execute actions when predefined conditions are met—reducing the need for intermediaries.

Trustless System: A system in which parties do not need to trust each other or a central authority because the system's protocol enforces honesty through consensus mechanisms and cryptography.

Zero-Knowledge Proofs (ZKPs): Cryptographic methods that allow one party to prove to another that a statement is true without revealing any information beyond the validity of the statement—used for privacy-preserving identity verification.

Chapter 6
AI-Driven Financial Technology:
Regulatory Challenges and Opportunities in the Digital Economy in Case Studies of EU and China

Poshan Yu
http://orcid.org/0000-0003-1069-3675
Soochow University, China

Jingwen Yan
http://orcid.org/0009-0008-2947-1750
Independent Researcher, China

ABSTRACT

The integration of AI into FinTech has transformed the digital economy, presenting innovation opportunities and regulatory challenges. Using CiteSpace, this study analyzes knowledge clusters in risk management, data privacy, and blockchain, revealing AI's progress in FinTech along with transparency and ethical concerns. Comparing EU and Chinese regulatory frameworks shows significant differences, with the EU's strict AI Act and China's thriving AI - FinTech scene. This emphasizes the need for flexible, globally - coordinated regulations to balance innovation and risk. Burst detection and timeline analysis uncover research gaps in AI applications in FinTech and blockchain - related transparency. The paper concludes by urging international collaboration and robust AI governance to fully realize AI - driven FinTech's potential in the digital economy.

DOI: 10.4018/979-8-3373-1509-6.ch006

I. INTRODUCTION

In recent years, the digital economy has been undergoing a revolutionary transformation, primarily propelled by the integration of Artificial Intelligence and Financial Technology. AI - driven FinTech represents a remarkable paradigm shift in the financial services domain, fundamentally altering the way financial services are designed, delivered, consumed, and regulated.

In the area of service delivery, AI has introduced a plethora of innovative applications. Algorithmic trading enables financial institutions to execute trades at high speeds, leveraging complex algorithms to analyze market trends and make split - second decisions (Dwivedi et al., 2021). Robo - advisors have democratized investment advice, providing personalized financial planning services to a broader spectrum of investors at a significantly lower cost. In fraud detection, AI - powered systems can analyze vast amounts of transactional data in real - time, identifying and flagging potentially fraudulent activities with a high degree of accuracy. Personalized banking services, powered by AI, offer customers tailored financial solutions based on their individual financial situations, spending habits, and goals.

These advancements have not only enhanced the efficiency of financial institutions but have also significantly improved accessibility to financial services. By leveraging AI, financial services can reach previously underserved populations, promoting financial inclusion. For example, in credit scoring, AI algorithms can analyze a wide range of data points, including non - traditional data sources such as social media behavior, online shopping patterns, and utility payment history. This comprehensive data analysis allows for a more accurate assessment of borrowers' creditworthiness, reducing the barriers that have historically limited access to credit for certain individuals and groups.

However, this rapid and far - reaching evolution of AI - driven FinTech also brings about a host of complex regulatory challenges. The digital economy, characterized by its borderless nature, real - time transactions, and the constant emergence of new technologies, demands regulatory frameworks that can adapt swiftly to technological changes. As AI continues to permeate every aspect of FinTech, it raises crucial questions regarding risk management. The complex algorithms used in AI - driven trading systems.

Data privacy is another major concern. With the extensive use of AI in FinTech, vast amounts of customer data, including sensitive financial information, personal identities, and transaction histories, are being collected and processed (Chen et al., 2024). Ethical governance is also at stake. The opacity of AI algorithms in many FinTech applications makes it difficult to understand how decisions are made (Dwivedi et al., 2021). In algorithmic trading, it can be extremely challenging to determine whether a trading loss is due to a flaw in the algorithm, market conditions, or other factors. This lack of transparency undermines trust in the financial system and raises questions about the fairness and reliability of AI - driven decision - making processes.

Furthermore, the adequacy of existing regulatory frameworks in dealing with these new challenges is being called into question. Traditional regulatory approaches often struggle to keep up with the rapid pace of technological change in AI - driven FinTech. There is an urgent need for adaptive and forward - thinking regulatory strategies (Rana & Daultani, 2023). These strategies should aim to harness the benefits of AI - driven FinTech, such as promoting innovation, enhancing financial inclusion, and improving economic efficiency, while effectively mitigating associated risks.

Understanding the regulatory challenges and opportunities of AI - driven FinTech is of paramount importance. It has far - reaching implications for the global economy, financial stability, and societal well - being. A well - regulated AI - driven FinTech ecosystem can contribute to the sustainable development of the digital economy, fostering innovation, and ensuring the protection of consumers and the integrity of the financial system.

II. LITERATURE REVIEW

The integration of AI into FinTech has revolutionized the digital economy, bringing about both significant opportunities and complex regulatory challenges. Many scholars have delved into various aspects of this integration, yet there remain gaps in our understanding.

1. AI - FinTech Integration and Its Implications

Scholars like Dwivedi et al. (2021) highlight that AI offers transformative potential across multiple industries, including finance. Zhang et al. (2021) illustrate how AI optimizes warehouse operations, and this same principle extends to its impact on FinTech decision - making. AI algorithms can analyze extensive financial data, enabling more precise risk assessment and customized financial planning in the financial sector.

Regarding blockchain - related FinTech, Ng, Ho & Wu, (2023) point out that the combination of AI and blockchain can enhance the security and transparency of financial transactions. Blockchain's decentralized ledger technology, when integrated with AI, provides an immutable record of financial activities, and AI can analyze blockchain data to detect irregularities.

2. Regulatory Challenges in AI - Driven FinTech

Data privacy is a crucial concern in AI - driven FinTech. Li, Shen, & Wen (2025) emphasize that in the digital age, data has become a crucial asset, and its improper management can have far - reaching consequences. The 2017 Equifax data leak serves as a prime example, where approximately 147 million consumers' personal information was compromised.

The lack of transparency and accountability in AI - based decision - making processes in FinTech is another significant challenge. Yogesh K. Dwivedi et al. (2021) note that the complexity of AI algorithms makes it difficult to understand how decisions are made.

3. Regulatory Frameworks and AI Governance

Laux et al. (2024) discuss the European Union's efforts to develop harmonized standards involving transparency, fairness, and accountability under the proposed Artificial Intelligence Act. They propose three pathways for standardization, with the "ethical disclosure by default" pathway aiming to shift ethical decision - making to local stakeholders.

In the context of China, while there is a lack of research on the regulatory frameworks in a comparative manner, studies on the general development of AI - related policies in China show that the government plays a crucial guiding role. Policies like the "Guidelines on Actively Promoting the 'Internet +' Initiative" and the "New Generation Artificial Intelligence Development Plan" have spurred enterprises to explore new business models and invest in AI research and development.

4. Research Gaps and the Current Study's Contribution

Previous individual discussions on the integration of AI into FinTech and its implications for the regulatory landscape in the digital economy have produced a wealth of empirical findings. However, most studies focus on either the technological advancements in AI - driven FinTech or the regulatory challenges in isolation, lacking a comprehensive and comparative analysis across different regions. There is a limited exploration of how regulatory frameworks vary across regions like the EU and China, and how they can be optimized in the face of rapid technological changes.

Another gap lies in the limited research on the emerging trends and knowledge clusters in AI - driven FinTech. There is a lack of in - depth exploration using advanced bibliometric analysis tools to map out the intellectual structure of the field.

This paper aims to fill these gaps. By using CiteSpace, a bibliometric analysis tool, we conduct co - citation and keyword analysis to identify knowledge clusters in areas such as risk management, data privacy, and blockchain. Our case studies of the EU and China offer in - depth insights into the regulatory challenges and opportunities associated with AI - driven FinTech, highlighting best practices and lessons learned.

This paper aims to explore how the integration of AI into FinTech has shaped regulatory challenges and opportunities in the digital economy. By focusing on case studies from the European Union (EU) and China, the study seeks to provide a comprehensive understanding of the regulatory landscape and identify emerging trends and future research directions. Consequently, this paper is designed to fill that gap by answering the following key questions.

The main research question guiding this study is: How has the integration of Artificial Intelligence into Financial Technology shaped regulatory challenges and opportunities in the digital economy?

To address this overarching question, the study will delve into the following sub - questions:

(1) What are the emerging trends and knowledge clusters in AI - driven FinTech research, particularly in areas such as data privacy and blockchain, as identified through CiteSpace co - citation and keyword analysis?
(2) How do regulatory frameworks and Al governance mechanisms vary across different regions and jurisdictions, particularly in the EU and China, and what insights can be drawn regarding stakeholder engagement and policy implementation?
(3) In light of the knowledge clusters in AI - driven FinTech research, how can regulators leverage the trends to foster innovation while maintaining regulatory compliance in the FinTech sector?

By examining case studies of regulatory sandboxes, the use of Regulatory Technology (RegTech), and international cooperation initiatives, we can identify best practices and potential pitfalls. Additionally, we will assess the effectiveness of current regulatory frameworks in promoting innovation while safeguarding financial stability, consumer protection, and market integrity. Through this analysis, we aim to offer practical recommendations for regulators to adapt to the rapid evolution of AI - driven FinTech, and develop regulatory strategies that strike a balance between innovation and compliance.

III. METHODOLOGY

1. Research Methods

CiteSpace, a powerful bibliometric analysis tool, enables the exploration of knowledge structures, research fronts, and emerging trends within a specific research field. By conducting co - citation and keyword analysis, CiteSpace maps the intellectual landscape of a research area. Co - citation analysis identifies pairs of documents that are frequently cited together, revealing the relationships between different research works. Keyword analysis, on the other hand, highlights the most prominent and emerging topics in the literature.

Case - study research involves an in - depth examination of specific cases to gain a detailed understanding of real - world phenomena (Yin, 2018). In this study, we selected the European Union and China as case study objects. The EU has implemented strict regulations, such as the AI Act and the General Data Protection Regulation (GDPR), to manage the risks associated with AI. China, on the other hand, has been actively promoting the development of AI - driven FinTech through various policies, aiming to enhance economic competitiveness and improve public services.

2. Data Collection

Consistently with recent systematic literature review studies, we searched through the WOS and Science Direct, to find papers written in English in the fields of "Economics, Econometrics and Finance" and "Business, Management and Accounting", and we restricted our search to post 2010 contributions. We extrapolated 1,520 papers overall.

Figure 1. Data selection and visualization approach

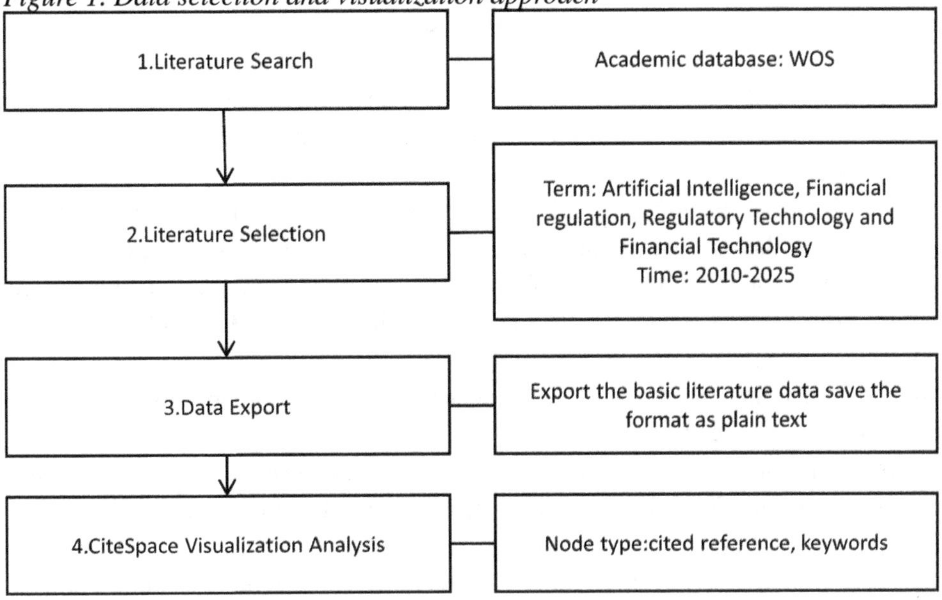

Source: WOS; Science Direct

To select meaningful results, we go through the process as shown in Figure 1. We processed the Title and Abstract of each contribution following the principle of reducing the probability of false negatives. We excluded all unrelated works concerned with other sectors (647), where "artificial intelligence" and "regulatory technology" were simply mentioned in the abstract or title but without relationship between the two. Those papers which are in medical sector studies (261) or covering generic sectors (129), as well as papers on the regulation of technology (47) and on regulation in general (48) are excluded.

To address these research questions, the study will employ a mixed-methods approach, combining bibliometric analysis with case study research. The use of CiteSpace will enable a systematic and data-driven exploration of the literature, providing a robust foundation for identifying trends, knowledge clusters, and research gaps. The case studies of the EU and China will offer in-depth insights into the regulatory challenges and opportunities associated with AI-driven FinTech, highlighting best practices and lessons learned.

3. CiteSpace Analysis of Emerging Trends and Knowledge Clusters in AI and Financial Regulation Research

In the field of AI and financial regulation research, the CiteSpace - generated keyword clustering map offers valuable insights into emerging trends and knowledge clusters.

As it is shown in Figure 2 and Figure 3, the cluster "#0 financial innovation" indicates that the application of AI in financial technology is driving continuous innovation. AI algorithms are revolutionizing financial services, enabling more accurate risk assessment, personalized financial planning, and efficient fraud detection. This trend is not only enhancing the competitiveness of financial institutions but also expanding the scope of financial services, making them more accessible to a wider range of consumers.

Regarding "#1 green innovation", it reflects the growing emphasis on sustainable development in the financial sector. AI is being used to develop green financial products, such as sustainable investment funds and carbon - footprint - based financial services. This aligns with the global trend towards environmental protection and sustainable finance.

The "fintech" and "artificial intelligence" clusters highlight the central role of these two concepts in the research field. Their continuous development and integration are reshaping the financial industry. Blockchain, which is closely related to fintech and AI, as shown in the "#3 fintech" and "#4 artificial intelligence" clusters, is enhancing the transparency and security of financial transactions. It provides an immutable ledger for recording financial data, reducing the risk of fraud and improving the efficiency of regulatory compliance.

However, these trends also give rise to knowledge clusters related to challenges. The "#2 financing constraints" cluster shows that while AI - driven fintech offers new financing opportunities, it also faces issues such as access to capital for startups in this field. The "#5 regulatory network" and "#10 bibliometric analysis" clusters emphasize the need for a comprehensive regulatory framework. As AI - driven fintech evolves rapidly, traditional regulatory methods are insufficient. There is a growing body of research, as indicated by bibliometric analysis, focusing on how to establish a regulatory network that can balance innovation and risk management.

In conclusion, the emerging trends in AI and financial regulation research show a strong drive towards innovation, sustainability, and technological integration. At the same time, knowledge clusters related to regulatory challenges and research methods are emerging, indicating the need for further exploration and study in these areas. This analysis provides a solid foundation for understanding the current research landscape and guiding future research directions in the field.

Figure 2. Keywords cluster in the field of "AI, FinTech & Regulation"

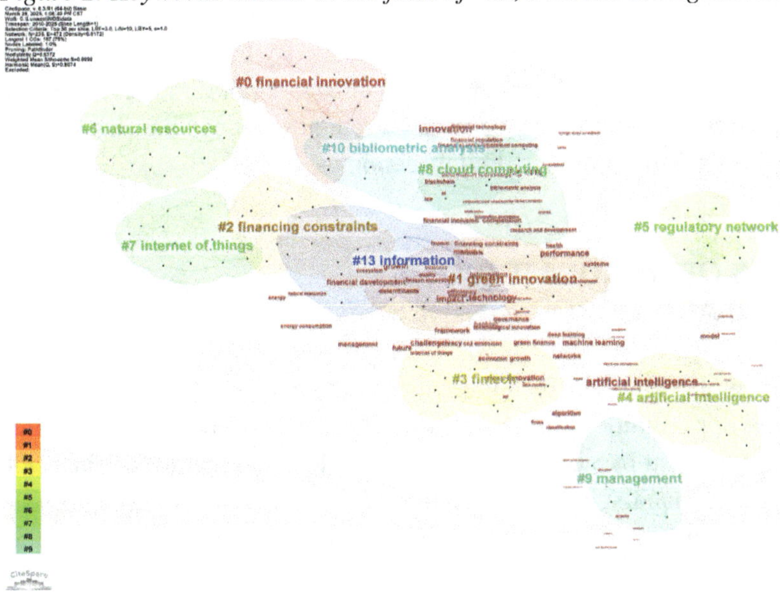

Source: ScienceDirect, 2010-2025

Figure 3. Keyword time diagram in the field of "AI, FinTech & Regulation"

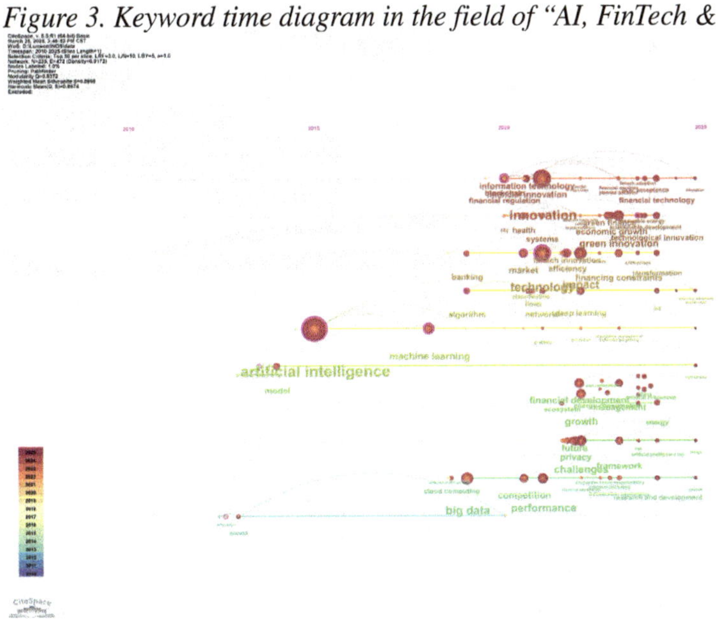

Source: Science Direct, 2010-2025

The findings of this study are expected to contribute to the academic literature on AI, FinTech, and regulation, while also providing practical recommendations for policymakers, regulators, and industry stakeholders. By fostering a deeper understanding of the regulatory landscape, the study aims to support the development of adaptive and globally coordinated regulatory strategies that can harness the full potential of AI-driven FinTech in the digital economy.

IV. OVERVIEW OF REGULATORY CHALLENGES AND OPPORTUNITIES

The integration of AI into FinTech presents a dual-edged sword. On one hand, it offers immense opportunities for innovation, such as enhanced decision-making processes, improved customer experiences, and the democratization of financial services. On the other hand, it introduces significant regulatory challenges, including issues related to transparency, accountability, data privacy, and the potential for systemic risks.

1. Opportunities

One of the most notable opportunities lies in the realm of decision - making processes. AI has the potential to revolutionize FinTech decision - making, much like its impact on optimizing warehouse operations (Zhang, Pee & Cui, 2021). In the financial sector, AI algorithms can analyze extensive financial data, enabling more precise risk assessment and customized financial planning. This ability to leverage diverse data sources allows for a more comprehensive understanding of borrowers, reducing the barriers that have historically limited access to credit for certain individuals and groups. Financial institutions can use AI - based models to assess the creditworthiness of borrowers more accurately, enabling them to make more informed lending decisions and extend credit to a broader customer base. This not only promotes financial inclusion but also helps to allocate capital more efficiently in the economy.

AI also has the potential to enhance the efficiency of financial fraud detection. AI - powered systems can analyze large volumes of transactional data in real - time, identifying patterns and anomalies that may indicate fraudulent activities (Dwivedi et al., 2021). These systems can continuously monitor transactions, looking for signs of unusual behavior, such as sudden large - value transactions, transactions from unfamiliar locations, or patterns that deviate from a customer's normal spending habits. By detecting fraud in real - time, financial institutions can take immediate action to prevent losses and protect their customers. AI - based fraud detection systems can automatically block suspicious transactions and alert customers and fraud prevention teams, reducing the financial impact of fraud on both individuals and the financial system as a whole.

In the investment management domain, AI - based robo - advisors can provide personalized investment advice at a lower cost, making investment services more accessible to retail investors. These robo - advisors can analyze an individual's financial goals, risk tolerance, and market trends to offer tailored investment strategies, democratizing the investment landscape. AI and machine learning are increasingly being used in various industries to optimize processes and make more informed decisions (Rana & Daultani, 2023). In the context of investment management, robo - advisors leverage these technologies to provide customized investment solutions that were previously only available to high - net - worth individuals. They can take into account a wide range of factors, such as an investor's age, financial goals, and risk tolerance, to create personalized investment portfolios. This not only makes investment advice more accessible but also helps retail investors to make more informed investment decisions, potentially leading to better financial outcomes.

In the realm of blockchain - related FinTech, the combination of AI and blockchain can enhance the security and transparency of financial transactions. Blockchain's decentralized ledger technology, when integrated with AI, can provide an immutable record of financial activities (Ng et al., 2023). AI can be used to analyze blockchain data, detect irregularities, and ensure the integrity of transactions. This is particularly important in supply chain finance, where blockchain - based AI systems can track the movement of goods and associated financial transactions, reducing the risk of fraud and improving the efficiency of the supply chain. The use of blockchain technology provides an immutable and transparent record of transactions, while AI algorithms can analyze this data to identify potential risks and ensure compliance with regulatory requirements. This technology can also streamline supply chain operations by automating processes such as payment settlements and inventory management, leading to cost savings and improved overall efficiency.

2. Challenges

However, this integration also brings about serious regulatory challenges. Data privacy is a paramount concern. With the extensive use of AI in FinTech, the collection and processing of vast amounts of customer data have become routine. In FinTech, AI systems often rely on large volumes of customer data, including sensitive financial information, personal identities, and transaction histories.As AI systems in FinTech rely heavily on vast amounts of customer data, including sensitive financial information, personal identities, and transaction histories, the mishandling of this data can have severe consequences. In the recent study, it is emphasized that data has become a crucial asset in the digital age, and its improper management can have far - reaching implications (Li et al., 2025). The sensitivity of financial data makes it a prime target for cyberattacks. As AI - driven FinTech systems become more prevalent, the risk of data breaches increases exponentially. Hackers may target these systems to obtain valuable customer information, which can be used for financial fraud, identity theft, or other malicious activities.

Another significant challenge is the lack of transparency and accountability in AI - based decision - making processes in FinTech. The complexity of AI algorithms makes it difficult to understand how decisions are made (Dwivedi et al., 2021). In algorithmic trading, it can be extremely challenging to determine whether a trading loss is due to a flaw in the algorithm, market conditions, or other factors. This lack of clarity undermines trust in the financial system. If investors cannot understand how investment decisions are being made by AI - powered systems, they may be reluctant to entrust their funds, which could have a negative impact on the growth and development of the FinTech industry. The opacity of AI algorithms is a major concern. In many cases, these algorithms are considered proprietary, and financial institutions are reluctant to disclose the details of how they work. This lack of transparency makes it difficult for regulators, investors, and customers to assess the fairness and reliability of the decisions made by these algorithms.

The potential for systemic risks is also a major issue in AI - driven FinTech. AI - driven FinTech can lead to increased interconnectedness in the financial system. As pointed out in previous study, the complex interactions between different AI - based components in the financial system can create a situation where a malfunction or exploitation of one AI - based trading algorithm could trigger a chain reaction that affects multiple financial institutions and markets (Sullivan & Fosso Wamba, 2024). High - frequency trading algorithms used by many financial institutions at the time contributed to rapid and extreme market movements, highlighting the need for better regulation of AI - driven trading systems. Such systemic risks can have far - reaching consequences, not only for the financial sector but also for the broader economy. A malfunctioning AI system in one financial institution could potentially cause a domino effect, leading to market instability and economic downturns.

Overall, traditional regulatory approaches are ill - equipped to address these new challenges. They are often struggling to keep up with the rapid pace of technological change in AI - driven FinTech. Based on the analysis of the policy in this paper, existing regulations haven't adequately cover new forms of AI - enabled financial products and services. There is an urgent need for proactive, flexible, and globally coordinated regulatory strategies. These strategies should aim to foster innovation while safeguarding public interest. This could involve the development of regulatory sandboxes, where new AI - driven FinTech products can be tested in a controlled environment, allowing regulators to balance innovation with risk management. Additionally, international cooperation is crucial to ensure that regulations are consistent across borders, preventing regulatory arbitrage and promoting a level playing field for businesses. By implementing such strategies, we can harness the potential of AI - driven FinTech without sacrificing security and privacy, ultimately promoting the sustainable development of the digital economy.

V. CASE STUDY OF KEY POLICIES BETWEEN EU AND CHINA FROM STAKEHOLDERS PERSPECTIVE

Artificial Intelligence has emerged as a transformative force in the digital economy, prompting both the EU and China to formulate a series of policies to harness its potential while addressing associated challenges. The regulatory landscapes for AI - driven FinTech in China and the EU reveal distinct yet converging trends, reflecting the global significance of AI in the digital economy. The Facebook data breach in March 2018, which compromised the personal information of approximately 87 million users, sent shockwaves through the global digital community and underscored the critical importance of data privacy and regulatory oversight in the digital age. This incident serves as a significant inflection point for analyzing and comparing the regulatory policies of the EU and China regarding AI - driven FinTech, as it spurred both regions to reevaluate and adapt their existing policy frameworks. By analyzing the key policies in pre - and post - Facebook data breach periods, we can more clearly discern the changes and continuities in the regulatory approaches of these two key regions.

1. Pre - Facebook Data Breach Policy Trends

1.1 Policy Trends in China

Before the Facebook data breach, Chinese policies were primarily focused on promoting the development of AI - driven industries and fostering innovation. Since 2015, initiatives such as the “Guidelines on Actively Promoting the 'Internet +' Initiative” encouraged enterprises to explore new business models. For example, companies in the e - commerce sector started leveraging AI for basic customer segmentation and recommendation systems. However, data privacy regulations were relatively less emphasized in the early stages. The focus was more on using AI to drive economic growth, with policies like the “New Generation Artificial Intelligence Development Plan” aiming to boost investment in AI research and development. These policies have had a positive impact on society as a whole. The development of AI - driven industries has created a large number of jobs, from

AI research and development to application - related service positions. Moreover, these policies helped enterprises in sectors such as manufacturing and logistics to start integrating AI - based technologies for process optimization, but there was limited specific regulation on data handling and privacy protection in the context of AI - driven FinTech.

1.2 Policy Trends in the EU

In the pre - Facebook data breach era, the EU had already started laying the groundwork for data - centric regulations. The General Data Protection Regulation (GDPR) was in the works and came into effect in 2018, just a few months after the Facebook incident. Even before its full implementation, the GDPR signaled the EU's growing emphasis on data privacy and protection. Additionally, policies related to AI were starting to emerge, with a focus on risk - assessment and ethical considerations for AI applications. Initiatives like Horizon 2020 were promoting research and innovation in AI, but also with an eye towards ensuring that new technologies were developed in a way that respected fundamental rights. Compared to China at this stage, the EU was more proactive in formulating comprehensive regulations related to data privacy, although the full impact of these regulations was yet to be seen.

2. Post - Facebook Data Breach Policy Changes

2.1 Policy Changes in China

The Facebook data breach brought data privacy to the forefront of the regulatory agenda in China. In response, the government began to introduce more targeted policies to strengthen data protection in the context of AI - driven FinTech. Policies such as the "Interim Measures for the Management of Generative AI Services" and the "National AI Industry Comprehensive Standardization System Construction Guidelines (2024 Edition)" started to incorporate more stringent data privacy requirements. For example, these policies required FinTech companies to obtain explicit user consent for data collection, implement better data security measures, and be more transparent about how customer data was used. Moreover, there was a greater emphasis on industry self - regulation, with the government encouraging industry associations to develop guidelines for data privacy and ethical AI use. This shift showed China's recognition of the importance of balancing innovation with safeguarding consumer data rights in the digital economy.

2.2 Policy Changes in the EU

After the Facebook data breach, the EU's commitment to data privacy and strict regulation was further solidified. The GDPR's full implementation led to a series of high - profile enforcement actions against companies that violated data protection rules, sending a strong message to the global business community. In the context of AI - driven FinTech, the EU continued to develop and refine its regulatory framework under the AI Act. This act emphasized risk - based classification of AI applications, ensuring that high - risk systems, especially those handling sensitive data, were subject to more rigorous scrutiny. The EU also increased its efforts to promote cross - border cooperation in data privacy enforcement, aiming to address the global nature of digital data flows. Compared to China, the EU's regulatory approach became even more stringent and comprehensive, with a greater focus on holding companies accountable for data privacy violations.

3. Differences and Similarities

One significant difference lies in their regulatory approaches. China focuses more on promoting innovation and industry development through integration and standardization. This is evident in its encouragement of AI applications in various sectors to drive economic growth and improve public services. For enterprises, this means more opportunities to explore new business models and expand their markets. In contrast, the EU emphasizes risk management and human - centric values. It prioritizes protecting individuals' rights and interests, which may require enterprises to invest more in compliance measures. Another difference is the level of centralization. China's policies are often issued at a national level, enabling unified implementation and resource allocation. This centralized approach can quickly mobilize resources for large - scale AI projects. In the EU, policy - making involves multiple institutions and member states, which may lead to more complex decision - making processes but also ensures diverse perspectives are considered.

Even after the Facebook data breach, the differences in regulatory approaches between China and the EU persisted. China still maintained a relatively more balanced approach between promoting innovation and strengthening regulation. Its policies aimed to provide clear guidelines for industry development while gradually enhancing data privacy protection. The centralized nature of policy - making in China allowed for quick implementation of new regulations across the country. In contrast, the EU remained committed to its human - centric and risk - management - focused approach. The multi - institutional and multi - member - state policy - making process in the EU continued to ensure diverse perspectives, but also sometimes led to slower regulatory responses compared to China.

Despite the differences, both regions share common goals. They aim to balance innovation with risk mitigation. China's focus on standardization and ethical governance and the EU's risk - based approach both seek to ensure the safe and sustainable development of AI. For enterprises, this means they need to comply with regulations while pursuing innovation. Additionally, both recognize the importance of AI in driving economic growth. China promotes AI - enabled industrial upgrading, while the EU invests in AI - related research and development to enhance its global competitiveness. This common goal also benefits consumers, as they can enjoy more innovative and reliable products and services, and is positive for society as a whole, promoting technological progress and social development.

Both regions also shared some similarities in their post - breach policy responses. They both recognized the heightened importance of data privacy in the digital economy and took steps to strengthen their regulatory frameworks. Both the EU and China aimed to build trust among consumers by protecting their data rights. Additionally, they continued to see AI - driven FinTech as a key driver of economic growth and innovation, and their policies reflected the need to support the development of this sector while managing associated risks.

VI. LEVERAGING TRENDS IN AI - DRIVEN FINTECH RESEARCH FOR INNOVATION AND COMPLIANCE

In response to the knowledge clusters identified in AI - driven FinTech research, regulators can adopt several multifaceted strategies to foster innovation while maintaining regulatory compliance within the FinTech sector. The ever - accelerating pace of AI development in the financial technology domain has not only unlocked unprecedented opportunities for innovation but has also introduced intricate challenges to the existing regulatory frameworks. Just as the Chinese digital economy, with the aid of technologies like blockchain, has transformed traditional economic models (Yu et al., 2022), regulators need to adapt to the unique characteristics of AI - driven FinTech. By capitalizing on emerging trends, regulators can construct an ecosystem that not only spurs the creation of innovative FinTech solutions but also upholds financial stability, safeguards consumer interests, and preserves market integrity (Dwivedi et al., 2021).

1. Policy - Making and Regulatory Sandboxes

1.1 Tailored Policy Frameworks

Regulators should meticulously develop policy frameworks that are exquisitely tailored to the unique characteristics of AI - driven FinTech. These frameworks must artfully strike a balance between flexibility to accommodate rapid technological advancements and clarity in providing well - defined compliance guidelines. Consider the highly dynamic field of algorithmic trading. As Dwivedi et al. (2021) astutely pointed out, the complexity of AI algorithms renders it extremely challenging to fathom the decision - making processes. Without proper disclosure mechanisms, market participants, including investors and regulators, may find it nearly impossible to understand how these algorithms operate.

Drawing a parallel with China's regulatory approach to blockchain in FinTech, where different application fields require specific regulatory responses (Yu et al., 2022), regulators of AI - driven algorithmic trading can enforce rules that require financial institutions to disclose the fundamental principles of their trading algorithms. This disclosure should encompass crucial details such as the types of data sources utilized, which could include alternative data like sentiment analysis from social media, the step - by - step decision - making logic, and the comprehensive risk management mechanisms in place. Understanding the key elements of trading algorithms through disclosure can help regulators identify potential risks and trends in the financial market (Rana & Daultani, 2023). By implementing such measures, transparency and accountability are significantly enhanced. Simultaneously, this approach still leaves ample room for innovation in algorithm design (Dwivedi et al., 2021). A well - crafted disclosure framework not only empowers regulators to conduct more effective market monitoring but also bolsters investor confidence, thereby ensuring the integrity of the financial markets.

In the context of AI - driven algorithmic trading, it's crucial to note that the lack of standardization in disclosure requirements can lead to regulatory gaps. As Laux et al. (2024) highlighted in the context of AI standardization in the European Union, standardization is a powerful regulatory tool. In the financial sector, establishing standardized disclosure requirements for trading algorithms could be a step forward. This would involve providing detailed documentation on how trading algorithms are developed, tested, and monitored, enabling regulators and market participants to better assess the potential risks associated with these algorithms.

1.2 Regulatory Sandboxes

Regulatory sandboxes have emerged as an invaluable tool for regulators to strike a harmonious balance between innovation and compliance. These sandboxes offer a controlled and secure environment where new AI - driven FinTech products and services can undergo rigorous testing. China's experience with the FinTech innovative regulation pilot, also known as the regulatory sandbox, provides a practical example (Yu et al., 2022). Launched by the People's Bank of China in 2019, this initiative has been implemented in 9 cities (areas) across the country. Regulators can then closely monitor the system's operation, evaluating its impact on consumers, the financial system's stability, and market competition. They can provide timely and constructive feedback to the startup, guiding it through an iterative process. As per established regulatory sandbox models, this iterative feedback loop helps the startup refine its product, ensuring that it meets regulatory requirements. Regulatory sandboxes can foster collaboration between regulators, FinTech startups, and other stakeholders (Rana & Daultani, 2023). By providing such testing grounds, regulators can encourage innovation without compromising the safety and soundness of the financial system (Dwivedi et al., 2021). Some forward - looking regulatory sandboxes even facilitate cross - border testing, enabling fintech companies to explore international markets while remaining under strict regulatory supervision.

When implementing regulatory sandboxes for AI - driven FinTech, regulators can learn from the EU's approach to standardization. Laux et al. (2024) discussed how the EU's standard - setting organizations (SSOs) play a role in developing standards for AI systems. In the context of regulatory sandboxes, SSOs could collaborate with regulators to develop standardized testing procedures and guidelines.

2. Promoting Stakeholder Engagement

2.1 Collaboration with Industry

Regulators should actively engage in collaboration with the FinTech industry to gain a deep understanding of the latest technological trends and innovation needs. This can be achieved through regular and structured dialogue, in - depth industry consultations, and collaborative joint research projects. By involving industry stakeholders in the regulatory process, regulators can develop more informed and effective policies that support innovation while ensuring compliance.

In China, the regulatory sandbox initiative encourages collaboration between financial institutions and high - tech enterprises. Most of the pilot projects in the regulatory sandbox are jointly applied by these two types of entities (Yu et al., 2022). Such collaboration mirrors the importance of stakeholder engagement in the development of effective policies, as seen in various industry - regulatory interactions (Dwivedi et al., 2021). Collaboration between regulators and the FinTech industry can promote international cooperation in the regulation of AI - driven FinTech.

Drawing from the EU's experience with standardization, Laux et al. (2024) emphasized the importance of stakeholder involvement in the standard - setting process. In the context of AI - driven FinTech, regulators can extend this principle to the development of regulatory policies. By promoting such inclusive stakeholder engagement, regulators can enhance the legitimacy and effectiveness of their regulatory policies.

2.2 Public Engagement

In addition to industry collaboration, regulators should also actively engage with the public to understand their concerns and expectations regarding AI - driven FinTech. This can help regulators develop policies that protect consumer interests and enhance public trust in the financial system. By involving the public in the regulatory process, regulators can ensure that the benefits of AI - driven FinTech are widely shared and that potential risks are properly addressed.

China's regulatory framework also emphasizes the protection of financial consumers. The administrative regulations focus on safeguarding consumers' rights, including the right to know, the right to free choice, and the right to property security (Yu et al., 2022). Public engagement is crucial for maintaining trust in the financial system, as demonstrated in previous studies on regulatory decision - making (Dwivedi et al., 2021). The lack of public engagement in the regulation of AI - driven FinTech can lead to ineffective regulations (Rana & Daultani, 2023).

To enhance public engagement, regulators can draw inspiration from the concept of ethical disclosure. Laux et al. (2024) suggested that providing the right information to local stakeholders can enable them to make informed normative decisions. In the context of AI - driven FinTech, regulators can ensure that the public has access to relevant information about the operation and impact of AI - based financial services. This could include publishing reports on the fairness and transparency of AI - powered lending platforms or organizing public awareness campaigns to educate consumers about the potential risks and benefits of AI - driven FinTech. By empowering the public with such information, regulators can foster a more informed and engaged citizenry, ultimately enhancing the effectiveness of regulatory policies.

3. International Cooperation

3.1 Harmonization of Regulatory Standards

Given the global nature of the digital economy, international cooperation is of utmost importance for regulators to effectively address the challenges and opportunities of AI - driven FinTech. Regulators from different countries and regions should collaborate to harmonize regulatory standards.Research showed that similar initiatives in other industries have demonstrated the positive impact of harmonized standards on global market operations (Wamba & Queiroz, 2020).

China, with its growing influence in the global digital economy, can contribute to the harmonization of regulatory standards for AI - driven FinTech. By aligning its regulatory sandbox practices with international standards, China can promote a more coordinated global regulatory environment (Yu et al., 2022). By harmonizing regulatory standards, regulators can reduce regulatory arbitrage and promote a level playing field for fintech companies operating across borders. This ensures that fintech companies are subject to consistent regulations regardless of their location, fostering fair competition in the global market. International cooperation can establish global centers for the regulation of AI - driven FinTech, similar to how the United States serves as a research hub for AI and ML - based supply chain literature (Rana & Daultani, 2023).

In the process of harmonizing regulatory standards for AI - driven FinTech, regulators can learn from the EU's experience with standardization. Laux et al. (2024) analyzed the challenges and potential pathways for standardization under the EU's Artificial Intelligence Act. They proposed that standard - setting organizations should avoid answering hard normative questions themselves and instead focus on creating standards that require "ethical disclosure by default." In the context of international cooperation for FinTech regulation, standard - setting bodies could adopt a similar approach.

3.2 Information Sharing and Joint Enforcement

Regulators should also establish robust mechanisms for information sharing and joint enforcement. This can help regulators detect and prevent cross - border financial crimes, such as cyber - fraud and money laundering. By working together, regulators can enhance their ability to monitor and regulate AI - driven FinTech activities in a global context. Information sharing and joint enforcement are essential for combating cross - border financial crimes, as emphasized in studies on international financial regulation (Dwivedi et al., 2021).

China's experience in promoting regional financial cooperation, such as in the Guangdong - Hong Kong - Macao Greater Bay Area, can serve as a model for international information sharing and joint enforcement. In this region, cross - border RMB payment and settlement projects have been piloted to enhance financial integration and security (Yu et al., 2022). In the case of a cross - border fintech fraud scheme, information sharing between regulators can lead to faster detection, more effective prosecution, and ultimately, the protection of the interests of consumers and the stability of the global financial system.

When establishing information sharing and joint enforcement mechanisms, regulators can consider incorporating the concept of ethical disclosure. Research proposed that ethical disclosure can shift ethical decision - making to local stakeholders (Laux et al.,2024)). In the context of cross - border FinTech regulation, regulators could require fintech companies to disclose information about their compliance with ethical and regulatory standards to relevant international regulatory bodies. This information could then be shared among regulators to facilitate joint enforcement actions.

REFERENCES

Chen, Y., Che, Z., & Wan, D. (2024). How does the financial technology innovation regulatory pilot influence financial regulation? *Finance Research Letters*, *69*, 106255. DOI: 10.1016/j.frl.2024.106255

Dwivedi, Y. K., Hughes, L., Ismagilova, E., Aarts, G., Coombs, C., Crick, T., Duan, Y., Dwivedi, R., Edwards, J., Eirug, A., Galanos, V., Ilavarasan, P. V., Janssen, M., Jones, P., Kar, A. K., Kizgin, H., Kronemann, B., Lal, B., Lucini, B., & Williams, M. D. (2021). Artificial Intelligence (AI): Multidisciplinary perspectives on emerging challenges, opportunities, and agenda for research, practice and policy. *International Journal of Information Management*, *57*, 101994. DOI: 10.1016/j.ijinfomgt.2019.08.002

Laux, J., Wachter, S., & Mittelstadt, B. (2024). Three pathways for standardisation and ethical disclosure by default under the European union artificial intelligence act. *Computer Law & Security Review*, *53*, 105957. DOI: 10.1016/j.clsr.2024.105957 PMID: 41322429

Li, W., Shen, Z., & Wen, X. (2025). Access to banking and corporate dividend policy: Evidence from the distribution of bank branches in China. *Research in International Business and Finance*, *75*, 102731. DOI: 10.1016/j.ribaf.2024.102731

Ng, S. C. H., Ho, G. T. S., & Wu, C. H. (2023). Blockchain-IIoT-big data aided process control and quality analytics. *International Journal of Production Economics*, *261*, 108871. DOI: 10.1016/j.ijpe.2023.108871

Rana, J., & Daultani, Y. (2023). Mapping the role and impact of artificial intelligence and machine learning applications in supply chain digital transformation: A bibliometric analysis. *Operations Management Research : Advancing Practice Through Research*, *16*(4), 1641–1666. DOI: 10.1007/s12063-022-00335-y

Sullivan, Y., & Fosso Wamba, S. (2024). Artificial intelligence and adaptive response to market changes: A strategy to enhance firm performance and innovation. *Journal of Business Research*, *174*, 114500. DOI: 10.1016/j.jbusres.2024.114500

Sullivan, Y., & Fosso Wamba, S. (2024). Artificial intelligence and adaptive response to market changes: A strategy to enhance firm performance and innovation. *Journal of Business Research*, *174*, 114500. DOI: 10.1016/j.jbusres.2024.114500

Yin, R. K. (2018). Case Study Research and Applications: Design and Methods (6th ed.). *SAGE Publications*,Inc.Retrieved from https://uk.sagepub.com/en-gb/eur/case-study-research-and-applications/book250150

Yu, P., Gong, R., & Sampat, M. (2022). Blockchain Technology in China's Digital Economy: Balancing Regulation and Innovation. In Global, I. G. I. (Ed.), *Blockchain Technology in China's Digital Economy: Balancing Regulation and Innovation* (pp. 132–157). IGI Global., DOI: 10.4018/978-1-7998-7927-5.ch007

Zhang, D., Pee, L. G., & Cui, L. (2021). Artificial intelligence in E-commerce fulfillment: A case study of resource orchestration at Alibaba's Smart Warehouse. *International Journal of Information Management*, *57*, 102304. DOI: 10.1016/j.ijinfomgt.2020.102304

Chapter 7
Integrating Cybersecurity Into Financial Planning:
Strategies for Protecting Client Assets in the Digital Age

Daniel Y. N. Tan
http://orcid.org/0009-0008-9177-0137
Golden Gate University, USA

ABSTRACT

This chapter covers incorporating cybersecurity into financial planning due to the increasing threats in the digital space. It explores fundamental risks such as phishing, ransomware, data breaches, as well as the client portal and planning software vulnerabilities. Focus is placed on proactive measures regarding asset protection encryption, multi-factor authentication, AI threat detection, and others. Cyberattacks case studies illustrate loss of organizational trust, reputation, and business, as well as compliance breaches. The responsibility of a financial planner as a proactive cybersecurity hygienist and other technologies like blockchain for cryptocurrencies as secure online transaction tools is also mentioned. Issues of compliance are analyzed with the policies set forth by GDPR and GLBA. This chapter highlights the need to cultivate a security culture that fosters enduring resilience while adapting to evolving threats, all while maintaining client trust.

INTRODUCTION

Digitization of financial services has automated wealth management for people and organizations using tools like robo-advisors, AI analytics, and blockchain. However, new technologies come with their own set of vulnerabilities and risks;

DOI: 10.4018/979-8-3373-1509-6.ch007

as such, cybersecurity measures are paramount in the protection of sensitive data. "Cybersecurity" is the practice of safeguarding systems, networks, and data from digital infiltration. With the changing business landscape, protecting sensitive data and controlling its access is not just an additional obligation; it is legal responsibility. Fiduciary duty encompasses safeguarding private information such as institutional phishing, ransomware, and data breaches targeting private and institutional information. Clients and institutions alike stand to lose incomparable value in assets, reputational trust, and legal standing within regulatory frameworks (Accenture, 2021).

Such considerations demonstrate why cybersecurity should be incorporated into the planning on finances. The use of advanced technologies in phishing has increased attackers' access points, resulting in breaches for as much as 70% of financial services firms in a single year (Deloitte, 2020). This corroborates Fang's (2024) observation about the growing explosion in, as well as the FinTech and DeFi paradigms, significantly worsening stress in the system and requiring active risk management. Concepts such as defense in depth (Nierengarten, 2006) and agency theory Uddin et al. (2020) neglect devoted caregivers who ethically ensure client privacy in client data and identity protective measures while implementing policies for access control through multi-level firewalls and intrusion detection. There is a gap in the literature regarding the insufficient outline of problems most financial planners face, such as client education and the rapid implementation of artificial intelligence systems for identifying and eliminating hostile infiltrators (Gupta et al., 2020).

Debates today also ignore the layers of regulations involved. Compliance, particularly for smaller firms, is still fractured at best, as Didenko (2020) notes with GDPR and GLBA. As Shaw et al. (2009) elaborate on behavioral finance theory, clients are overexposed to social engineering techniques due to biases. Planners are left with the gap of combining technical solutions with human-centric education.

Compiling industry wisdom, regulations, and available research, this chapter fills current gaps by proposing actionable steps for enhancing trust and resilience in the digitized world of finance. Alongside technological infrastructures, clients need to be empowered through education, making cybersecurity relevant to financial planning.

Background

Recent technological advancements have facilitated a drastic change in the global economy within the past few years. The emergence of digital tools and assets like cryptocurrencies and blockchain technology has changed the traditional approaches to financial management by both businesses and individuals. At the same time, these advancements pave the way towards new hurdles, especially in the field of cyber security (Fang, 2024).

Problem Statement

As technology becomes more pervasive in financial planning, wealth managers and financial advisors now must worry about the security and protection of the assets of their clients. The ever-evolving complexity of cyber criminals, who are perpetually attempting to infiltrate financial institutions and obtain confidential information, pose a danger to the financial health of clients.

Objectives

This chapter seeks to analyze the strategies and best approaches to incorporating cybersecurity into the financial planning process. The guiding purposes are to:

- Determine the most critical cybersecurity threats that impact the effectiveness of the financial planning process.
- Investigate the various protective measures and strategies that can reduce the threat level.
- Investigate the impact of technology on the protection of clients' assets.
- Evaluate the issues concerning cybersecurity and financial planning in relation to regulation and compliance.
- Recommend how clients can be taught best practices in cybersecurity.

LITERATURE REVIEW

Overview of Cybersecurity in Financial Services

The financial services sector has become one of the most sought-after sectors for cyberattacks because of the sensitive data involved and the continuing use of technology. Phishing attacks and advanced persistent threats are just some of the more common techniques utilized by cybercriminals. Now more than ever, wealth management firms and other financial institutions must ensure that their cybersecurity infrastructure is solid enough to defend against attacks and protect client assets. Accenture (2021) estimates that firms in the financial services sector suffer from cyberattacks daily, 300 percent more than businesses from other sectors. The rapid growth of FinTech and digital banking has only added to the problem and made it more important for financial planners and clients alike. Some trends shaping the industry include the use of AI technology to anticipate and mitigate risks, blockchain technology for enhanced security, or robust legal policies directed against consumer data negligence. Yet even with these changes, financial planners trying to protect client assets will continue to struggle due to the relentless sophistication of modern-day cybercriminals.

Financial Planning in the Digital Age

The modernization of technology in financial planning has changed how advisors and clients work together and how portfolios are managed. The growth of robo-advisors along with the adoption of cloud-based services have Simplified financial planning, enabling instant access to client information and custom-tailored investment plans. At the same time, these changes pose additional risks, since financial planners must face potential data breaches, cyberattacks, and even affiliate fraud. For instance, Deloitte (2020) shows that 70% of the companies that provided financial planning services had suffered at least one case of cybersecurity violation during the previous year, and most of these cases were caused by phishing attempts. Financial planners have an obligation to safeguard sensitive information about the clients, including their personal financial details, investments, and even their retirement plans form the wrath of cyber criminals.

These risks have only increased with the shift to remote work during the COVID-19 pandemic, as both financial planners and clients depend more on cloud-based systems and unsecured networks. This further highlights the need to mitigate these risks through better cybersecurity protection to sensitive financial data and maintain trust.

In the digital era of financial planning, protecting sensitive data with multi-factor authentication, encryption, and regular software updates has become essential.

Theoretical Frameworks

From a theoretical standpoint, the integration of cybersecurity in financial planning can be approached in different ways. A strategy that applies perfectly in this case is the "defense in depth" approach - it revolves around the addition of various protective measures to lower the chance of an attack succeeding in the specific area of financial planning (Nierengarten, 2006). "Risk management" is another important concept since it states that financial planners must assess and control potential cybersecurity risks to clients' assets. As an example, risk management theory functions based on recognizing, evaluating, and mitigating risks in relation to the organizational goals set to achieve the overarching goal. In addition, "agency theory" matters here as it shows that financial planners have a responsibility to their clients as fiduciaries which includes protecting them from cyberattacks on their finances. (Uddin et al., 2020).

This includes performing periodic risk assessments, taking preventive measures, and designing response actions in case of an incident. Here, Behavioral Finance theory comes into play as it deals with the cognitive aspects of decision-making as it is made in the specific context. This is necessary in teaching clients regarding the best practices of cybersecurity and making sure that they participate in the security process.

Cybersecurity risks pose unique challenges to financial planners and their clients. As financial planners, they must devise communication plans that promote secure behaviors from their clients. For example, understanding the value of an asset may motivate a client to implement multi-factor authentication (MFA).

Gaps in Existing Research

Though the literature is insightful on the importance of cybersecurity in the financial services sector, there are still gaps that this chapter seeks to fill.

To start off, the existing literature mainly focuses on the cyber security concerns of institutions such as banks and investment companies, while neglecting the worries of financial planning firms. Particularly, there is a lack of literature concerning the financial planners' struggles within the scope of client education and communication. Additionally, there seems to be no literature detailing how AI and blockchain and other emerging technologies are shifting the cyber security landscape of financial planning services. Finally, financial planners need more ideas on how to protect their clients' assets.

This chapter will try to address these gaps by developing a systematic approach integrating cyber security into financial planning using recent academic research and industry works.

CYBERSECURITY RISKS IN FINANCIAL PLANNING

Types of Cyber Threats

While the convenience and efficiency that comes with financial planning in the digital age is unmatched, cyber threats are more prevalent than ever. Financial planners must understand these threats to guard their clients' financial interests. This section covers different breaches including, but not limited to phishing, ransomware, data breaches, and malware, alongside their real-world implications.

Phishing

Phishing scams are defined as fraudulent attempts to extract personally identifiable information, for example, account names, passwords, and credit card information. "Phishers" usually send emails or text messages posing as valid companies. These messages may lead victims to spoof sites pretending to be real login pages, where users are tricked into submitting their credentials. A notable example is in 2016, when phishers targeted the Democratic National Committee and used sophisticated phishing techniques to obtain sensitive emails. Chang (2022) explains how attempts at phishing use the logos of known companies and often sophisticated replicas of the original site, making it more difficult to detect. When offering financial services, phishing scams can lead to accessing clients' transactions and executing fraudulent activities using stolen credentials.

Ransomware

In a ransomware attack, the motive of the attacker would be to hold a victim's information hostage behind encryption and walls and ask for a ransom fee to bring it back. Frequently, these attackers break into systems through a phishing email sent with a particular intent or through an infected web page. If the user system is compromised and the ransomware successfully implanted, it goes ahead to encrypt and render useless to the user files that are crucial.

Consistently, the attackers will ask for payment to be made, in some cases, using cryptocurrency while in return offering the decryption key. An example of this is the attack by supposed North Korean hackers in 2017 concerning the WannaCry ransomware where they managed to infect a couple of institutions and international corporations, colleges, public health facilities and state agencies. For financial service providers, a ransomware invasion can paralyze the processes since operations as in accessing customer information will be impossible amid huge financial losses if the ransom is disbursed.

Data Breaches

Data breaches are defined as the unauthorized access of sensitive information including client financial records, personally identifiable information (PII), and intellectual property. Weak security measures, insider threats, and advanced hacking tactics all contribute to data breaches and culminate in devastating vulnerabilities for organizations. For instance, the 2013 Target breach in which hackers absconded with millions of customer credit cards illustrates the catastrophic impact this could have on a data holder. In the context of financial management, breaches could lead to compromised protected information including sensitive PII, financial records, and investment strategies which in turn precipitate identity theft and significant pecuniary loss or reputational harm to the firm. Hermawan et al. (2023) have highlighted how such breaches have irrevocable consequences for the integrity and confidentiality of sensitive financial data.

Malware

Malware describes software that is created for the particular purpose of causing damage or disabling computer systems. This malicious software can exist in the format of viruses, worms, and even Trojans. Malware can be disseminated via infectious attachments, compromised software, or infected web pages. One notable example is the ILOVEYOU malware from the year 2000 which distributed itself via email attachments and caused extensive damage to computer systems globally. Malware infections can steal private information, disrupt business processes, and give unconsented access to the systems to attackers, therefore being a huge risk to planning financial firms. Fang discusses these issues citing "sensitive data can undercut trustworthiness as a result of a malicious information attack that might lead to financial loss and damage reputation" (Fang, 2024).

Other Threats

Apart from these broader threats, financial planners need to pay attention to social engineering, denial-of-service attacks, and insider threats as other emerging risks. Social engineering is the process of obtaining confidential information by infiltrating social circles and convincing individuals to take actions that would compromise security. Systems are rendered useless to bona fide users through traffic bombardment or denial-of-service attacks. These threats come from within the organization from actors such as employees and contractors who engage in bad faith attempts to disrupt the organization's goals. Cristea (2020) mentions several security threats, including phishing, electronic fraud, and impersonations.

Safeguarding client assets requires attention on multiple fronts, integrating effective technological security with employee training and client education. Financial planners need to understand the changes in the risks and use proper methods to guard against them.

Vulnerabilities in Financial Planning Tools

The rise of digital infrastructure has led to improvement in efficiency and access when it comes to financial planning. However, client data and assets are put at risk because of such systems. In this sub-section, the focus is placed on the risks associated with the most popular financial planning tools, which include digital platforms, client portals, communication tools, and financial planning software.

Digital Platforms

Within the context of digital platforms, a variety of services pertinent to financial planning are integrated, such as trading platforms, robo-advisors, and portfolio management systems. Client sensitive information like confidential financial data, investment plans, and personally identifiable information are often kept on file. There can be potential weaknesses in their security from insufficient security measures, bugs in the application, or weak data encryption. For example, a trading platform may not adequately protect client accounts from being accessed by unauthorized users who wish to execute fraudulent trades. Kakareka (2009) articulates how vulnerabilities can arise from knowledge bases and systems that have not been updated for a long time. Lowered exposure to potential risks can be attained through more thorough security checks and system updates.

Client Portals

With the help of client portals, clients can now easily access their financial statements, investment portfolios, and even communicate with their financial planner online. Portals like these are extremely useful; however, they are prone to numerous hacking attempts if sufficient security measures are not implemented. The lack of multi-factor authentication, weak passwords, and insecure communication channels can place clients' information at the mercy of hackers. Garg et al. (2022) explain how an imbalance of data coupled with noisy historical data creates a risk that is too often taken advantage of. Enhanced privacy for clients would require the incorporation of improved multi-factor authentication and more robust encryption protocols.

Communication Tools

Financial planners employ various tools such as email, instant messaging, and video conferencing software to interface with clients. All these tools are susceptible to phishing, malware, and eavesdropping attacks. An example of this would be an attacker sending a phishing email to a client and upon opening it, they unknowingly expose their credentials which allow the attacker to take control of the client's financial accounts. Chang (2022) points out that bad actors have become more advanced in conducting phishing attacks. With these phishing attacks, the employment of secure communication systems, establishing strong phishing education for clients, and securing emails helps mitigate these attacks tremendously.

Financial Planning Software

With software tailored for financial planning, planners can evaluate client data, formulate financial strategies, and manage investments. Poorly safeguarded software could jeopardize sensitive client data if exposed to a breach. Weak security protocols, inadequate software protection, and ineffective access controls could permit unauthorized access to sensitive data. Often, as pointed out by Neal (2020), advisors do not utilize the complete functionalities of financial planning software, which enables them to overlook important aspects—including protective measures. Regular updates, complicated passwords, and stringent access restrictions are essential for protecting financial planning software.

Other Vulnerabilities

Besides these instruments, other threats could arise from the lack of basic cybersecurity hygiene of financial planning companies, such as inadequate internal controls, insufficient employee training, and weak incident response planning. Zhang et al. (2020) delineates the vulnerabilities of information systems. To mitigate these risks, companies should implement comprehensive cybersecurity policies alongside training and strategic incident response plans.

To eliminate these gaps, a mix of sophisticated security techniques alongside client and employee education is required. In the scope of financial planning, asset protection and maintaining client trust places cybersecurity on the critical path. Feyan et al. (2023) highlighted the importance of protective measures regarding data and cybersecurity.

Impact on Clients and Financial Planners

In the world of cybersecurity, data breaches can have severe ramifications not only for clients but also for financial planners. These incidents extend far beyond the financial implications, as they undermine the trust, reputation, and overall well-being of the entire financial planning industry. This sub-section examines the various profound damage resulting from breaches of cybersecurity, which include, but are not limited to, diminished trust, economic damage, reputational damage, regulatory penalties, and emotional distress.

Loss of Trust

The erosion of trust in a financial firm and the planner is perhaps the most detrimental consequence of the breach of cybersecurity. The moment a client's private data is exposed, there is an almost certain likelihood that the client will become skeptical about the planner's skills and ability to protect their sensitive information.

Such erosion of trust may translate to client attrition, negative word of mouth, and difficulty in obtaining new clientele. Kitces (2014) elaborates on the significance of communications by the advisor as well as their trustworthiness in the eyes of the client. Restoring trust after a breach is a long, painstaking process that requires transparent communication, proactive outreach, and clear evidence of enhanced security measures. Asaad (2020) stresses the responsibility of the financial advisors to offer insight as well as strengthen confidence among investors.

Financial Damage

Security breaches and other cybersecurity incidents inflict devastating monetary losses on clients and financial planners. Clients can incur direct financial loss from the theft of money, unauthorized access to funds, and identity theft. On the other hand, financial planners face losses in conducting investigations into the breach, implementing remedial measures, offering credit monitoring services to affected clients, and litigating legal action (Fox & Bartholomae, 2020). As caretakers of household finances, planners must provide enduring explanations when unanticipated financial shocks disrupt clients' budgets. Partial financial compensation is offered through cybersecurity insurance policies. However, these policies do not adequately reimburse lost reputation, trust, and other non-tangible damages.

Reputational Risks

A cybersecurity breach can critically damage the standing reputation of a financial planning firm. Public backlashes may arise, especially with social media, which combined with regulatory scrutiny, poses challenges in attracting potential clients. As Fang (2024) highlights, the overriding reputation in the industry makes it easier to gain and retain clientele for fintech firms. It further hampers the firm's capacity to secure partnerships, attract investors, or sustain professional certifications. Restoring a damaged reputation necessitates employing a strategy based on open dialogue, active engagement with stakeholders, and verifiable enhancement of security measures.

Regulatory Scrutiny

Like other businesses, financial planning firms need to comply with several regulations focused on protecting data security and data privacy. Any breach of cybersecurity may invite some form of regulatory scrutiny along with measures which may include monetary fines and lawsuits. The authors Payne & Asebedo (2019) mention that financial planners actively participate in making recommendations because of client assessments. Avoiding regulatory risks brings the necessity of protective compliance with laws such as GDPR and CCPA. To avoid regulatory penalties, financial planners must ensure that the claim of compliance can be substantiated through demonstration of active control mechanisms such as structured security attainment measures, scheduled risk assessments, and defined incident response plans.

Psychological Distress

Both clients and financial planners can suffer immense emotional trauma due to a cybersecurity breach. Clients often feel anxious, stressed, and vulnerable after their financial information is compromised. Financial planners often feel the pressure of resolving the breach, managing emotional client concerns, and making amends for the unfortunate incident. Wilcox (2024) focuses on the need to leverage technology in providing support to clients during uncertain periods. Setting up adequate resources such as counseling services and identity protection can lessen the negative impact of a cybersecurity breach distress for an individual.

To lessen the consequences of cybersecurity breaches, an anticipatory and holistic strategy is paramount. Financial advisors must weave cybersecurity into their practice management, safeguard systems and websites, inform clients about safe online practices, and formulate an incident response strategy. These efforts will enable financial planners to better serve their clients, safeguard their firms, and uphold the reputation of the financial planning profession.

CASE STUDIES

Incidents of cyberattacks on financial planning firms serve as troubling case studies reflecting the consequences of poor cybersecurity practices. Analysis of these case studies is helpful in improving security measures and safeguarding client assets. Though most breaches involving private details of clients are kept confidential, there are published accounts involving firms and other general industry trends that reveal the attacks and their effects.

Case Study 1: Data Breach at a Wealth Management Firm

One instance involves a reputable wealth management firm which was breached after an employee fell victim to a phishing scam. The cybercriminal accessed the firm's network with the captured account and collected massive amounts of sensitive data including client documents, financial information, and personal identifiable information. The firm suffered financial losses from legal proceedings, regulatory fines, and costs associated with providing credit monitoring services to clients affected by the breach. The firm's reputation also suffered significantly because of the breach, causing them to lose clients and making it difficult to acquire new ones. This case illustrates the importance of providing comprehensive training to employees at all levels of the organization and implementing robust email filtering systems to defend against phishing attacks.

Case Study 2: Ransomware Attack on a Financial Planning Practice

A small-scale financial planning office fell victim to a ransomware attack which led to the encryption of client data and systems. The cybercriminals demanded ransom for the decryption key claiming that they would release sensitive information if their demands were not met. The firm restored data from backups rather than complying with the directives, however, the operations were disrupted for several days. The attack resulted in a monetary loss and proved to be inconvenient for the clients. This case highlights the significance of periodic data backups and plans of action to counter the impacts of ransomware. LAWPRO (2025) cite a case study that included law firms acknowledging the possible disruption, litigation consequences, and incidents of such nature. While not directly pertaining to cyber security, it demonstrates the necessity of the concept of preparedness and risk management.

Case Study 3: Exploitation of a Vulnerability in Financial Planning Software

Through a cyberattack which exploited a popular financial planning software application, malicious actors retrieved sensitive financial data including account details, transaction summaries, and investment plans from clients at multiple financial planning firms. The clients affected suffered monetary losses while the vendor and the financial planning firms using the compromised software suffered immense reputational damage. This case demonstrates the importance of maintaining software updates and applying relevant patches for vulnerability management. Most advisors, as Neal (2020) discusses, struggle with software upkeep and actively failsafe configuration in the context of cyber defense.

Case Study 4: Insider Threat at a Financial Advisory Firm

An employee within a finance consultancy was discovered to have been embezzling money from various clients for a significant period. The staff member exploited inadequate supervision and weak internal controls to transfer money from client-specific accounts to their personal account. The embezzlement caused substantial financial losses to the clients while simultaneously tainting the reputation of the consulting firm. This case illustrates the potential threat of insiders and the lack of internal control, constant review and appraisal, and employee scrutiny.

Lessons Learned

The examined case studies demonstrate the range of cybersecurity threats facing financial planning firms along with the possible disastrous consequences that inadequate security can cause. They underscore the importance of comprehensive approaches to cybersecurity which encompass robust technological defenses, staff training, client education, and effective incident management strategies. Cristea (2020) examines various frameworks of security concern and makes a case for proactive adjustment. Financial planners can learn from these experiences and enhance their cybersecurity measures to better defend their clients' assets in cyberspace.

STRATEGIES FOR MITIGATING CYBERSECURITY RISKS

Proactive Measures: Building a Robust Cybersecurity Framework

In today's world, safeguarding a client's assets necessitates that financial planners adopt a proactive and multi-faceted strategy concerning cybersecurity. This section addresses some of the most critical proactive measures like encryption, multi-factor authentication, secure client portals, intrusion prevention systems, vulnerability scanning, and penetration testing. The implementation of these proactive measures will significantly reduce the risks associated with cyberattacks and their devastating consequences.

Encryption: Safeguarding Data Confidentiality

Encryption is one of the basic protective measures ensuring information confidentiality by rendering the information unreadable. Encryption guarantees that even stolen data does not make sense to unauthorized individuals. Helpful strategies to mitigate data breaches include encryption as discussed by Cerrato (2016). Financial planners should encrypt sensitive client data both in motion and at rest. Information transiting over the internet or email is termed as data in motion and should be secured with TLS/SSL and similar tools. Additionally, information stored on servers on hard drives, referred to as data at rest, should be protected with high-grade encryption schemes.

Multi-Factor Authentication: Strengthening Access Control

MFA enhances security by requiring users to authenticate with more than one method before accessing a system or account. E-finance consumers face authentication and protection issues as elaborated by Shin et al. (2009), emphasizing the cost versus the convenience balance. MFA usually integrates something the user knows (password), something the user has (security token), or something the user is (biometric fingerprint), and not just a password. The implementation of MFA can greatly mitigate the risk of unauthorized access, even when passwords are exposed.

Secure Client Portals: Facilitating Secure Communication and Data Exchange

Financial planners and clients can interact with and exchange sensitive information over secure client portals. MFA access restrictions and encryption to uphold data confidentiality are necessary protective measures. Clark (2024) notes the robust technology needed in the ongoing battle against cybercrime. In addition to the functions mentioned, secure client portals also enable confidential document sharing and e-signature capabilities.

Intrusion Detection Systems: Monitoring for Suspicious Activity

Intrusion detection systems (IDS) supervise incoming and outgoing traffic on a network looking for any activity that can be classified as suspicious, which may indicate a possible cyber-attack. An attempt, whether it be unauthorized access, malicious software, or any such infections, can be detected, and these systems will notify the system administrators of the potential danger. Salinger (2022) covers those cybersecurity practices concerning advisory firms, particularly the employing of managed service providers for constant assistance and monitoring. If potential problems are dealt with in a timely manner, financial planners can minimize the effects of such cyber-attacks.

Vulnerability Scanning: Identifying and Addressing Security Weaknesses

Vulnerability scanning tools assist in automatically detecting security gaps in the system and software. Kakareka (2009) summarizes the work done in vulnerability assessment. These tools can find known system vulnerabilities, like software that is out of date or systems that are misconfigured and suggest how to solve them. Routine vulnerability scans make it easier for financial planners to anticipate threats and support a strong defensive cybersecurity posture. Zhang et al. (2020) outlines a comprehensive framework focused on analyzing information system vulnerabilities.

Penetration Testing: Simulating Real-World Attacks

Through conducting ethical hacking in the form of penetration tests, a company can devise a plan regarding how they would handle a cyberattack in real life to exploit vulnerabilities within the systems as well as the security controls. Salinger (2022) cites this as one of the uses of penetration testing within the cybersecurity programs. An ethical hacker, also referred to as a hacker or a penetration tester, tries different ways of circumventing a company's defenses to obtain useful information concerning security weaknesses within the system of the stakeholders. Routine penetration testing enables financial planners to mitigate potential risks and proactively address system vulnerabilities before they can be exploited by malicious actors.

Proactive measures of this nature allow financial planners to devise a comprehensive cybersecurity framework that protects client assets and enhances overall security posture. The protection of sensitive information and maintaining the trust of clients necessitates a proactive approach, which is essential to the sustainability of the financial planning industry in the long term.

Regular Audits and Risk Assessments: A Foundation for Cybersecurity

Like other businesses, financial planners must always perform regular audits and risk assessments in relation to cybersecurity concerns. With these audits and assessments, firms can identify weaknesses, estimate risks, and take appropriate steps to fortify protective measures concerning clients. This section outlines the importance of regular audits in conjunction with risk evaluation, managing exposures, incident response, business continuity planning, recovery, and business impact analysis.

Regular Audits: Evaluating Security Posture

Regular security audits evaluate deeply the cybersecurity posture of a firm. Internal personnel or external cybersecurity professionals can perform these audits. Khudhair et al. (2019) highlights the logic of audit in the case of sound risk management and internal control. A complete audit assesses various elements of security like network infrastructure, software applications, access control, data protection, security policies in place, and the security audit practices. Understanding the audit results helps expose gaps and weaknesses, which financial planners can mitigate to strengthen the firm's posture.

Comprehensive Risk Assessments: Identifying and Prioritizing Threats

Thorough risk assessments include examining various risks in cybersecurity, estimating their likelihood as well as impact, and prioritizing them to determine appropriate responses. SEC (2022) comments on the proposed set of rules for the managed cybersecurity risks for registered funds and advisers, which includes an obligation to conduct risk assessments. Risk assessments must be scoped this way due to the size and nature of different industries, the clientele, and the nature of the data involved, some of which is sensitive. Understanding threat vectors through a detailed risk assessment allows financial planners to make informed decisions about resource allocation and prioritization of mitigation strategies. Didenko (2020) reviewed financial sector cybersecurity regulation, reporting progress in risk assessment effectiveness.

Vulnerability Management: Addressing Security Weaknesses

Vulnerability management pertains to the processes of discovering, evaluating, and rectifying security weaknesses found within systems and software applications. Garg et al. (2022) centers his attention on predicting vulnerabilities based on data from the past. As discussed previously, consistent vulnerability scanning is vital in the overall lifecycle of managing vulnerabilities. An organization's workflows and processes must be arranged in order of the identified vulnerabilities by priority and risk level. Financial planners need to act on relevant patches, updates, and other changes that mitigate vulnerabilities that need to be resolved immediately.

Incident Response Planning: Preparing for Security Incidents

Even with active efforts to enhance security, incidents may still occur. A clearly defined strategy describes the way responses will be handled for cyberattacks or data breaches. Nierengarten (2006) talks about a plan for wealth management firms aimed at minimizing the risk of cyber threats, highlighting the importance of being able to respond effectively to incidents. Procedures for recognizing and containing the incident, alerting relevant constituents, rebuilding systems and data, and performing a post-incident review to prepare systems for future attacks should also be included. The incident response plan must be periodically tested and updated to ensure effectiveness.

Business Continuity and Disaster Recovery Planning: Ensuring Business Resilience

Business continuity and disaster recovery planning policies enable businesses to continue operating even during extreme disruption scenarios like natural disasters or cyberattacks. SEC (2022) cites that the review of BCDR policies is part of SEC's proposed risk management framework. These procedures must include backing up and restoring data, minimizing the interruption of key business processes, and communicating with clients or other stakeholders during the business disruption. Effective BCDR architectural plans enable financial planners to minimize operational downtime, safeguard client assets, and sustain business operations during unforeseen events.

BCDR planning, along with periodic audits and detailed and comprehensive risk assessments, increase the security posture of financial planners. These policies enable firms to identify and rectify vulnerabilities, mitigate potential risks, and enhance continuity of operations in the event of cybersecurity attacks or other disruptive incidents. This is essential to protect clients' assets, maintain client trust, and practise sustainable financial planning over time.

Incident Response Planning: A Crucial Element of Cybersecurity

The planning and investment strategies of an organization should incorporate the incident response strategy as it is critical to the overall cybersecurity framework. Preventative measures, no matter how robust, will always have the risk of breach for every enterprise. The right incident response plan allows enterprises to mitigate cyberattacks appropriately and expeditiously. These measures can minimize damage while empowering the business to continue operations. This segment concentrates on examining the constituent elements of incident response planning which comprise preparation, detection and analysis, containment, eradication and recovery, as well as post-incident activity.

Preparation: Building a Solid Foundation

Being ready for an attack is the guiding principle for effective response to an incident. This covers planning, preparing a response team, and establishing effective communication. For incident response preparation, NIST (2013) provides a thorough, almost encyclopedic account of computer security incident handling, giving a single incident's pre and post response analysis. There need to be defined policies on roles, responsibilities and actions required for each level of security breach for the incident response plan. Team readiness is achieved through structured training and simulations. The text highlights the issue of incident response as well as the relevance of the skill sets needed (Shinde & Kulkarni, 2021).

Detection and Analysis: Identifying and Assessing Security Incidents

To mitigate the detrimental effects that come with security incidents, timely detection is essential. It includes system checks for abnormal activity, revision of security logs, and utilizing intrusion detection systems. Both detection and action controls of cyberattack response should be robust, as stated by Blum (2020). Following the identification of a possible incident, an extensive analysis of the incident, impact, and scope must be conducted. This helps in deciding an effective response action plan.

Containment: Limiting the Damage

Containment focuses on preventing further propagation and the effects of a given security incident. This may include system isolation, account suspension, and stopping harmful traffic. Governments' policies and strategies to improve security in cyberspace, as described by Gurkaynak et al. (2013), emphasize the critical need for rapid response and containment. These measures must be taken, with little delay, and in a manner that is clear-cut and devoid of hesitation, in order to lessen the collateral damage and loss of information.

Eradication and Recovery: Restoring Systems and Data

Eradication includes eliminating the causes of security incidents such as malware or an account that has been compromised. Shostack & Dykstra (2024) discuss handling pandemic-scale cyber threats and they are emphasizing the need of developing doctrine and skill sets pertaining to cyber threats. Recovery involves restoring the systems and data to their condition prior to the incident. This may require the use of backup copies, the reconstruction of compromised systems, and applying patches. The restored systems and data also require validation and verification testing to confirm they are intact and secure.

Post-Incident Activity: Learning from Experience

As Nist (2013) noted, post-incident analysis is equally as important as the prevention strategy itself. This is aimed at understanding the security breach and outlining how to respond better in future cases. It encompasses evaluation of the primary underlying cause of the incident, the efficacy of the executed response, and the possible refinement options. All conclusions reached from the analysis should be used to revise the incident response framework, improve overall preventive strategies as well as augment the related training programs.

The training of an organization's employees and having a structured incident response plan equips financial planners to mitigate the impact of cyberattacks. A well drafted and implemented incident response plan minimizes damage from security breaches, mitigates loss of important client data, and preserves business continuity. An incident response geared towards providing comprehensive solutions to financial planners and their clients while at the same time anticipating potential risks to the industry demonstrates responsible client care as well as protective stewardship of the financial planning profession.

Role of Financial Planners: Champions of Cybersecurity

Cybersecurity is part of security awareness that must be integrated into the culture of financial planners. In addition to protective actions, financial planners are obligated to warn the clients regarding the risks so that they can protect their financial information. This portion elaborates on the diverse roles of financial planners concerning cyber security advocacy like client training, security culture promotion, security information dissemination, password and multi-layered authentication advocacy, environmental monitoring, and anticipating emerging threats.

Client Education: Empowering Clients to Protect Themselves

Financial planners need to actively inform clients about the potential cyber risks and best practices to avert them. As Camillo (2017) points out, cybersecurity is critical in the modern world, especially for companies that work with sensitive information. Such education may concern phishing scams, malware, social engineering, and password safety. Financial planners can equip their clients with materials in the form of checklists and guides so they can apply them to enhance their cybersecurity at home and their workplace. Kitces (2014) describes advisor communication in the context of client segmentation for education about cybersecurity which could be customized for specific clientele.

Fostering a Security-Conscious Culture: Leading by Example

Financial planners must foster a culture of security awareness and attitude within their own firms. Kaplan (2016) discusses cybersecurity in the context of business advantage and emphasizes the need for integration of policies and practices into business strategies. This includes establishing reasonable and adequate security policies, training all employees on security awareness on a regular basis, and encouraging correct data handling. Adopting such strong policies allows financial planners to demonstrate a cybersecurity commitment, enabling them to lead clients toward embracing security deeply.

Promoting Secure Communication Practices: Protecting Client Data in Transit

Communication methods focusing on security should be applied by financial planners for protecting client information whilst in transit. Shin et al (2009) discusses concerns surrounding the identity authentication and security of e-finance users, paying particular attention to the role of communication systems. Such practices include the application of encrypted email, secure file transfer protocols, as well as Virtual Private Networks (VPNs) for remote access of client information. As far as clients are concerned, financial planners should also alert them about the risks involved in performing financial transactions over public networks.

Advocating for Strong Passwords and Multi-Factor Authentication: Enhancing Account Security

As multi-factor authentication (MFA) enabled security measures strengthen protections, advisors should discuss with clients the importance of multi-layered security, recommending the creation of individual strong passwords for every financial account alongside enabling MFA wherever possible. Landwehr (2008) emphatically outlines the importance of intricate passwords as fundamental security measures. Through mobile applications, MFA adds another layer of verification requiring a code. Financial planners can walk clients through the process while explaining the benefits of enabling multi-factor authentication.

Staying Informed about Evolving Threats: Adapting to the Changing Landscape

As new threats appear, the cybersecurity landscape changes and becomes more sophisticated. Financial planners need to remain alert to new threats to client assets to protect them. Cristea (2020) discusses security threats in the global and national context and emphasizes the imperative to constantly stay aware of changing threats. Some of the approaches include subscribing to security bulletins, attending industry conferences, or even undergoing cybersecurity training. Financial planners can change their security practices and inform clients of new threats by remaining educated.

By taking the time to teach clients, and create a culture of security, and secure behavior, financial planners can effectively improve the cybersecurity posture of their clients and their firms. In the modern era where everything is digitized, proactivity and comprehensiveness in approach is quintessential as it helps in cultivating trust among clients, safeguarding client wealth, and guaranteeing the future of financial planning practice.

LEVERAGING TECHNOLOGY FOR ENHANCED SECURITY

AI and Machine Learning: Bolstering Cybersecurity Defenses

Cyber threats are continuously evolving, but so are the methods of addressing them, and among the most powerful methods is the use of artificial intelligence and machine learning. These technologies have the capability to process huge amounts of data, detect certain behaviors that may suggest threatening activity, and even act automatically in case of a security breach. The role of AI and ML in improving cybersecurity for financial planning is multi-dimensional and includes, but is not limited to, threat detection, prevention of fraud, vulnerability assessment and management, incident response, and proactive personal security.

Threat Detection: Identifying Malicious Activity in Real-Time

Through examining system logs, network interactions, and various data sources, AI and ML techniques can assist in identifying potential cyberattack attempts. EMarketer (2023) analyses the applications of AI in the finance sector, citing examples that include automating fraud detection and cyber defense systems. Besides unlocking new threats, these algorithms can uncover known malicious malware patterns and reveal novel ones through behavior-based identification. Immediate recognition of threats facilitates prompt response actions from financial security planners to curtail obtrusive damage during breaches. Gupta et al. (2020) discusses the application of AI in malware analysis, restating the necessity of AI for perimeter cybersecurity.

Fraud Prevention: Protecting Client Assets from Financial Crimes

The role of AI and ML in combating identity theft, account takeover and even unauthorized transactions is crucial within fraud detection technology. Kurshan et al. (2024) observe the use of AI in detecting financial fraud and discuss the proactive AI countermeasures to problems posed by evolving AI technologies needing attention. These technologies analyze transaction records, identify dubious behavior, and flag them as suspicious for further examination. With the adoption of AI-based fraud prevention systems, financial planners are now able to protect their clients' accounts from financial crimes.

Vulnerability Management: Proactively Identifying and Addressing Security Weaknesses

AI and ML technologies help to automate vulnerability discovery and risk prioritization in software and systems. Garg et al. (2022) analyzes the use of AI-based predictive security and proactive defense dealing with the security issues by forecasting events utilizing historical data. As part of these techniques, code is examined for potential vulnerabilities and the likelihood of the vulnerability being exploited is predicted. Financial planners, for example, strengthen the defenses in place to mitigate the impact of successful cyber-attacks. In Kakareka 2009, we can find a broad description of an overview of a particular process of vulnerability assessment, which is known as one of the basic processes in cybersecurity.

Incident Response: Automating and Accelerating Response Efforts

AI and machine learning can enhance the accuracy and speed of an organization's incident response activities, allowing financial planners to react to incidents in a timely manner. As Khan et al. (2022a) study the effect of AI on finance, especially on its cybersecurity dimension, they note that AI can process alerts, diagnosing incidents, and implementing mitigation measures autonomously. The use of automated response procedures enables significantly faster recovery from the impact of a security incident and a return to 'normal' operational activities.

Personalized Security: Tailoring Security Measures to Individual Client Needs

AI and ML can be utilized to tailor security controls to individual clients and their risk profiles. Giorgio Barba Navaretti et al. (2023) cites the example of AI taking over chores like household budget management, highlighting the potential of AI-driven personalized financial advising. This method can customize both security and training instructions to the specific needs and risk tolerance of each client. Customized security helps streamline the cybersecurity processes while minimizing the burden on the client experience.

Through the appropriate application of AI and ML, financial planners can dramatically enhance their cyber defense posture and protect client assets during this technological era. Automated threat prevention, detection, responsive incident management, and bespoke shielding are some of the protective measures these technologies provide. As AI and ML technologies are enhanced, they will become increasingly important in safeguarding the financial planning profession from the rapidly growing risk of cyberattacks and other hostile attacks.

Blockchain Technology: Enhancing Security and Transparency

Leveraging blockchain technology for financial planning is critical because of the automation, decentralization, and immutability features of blockchain which create self-operating, transparent, and secure registries. With blockchain, financial planners can record every interaction without the possibility of modification, which increases data integrity while enabling the automation of numerous functions. This subsection outlines the use of blockchain technology towards improving cybersecurity in financial planning, including secure transactions and data storage, heightened data integrity, identity and client onboarding verification, and compliance with set regulations.

Secure Transactions and Data Storage: Protecting Sensitive Information

The combination of blockchain's decentralized framework and its encryption methods forms a secure infrastructure for the storage and transfer of sensitive financial information. Fang (2024) states that blockchain can accurately be described as a distributed transactional database in which information is verified by nodes and transferred using public keys, therefore guaranteeing a secure exchange of information. Feng et al. (2024) explains the importance of fundamental cryptography in symmetric and asymmetric encryption and how they secure and maintain trust in blockchain systems. Financial planners can mitigate these risks by storing sensitive client data on a distributed ledger which minimizes the potential for data breaches. With each transaction stored on the blockchain, an individual and verifiable trail is established.

Enhanced Data Integrity: Ensuring Trust and Transparency

Blockchain's immutability ensures that once data is recorded on the blockchain, it cannot be altered or deleted without detection. Maity (2022) explains the concept of hashing, where any change in data results in completely different hash, ensuring data integrity. This feature enhances data integrity and builds trust between financial planners and their clients. Clients can be confident that their financial information is accurate and has not been tampered with. This transparency can also facilitate audits and regulatory compliance.

Secure Client Onboarding and Identity Verification: Streamlining Processes

As stated in Stafford & Treiblmaier (2020)), blockchain technology can secure medical records and could be expanded to other fields which need protected information. Blockchain could also optimize the client onboarding and identity verification processes, making them more secure. By keeping client identity information on the blockchain, financial planners can effectively remove the need for redundant verification processes. As a result, onboarding time is reduced, and the client's experience is enhanced. Moreover, identity verification with the help of blockchain technology can reduce fraud and identity theft.

Improved Regulatory Compliance: Meeting Evolving Requirements

According to Didenko (2020), regulation in cyber security on financial services and the prospects of legal integration were discussed. Blockchain helps with data privacy protection compliance, especially in the security infrastructure pertaining to safeguarding sensitive information. Blockchain aids in automating compliance processes and audits which decreases the burden on attestation and verification processes. Moreover, the inability to alter blockchain records aids in complying proof requirements for data storage policies.

Challenges and Future Directions: Navigating the Evolving Landscape

Though blockchain has great benefits for improving cybersecurity in financial planning, there are hurdles that must be overcome. Thelittlewonder (2018) cites the ever-present 51% attack as a possible weakness in blockchain networks. Other vulnerabilities lie within a system's scalability, interoperability, and lack of proper regulation. Fahmideh et. al. (2022) undertakes a research article which identifies novelties in the field of blockchain and explains the processes of invention and development taking place in this area of study. It is reasonable to expect that the possibilities of using blockchain technology in financial planning will continue to grow as technology advances and legislation changes.

Financial planners can improve security, transparency, and efficiency of various operations by making use of blockchain technologies' distinct features. Even though certain difficulties still exist, it is obvious that blockchain has potential advantages for the financial planning sector. As these innovations evolve, blockchain technology will have greater prominence in protecting client assets and cultivating trust in the digital era.

Emerging Tools and Platforms: Navigating the Evolving Cybersecurity Landscape

The ever-evolving landscape in cybersecurity, coupled with new threats and vulnerabilities, increases the need to monitor emerging technologies for financial planners. This chapter analyses the modern security advancements in financial planning, focusing on threat detection and prevention systems, cloud security, multi-authentication and biometric identification, and cyber training tools.

Advanced Threat Detection and Prevention Platforms: Proactive Security Posture

Sophisticated AI or ML algorithms bundled within advanced threat detection and prevention platforms intercept real-time cyber threats. As stated in Umoga & Oluwademilade (2024), new cybersecurity risks associated with financial technologies have been noted which call for stronger measures to be put in place. Comprehensive analysis of an organization's data procured from multiple sources like graphing network traffic, inspecting system logs, and endpoints helps in identifying any form of irregularities or mal intentions. They assist in incident response automation which enables the reduction of overall damage inflicted by security breaches. Shinde &

Kulkarni (2021) provide a detailed overview of an important consideration in cybersecurity management: incident response.

Cloud-Based Security Solutions: Enhanced Scalability and Flexibility

Unlike traditional on-premises security infrastructure, cloud-based security solutions allow for greater scalability, flexibility, and cost optimization. Kaplan (2016) discusses cloud-based solutions in the wider context of cybersecurity for commercial advantage and emphasizes the merits of these solutions. A vast array of security services can be provided, such as firewall protection, intrusion detection and prevention, data loss prevention, as well as security information and event management. Cloud-based solutions also ease the management of security controls, which lessens the workload for IT personnel.

Multi-Factor Authentication and Biometric Security: Strengthening Access Controls

MFA and biometric methods of security enable identity verification through multiple factors, thus adding a further protective layer. Shin et al. (2009) consider e-finance consumers' authentication and protective issues, noting that the balance between security and usability is crucial. Typical forms of MFA consist of something the user knows (password), something the user possesses (security token), and something the user is (biometric information such as a fingerprint). Biometric methods of security check identity by using distinguishing features of a person such as fingerprints, facial geometry, or voice prints. Even if passwords are breached, these measures help minimize the chances of unauthorized access considerably.

Cybersecurity Awareness Training Platforms: Empowering Employees and Clients

Cybersecurity awareness training platforms teach employees and clients about potential threats and assist them in avoiding such dangers. Reducing the risk of data breaches involves training employees and making them more aware, as indicated by Cerrato (2016). Illustrative training modules alongside simulations and assessments can be provided by these platforms to strengthen safe online behavior and crucial concept retention. Providing awareness concerning cybersecurity risks to employees and clients aids in diminishing the attempt to launch cyberattacks. Guidance on SEC cybersecurity compliance for advisors is discussed by Clark (2024).

Innovations in Cybersecurity for Financial Planning: Staying Ahead of the Curve

Multiple cutting-edge tools and platforms are being developed for the financial planning sector. Wilcox (2024) explores the use of technology during client management in turbulent periods, accentuating cybersecurity issues. These tools assist financial planners in evaluating their cybersecurity posture, identifying potential weaknesses, and implementing security measures. They also offer automatic compliance reporting and risk management services. SEC (2022) analyses the SEC's proposed cybersecurity risk management regime for registered funds and advisors.

Keeping pace with new tools and platforms aids financial planners in modernizing asset protection in contemporary cybersecurity. Such innovations lend themselves to powerful threat detection and prevention, access control enforcement, and cybersecurity training for personnel and clients. Financial planners specializing in this area require adaptable strategies and proactivity as the cybersecurity landscape evolves.

REGULATORY AND COMPLIANCE CONSIDERATIONS

Overview of Cybersecurity Regulations: Navigating the Legal Landscape

Financial planners work in a highly regulated environment, and cybersecurity is no exception. Compliance with the numerous rules and frameworks which govern the collection, storage, and use of client information requires financial planners to have appropriate cybersecurity measures in place. This subsection outlines basic cybersecurity regulations like GDPR, CCPA, GLBA, and other relevant frameworks. These regulations are important in ensuring that sensitive client information is sufficiently protected, compliance obligations are met, and potential legal liabilities are avoided.

General Data Protection Regulation: Protecting Personal Data of EU Citizens

The GDPR is a comprehensive data protection regulation enacted by the European Union. Organizations collecting, processing, and even storing the personal data of EU citizens are required to meet specific criteria irrespective of where they are domiciled. Zhu & Song (2022) stresses further the necessity of safeguarding one's privacy and personal information in the digital e-commerce context and the overall frame of data protection. Essential components of GDPR encompass the rights of data subjects (example, access, modification, deletion), the obligation to inform about breaches of data, and the requirement of data minimization principles. Financial planners who handle the data of EU citizens must comply with the GDPR, even if they are not based in the EU.

California Consumer Privacy Act: Protecting the Privacy Rights of California Residents

The CCPA is a state-level data privacy law in California that grants consumers significant control over their personal information. It provides California residents with the right to know what personal information businesses collect about them, the right to delete their personal information, and the right to opt-out of the sale of their personal information. Sadoian (2024) lists the CCPA among the main cybersecurity regulations that apply to the financial services industry. Financial planners who serve clients in California must comply with the CCPA.

Gramm-Leach-Bliley Act: Safeguarding Financial Information

The GLBA or Gramm-Leach-Bliley Act of 1999 is a federal law in the United States that governs the treatment of consumers' personal information by financial institutions. Sadoian (2024) identifies this act as critical legislation protecting the financial privacy of consumers. It obliges particular businesses to take appropriate steps to avert non-public information's access and protection breaches. Financial planners are bound by the provisions of GLBA and are therefore required to undertake relevant cybersecurity measures to secure clients' sensitive financial information. The SEC (2024) provides a summary of cybersecurity laws concerning financial services which incorporate the GLBA.

Other Relevant Frameworks: Industry Standards and Best Practices

Alongside specialized guidelines, many industry standards and best practices have developed policies related to cybersecurity for financial planners. For instance, the NIST Cybersecurity Framework provides a structured approach to mitigating cybersecurity threats. According to NIST (2013), there is a provided computer security incident handling guide from NIST. The ISO 27001 standard outlines the requirements needed for the establishment, implementation, maintenance, and continual improvement of an information security management system. In Carello et al. (2023), the authors explain the systematization of cybersecurity regulations, standards, and guidelines in the healthcare sector, although the insights provided can be viewed from other sectors' viewpoints. Integrating these frameworks will assist financial planners in improving their cybersecurity capabilities while affirming their commitment to safeguarding client information. In the work of Fischer (2014), federal laws regarding cybersecurity are presented.

Navigating the Complex Regulatory Landscape: Staying Informed and Compliant

Laws concerning cybersecurity are complex and change constantly. Financial planners must be acquainted with new policies, changes to policies, and emerging best practices that are introduced. Mantelero et al. (2021) analyzes regulations on cyber security, along with the steps taken by the government to strengthen the country's cyber security. Consulting a lawyer who specializes in cybersecurity can assist financial planners to follow the cyber laws in place so as not to breach compliance. If constant rules in relation to cyber security are followed, client assets will be secured, trust will be built, and legal as well as reputational harm can be adverted.

Compliance Challenges for Financial Planners: Balancing Security and Efficiency

For financial planners, installing advanced cybersecurity measures while staying efficient during operations can be quite a task. It requires detailed planning, distribution of the right resources, and readiness to make changes. Striking a balance between security measures and the growing demand of modern business is challenging. This sub-section focuses on important compliance challenges which financial planners deal with the conflict between security and operational efficiency, third-party vulnerability management, emerging compliance regulations, and the adoption of a cybersecurity culture.

Balancing Security with Operational Efficiency: A Delicate Balancing Act

Strict measures of security may at times impede operational efficiency. For example, overly advanced authentication procedures can delay client transactions and interactions. Shin et al. (2009) does explain the cost versus ease-of-use dichotomy regarding authentication and protection for e-finance consumers. In the same manner, extreme data encryption can also reduce system performance. This means that the financial planners must try to achieve a balance between security and efficiency, accomplishing robust security measures while minimally impeding business operations. Clark (2024) discusses, the Automated AI for Compliance supports automatic tracking and reporting of compliance which helps improve other processes by shifting attention to more important tasks.

Managing Third-Party Risks: Securing the Supply Chain

To optimize resources, most financial planners outsource activities such as cloud storage, software application, and IT support. This dependency poses cybersecurity vulnerabilities because any flaws in a third party can jeopardize a financial planner's sensitive information. Feyen et al. (2023) identifies operational and cyber security threats to financial services due to the use of external cloud storage and data supply vendors. Careful selection of vetted third parties and appropriate security protections is necessary for contract management.

Addressing Evolving Regulatory Requirements: Staying Ahead of the Curve

Professionals in the financial planning sector are required to modify their compliance methodologies perpetually due to advancements in cybersecurity requirements. New policies may entail supplementary components such as a tightened schedule for data breach notification or stronger safeguards for sensitive information. Remolina (2019) examines the regulatory concerns of open finance and data sharing within the context of the broader financial industry. Financial planners need to track changes in legislation, evaluate their implications, and make operational changes necessary for sustained compliance. Gellert (2020) offers a comprehensive account of regulation and compliance, which includes an analysis of Reg BI and regulators' interventions. Being up to date with regulatory changes is necessary to avoid sanctions and protect the organization's image

Fostering a Culture of Cybersecurity Awareness: Empowering Employees and Clients

Fostering a culture of cybersecurity awareness is foundational to compliance. Everyone as well as employees in the organization need to appreciate the significance of cybersecurity and protecting sensitive data. Schwartz (1991) points out the lack of oversight within the financial planning profession and the troubles it is likely to encounter, calling for circumspection. Financial planners ought to offer regular cybersecurity best practices training to both employees and clients covering issues like password management, phishing, and internet browsing. Yeske & Buie (2014) capture the essence of trust and commitment in financial planning engagements and client relationships as important aspects of communication. Developing a culture of cybersecurity awareness has the potential to decrease the risk of human error and improve the security posture.

Overcoming Compliance Challenges: A Proactive Approach

Proactive and holistic measures are necessary in overcoming the compliance difficulties in cybersecurity. Financial planners must first address the cybersecurity needs of their businesses by acquiring the right technologies and training, as well as company policies, set procedures, and document workflows. Open banking has promise but also poses risks — security is one of the most important ones, especially when it comes to sharing sensitive information (Giorgio Barba Navaretti et al., 2023). Regular assessments of system security, vulnerability, and penetration testing can help pinpoint possible threats and weaknesses to systems. Through proactive strategy formulation, financial planners can resolve compliance concerns, optimize client asset protection, and sustain confidence in an agile digital environment.

Best Practices for Regulatory Adherence: A Proactive and Adaptive Approach

A financial planner's scope of practice is broad due to the compliance requirements of cybersecurity. This implies that they must stay current with active regulations, take appropriate compliant measures, foster compliance with a climate of cybersecurity at all organizational levels, and assess security practices and policies as part of a continual improvement process. This subsection analyzes the fundamentals of compliance as law and the underpinning regulatory policies so that financial planners can protect their client's assets and trust.

Staying Updated and Implementing Compliant Measures: A Continuous Process

As technology improves and new risks emerge, cybersecurity regulations are revised. Consequently, financial planners need to track legislative changes and updates. Subscribing to relevant newsletters, attending specific conferences, and consulting legal specialists in cybersecurity help financial planners stay informed with the regulations. Zhu & Song (2022) focuses on regulation and compliance and highlights the gaps in being uninformed about e-commerce regulations. After understanding the regulations, business policies, as well as compliance processes need to be framed. This entails modification of internal policies and procedures, the purchase of new equipment, and additional employee training.

Conducting Regular Risk Assessments: Identifying Vulnerabilities and Implementing Safeguards

Regular and structured risk assessment is critical for understanding an entity's vulnerabilities and managing risks appropriately. A thorough risk assessment will evaluate a financial planner's activities from data storage, through network security, to access control and incident response systems. One of the fundamental components of cybersecurity risk management is vulnerability assessment which, as Kakareka (2009) notes, is one of the basic parts. After the risk assessment, it is very important that financial planners implement appropriate security controls to eliminate the identified vulnerabilities. These may involve more stringent password protection, multi-step authentication processes, encryption of sensitive data, firewall installation, or employing intrusion detection systems.

Developing and Implementing Comprehensive Cybersecurity Policies and Procedures: Establishing a Framework for Security

A firm's information security policies must be crafted in such a manner that they enable compliance with the relevant legislation. Such policies should contain at a minimum how the financial planner dealt with cybersecurity issues, assign responsibilities, outline protocols for incident response, data breach notification, and other security actions as may be deemed necessary. The policies and procedures described in this document aim to enable the management of access rights within the financial services industry. They present fundamental policies defining their procedures and importance to protecting sensitive information. The policies and procedures established need to be reviewed and modified periodically to incorporate changes in the regulatory environment and meet best practices (Banoczi et al., 2017).

Fostering a Culture of Cybersecurity Awareness: Empowering Employees and Clients

Fostering a proactive culture of cybersecurity awareness is key in ensuring compliance. All employees and clients need to appreciate the degree of cybersecurity sensitivity and their responsibility regarding information protection. Financial planners should ensure that employees are trained on cybersecurity issues, including strong password management, phishing awareness, and safe browsing practices. Nierengarten (2006) has developed a plan for wealth managers on how to reduce the risks of cyberthreats, which incorporates education and cognitive awareness as fundamental components. Equally important is the client education. Financial planners ought to inform them concerning the business firm's cybersecurity and advise them on appropriate online activities while encouraging them to inform the firm in case they notice anything unusual.

Partnering with Cybersecurity Experts: Leveraging Specialized Expertise

Cybersecurity is a broad and highly dynamic domain, which poses a challenge for most financial planners since they may lack the requisite human capital to effectively manage every facet of cybersecurity. Engaging the services of cybersecurity professionals offers the needed knowledge and tools. McCarthy (2021) elaborates on the responsibilities of advisors around cybersecurity and their possible gains from engaging professionals. Using the help of a cybersecurity consultant can be beneficial during risk analysis, the drafting of policies, planning for incidents of responses, and other crucial security operational activities. At the same time, they can help with other activities that need supervision, aiding the financial planners to deal with issues of new threats and changes in policies that need compliance with. If these approaches are embraced, financial planners will improve their cybersecurity defenses while ensuring compliance with regulations and client protection in an increasingly digital world.

EDUCATING CLIENTS ON CYBERSECURITY

Importance of Client Awareness: Empowering Clients in the Digital Age

With the rise of new technology, awareness is the key to protecting clients' assets from financial cyberattacks. An educated client will always be the first line of defense. This section focuses on the importance of client awareness and its impacts toward preventing cyber-attacks, shared responsibility and trust in the client planner relationship.

Preventing Cyberattacks Through Education: Empowering Clients to Identify and Avoid Threats

Client education is essential in mitigating the probability of cyber-attacks. Clients who understand the dangers of online cyber threats and the safe practices that they can adopt are able to protect themselves from the attacks. Camillo (2017) provides a myriad of illustrative cases relating to cybersecurity while underscoring the essential role of education in averting risk. Educators need to teach clients the basics of phishing and social engineering as well as more advanced techniques such as malware and adware infiltration.

As illustrated by Shaw et al. (2009), the impact of an individual's actions regarding information technology safety within an organization and the broader society emphasizes the value of security awareness training. Users need to identify destructive emails, websites, and messages, and learn not to activate dangerous links or attachments. Users should also be taught about protecting passwords, updating software, and the security risks associated with public Wi-Fi.

Fostering Shared Responsibility: Building a Collaborative Security Approach

Mitigating cyber threats is not solely the responsibility of the financial planner; it also involves the client's part. Salinger (2022) discusses what advisory firms should be doing on cybersecurity, and security communication with the client is equally important. Cybersecurity education fosters shared responsibility among clients and financial planners. Clients who are educated regarding the importance of cyber security are more likely to practice safe online behaviors and report suspicious activity to their financial planner. This collaborative approach enhances the collective security posture while lowering the risk of successful cyberattacks.

Building Trust and Transparency: Strengthening the Client-Planner Relationship

Trust can be fostered and the client-planner relationship strengthened by open communication and disclosure of cybersecurity measures. Kitces (2014) highlights the importance of security in fintech and the firm's need to focus on cybersecurity to build trust with clients. Clients whose financial planners take the time to explain their cybersecurity policies are likely to feel that their assets are adequately protected. Financial planners need to inform clients about their policies and procedures, take proactive measures to protect client information, and invite clients to have discussions where they can voice concerns. Such discussions enhance trust and solidify the client-planner relationship.

Empowering Clients to Take Control of Their Cybersecurity: Promoting Proactive Security Habits

Client education actively empowers clients to control their own cybersecurity. Khan et al. (2022b) discusses cybersecurity in conjunction with risky behaviors and highlights proactive cybersecurity awareness behavior as an important aspect.

Client education materials should offer instructions on strong password creation, online account security checklists, reputable antivirus software recommendations, and generalized resources that can help enhance clients' cybersecurity practices. Furthermore, clients should be encouraged to actively reconcile their financial accounts to identify discrepancies and report any suspected unauthorized activity without delay.

Promoting Ongoing Education and Awareness: Adapting to Evolving Threats

Client education ought to be done on a continual basis because technological risks evolving due to cybersecurity dangers. To keep up with new threats and advanced techniques, financial planners need to revise their client education materials on a regular basis. "Shostack & Dykstra (2024) discuss how we manage cyber threats at a pandemic level and stresses the point of continuous change and learning." To highlight important details and address client queries, planners should hold regular cybersecurity awareness lectures or webinars. Planners can help equip their clients to outsmart emerging threats and protect their financial assets by encouraging education and awareness regarding ever evolving dangers in the modern world.

Guidelines for Secure Online Behavior: Practical Tips for Protecting Client Assets

Secure online behavior is essential to maintaining one's financial security in a globalized world. Clients must practice safe online behavior alongside being on guard against possible cyber-attacks. This subsection outlines practical recommendations that allow clients to actively participate in protecting their personal financial data and work with financial planners to establish an appropriate financial security environment.

Strong Password Management: The Foundation of Online Security

The first line of defense against inappropriate access to accounts is the use of strong passwords. Clients ought to use strong passwords, which are unique to each financial account so that pet names or dates of birth won't make those passwords to be easily guessed. (6 Tips to Tighten Your Online Security, 2022) also has helpful suggestions for security, such as changing passwords frequently and using more complicated ones. Passwords should be at least 12 characters long and include a combination of uppercase and lowercase letters, numbers, and symbols. Using a password manager can help clients generate and store strong passwords securely.

Barnett (2021) discusses security tips in the context of online dating, emphasizing the importance of strong passwords and threat modeling. Clients should also enable multi-factor authentication whenever possible, adding an extra layer of security to their accounts.

Phishing Awareness: Identifying and Avoiding Malicious Emails and Messages

Phishing attacks are a common tactic used by cybercriminals to steal sensitive information. Appropriate instruction needs to be provided regarding the recognition of phishing emails or messages that may come from fake or pseudo sources, for example, banking institutions. Issac et al. (2014) examines phishing attacks and countermeasures and offers relevant analysis on the issue. These messages may contain urgent requests asking for personal information, unexpected attachments, or links leading to phishing websites. Clients need to verify the authenticity of any unsolicited messages or emails containing personally identifiable information and refrain from clicking links or opening files sent by unknown individuals. Hong (2012) describes phishing attacks and argues that users must be informed and educated about such techniques. Such users need to verify the sender by calling the institution directly using the number found on their official website or a trusted source.

Safe Internet Browsing Habits: Protecting Against Malware and Other Threats

Appropriate internet browsing etiquette needs to be observed to mitigate risks posed by malware and other cyber threats. Customers should not visit suspicious websites, especially those containing pirated or other illicit content. Gupta et al (2020) provides an example of AI malware analysis illustrating the incredible and sophisticated growth of malware and the urgent need for strong protective measures. Customers should also not click on or engage with pop-up ads or links from pages they do not recognize. The employment of a reputable antispyware and antivirus program, as suggested by Nist (2013), serves as a good defense against infections. In the computer security incident handling guide, Nist (2013) provides pertinent information and guides on the response to and management of security incidents. Clients must also remember that software updates often come with security fixes for known problems, thus keeping software up to date is necessary in safeguarding the system.

Securing Devices and Networks: Protecting Against Unauthorized Access

Unauthorized access to financial information is a significant concern, which is why device and network security measures must be put in place. Safeguarding personal passwords for computers, smartphones, and other devices is essential for the clients. According to Seltzer and Lewek (2019), data security measures include the use of a virtual private network which adds an additional layer of protection. Clients should ensure that firewalls and intrusion detection systems are also configured on their home networks. Clients need to be especially cautious when using public Wi-Fi, particularly when trying to access sensitive financial information. Tuliao (2014) outlines how one can safely shop online and includes using strong passwords and avoiding questionable websites as some of the pointers. While the use of public Wi-Fi is somewhat of a security challenge, a VPN can mask the client's location, thus allowing for added security measures.

Regularly Monitoring Accounts: Detecting and Responding to Suspicious Activity

Regular monitoring of financial accounts is crucial as it allows for timely response to any suspicious activities. Clients need to review their account statements and transaction logs for any abnormal activities. Cerrato (2016) addresses the need for constant observation and immediate identification of activities, while arguing about reducing the possibility of a breach. In addition to this, clients ought to monitor their credit reports for unknown accounts or inquiries. If clients observe anything unusual, they must contact their financial planner and the relevant financial institution immediately. Shinde & Kulkarni (2021) highlighted incidents regarding security breaches and emphasized the proactive approach towards minimizing the impact of such incidents. Following these guidelines will allow clients to reduce the risk of having to deal with cyber sabotage in the current digital landscape.

Building Trust Through Transparency: Communicating Cybersecurity Efforts to Clients

Trust plays an integral role in the client-planner relationship and often constitutes the foundation of the financial industry. Trust can be easily compromised and thus making cybersecurity policies known become important. This paragraph explains the need to communicate and demonstrate trust in relation to asset security.

Communicating Cybersecurity Efforts to Clients: Fostering Understanding and Confidence

Financial planners also must communicate with their clients regarding the measures being taken to secure sensitive data. Alleviating stress and promoting trust can be accomplished through effective communication of cybersecurity measures. Clark (2024) analyzes the compliance of the SEC concerning cybersecurity and claims that training and professional development of staff is crucial. Financial planners ought to explain the policies and procedures concerning information technology security as policies on documentation and plain language without any jargon. Such explanations should include the use of encryption, firewalls, and intrusion detection system. Xi (2024) discusses the provision of financial service compliance solutions and highlights the importance of regulatory compliance for credibility. Providing regular and proactive updates about changes in security policies, as well as incidents that require immediate attention helps reinforce the organization's commitment to transparency.

Demonstrating Commitment to Security: Building a Culture of Trust

Being transparent includes more than communicating security protocols; it also means showing active commitment to cybersecurity practices on all levels of the financial planning practice. Policy-based financial planning as discussed by Yeske & Buie (2014), brings into consideration trust and relationship management with clients. It is prudent for financial planners to deploy advanced and up-to-date technology defensive mechanisms. Blum (2020) articulates building resilience through detection, response and recovery emphasizing proactive defensive security. Conducting routine security checks and vulnerability scans have the potential of identifying weaknesses and mitigating them. Taking part in professional cybersecurity initiatives and obtaining industry-recognized certifications goes a long way in demonstrating commitment to best practices.

Fostering a Collaborative Approach to Security: Empowering Clients to Participate

The planner cannot single-handedly shoulder the burden of cybersecurity; it is a joint responsibility with the client. Nierengarten (2006) offers a plan for wealth managers to reduce cyber threats, emphasizing the importance of client communication and collaboration. Advisors should counsel clients on their responsibility for maintaining overall security, which includes selecting robust passwords, evading phishing attempts, and reporting abnormal behavior. Trottman-Adewumi et al. (2017)

elaborates on protection policy, underscoring the need for clients and advisors to work together to balance security risks. Open dialogues and feedback systems can help clients become more involved in safeguarding their financial assets.

Addressing Client Concerns and Questions: Building Open Communication Channels

Clients might have specific questions regarding cybersecurity and financial planners should leave space for open communication with clients to attend to such matters. Castle et al. (2016) has noted the role of communication in financial planning stating the facilitative nature of client's comfort in discussing their issues pertaining to finances. Clients can be encouraged to contact an organization by offering them simple and direct means of communication like providing an email or a phone number for questions regarding security issues.

A financial planner should answer all questions posed by the client timely and in detail, proving their dedication to transparency and responsiveness. (Does My Smart Device Provider care about My Privacy? Investigating Trust Factors and User Attitudes in IoT Systems, n.d.) investigates the trust components of IoT systems and discusses the need for clear and two-way communication as a necessary factor for user trust. Financial planners can gain and strengthen the trust of their clients by answering their concerns and questions truthfully, and in return, improve the client-planner relationship.

Emphasizing the Importance of Data Privacy: Protecting Client Information

With the growth of technology, clients are continually worried about the security of their private financial information and, thus, financial planners must highlight their dedication to confidentiality. Open banking and data sharing comes with ample advantages; however, it raises questions regarding the security and privacy of such information Giorgio Barba Navaretti et al. (2023). Financial planners need to tell their clients what their data protection and privacy policies are and the different client-data processes that take place.

Remolina (2019) addresses the risks of open finance relating to data protection and compliance. Furthermore, they must comply with data protection legislation such as GDPR or CCPA and ensure that their data handling practices are secure and nonopaque. Financial planners can foster client trust and illustrate their commitment to asset protection in the digital realm by demonstrating privacy efforts and clear communications regarding data management.

CONCLUSION

The integration of cybersecurity into financial planning is no longer optional but a fundamental necessity in safeguarding client assets and maintaining trust in an increasingly digital world. As technological advancements such as AI, blockchain, and cloud-based platforms redefine financial services, they simultaneously introduce complex vulnerabilities—from phishing and ransomware to insider threats and data breaches. This chapter underscores that mitigating these risks requires a multi-layered approach, combining robust technological defenses (e.g., encryption, multi-factor authentication) with proactive client education and regulatory adherence.

Case studies of ransomware attacks and data breaches illustrate the catastrophic consequences of inadequate security measures, including financial losses, reputational damage, and regulatory penalties. These examples reinforce the urgency of adopting frameworks like defense in depth and risk management theory to build resilient systems. Equally critical is the role of financial planners as educators, empowering clients to recognize threats like social engineering and adopt secure practices.

Regulatory frameworks such as GDPR and GLBA provide essential guidelines, yet compliance demands continuous adaptation to evolving threats and technologies. Emerging tools like AI-driven threat detection and blockchain-based transparency offer promising solutions but require ongoing investment and expertise.

Ultimately, the future of cybersecurity in financial planning hinges on collaboration—between planners, clients, regulators, and technologists—to stay ahead of cybercriminals. By prioritizing transparency, fostering a culture of security awareness, and leveraging innovation, the financial sector can navigate the digital age with resilience, ensuring that trust remains the cornerstone of client-advisor relationships. As cyber threats evolve, so must strategies to combat them, ensuring that financial planning remains both progressive and secure.

REFERENCES

Asaad, C. T. (2020). Investor confidence: Are you your own worst enemy? In Financial Planning Review (Vol. 3, Issue 2). Wiley. https://doi.org/DOI: 10.1002/cfp2.1092

Banoczi, J., Edwards, S., Irrechukwu, C., Klosterman, J., Perper, H., Prince, S., Symington, S., & Wynne, D. (2017, August 31). Access Rights Management for the Financial Services Sector. Nist.gov. https://csrc.nist.gov/publications/detail/sp/1800-9/draft

Barnett, D. (2021, June 7). Security Tips for Online LGBTQ+ Dating. Electronic Frontier Foundation. https://www.eff.org/deeplinks/2021/06/security-tips-online-lgbtq-dating

Blum, D. (2020). Institute Resilience Through Detection, Response, and Recovery. Rational Cybersecurity for Business, 259–295. https://doi.org/DOI: 10.1007/978-1-4842-5952-8_9

Camillo, M. (2017). Cybersecurity: Risks and management of risks for global banks and financial institutions. https://doi.org/DOI: 10.69554/EPYV4777

Carello, M. P., Spaccamela, A. M., Querzoni, L., & Angelini, M. (2023). A Systematization of Cybersecurity Regulations, Standards and Guidelines for the Healthcare Sector. ArXiv (Cornell University). https://doi.org//arxiv.2304.14955DOI: 10.48550

Castle, S., Pervaiz, F., Weld, G., Roesner, F., & Anderson, R. (2016). Let's Talk Money. https://doi.org/DOI: 10.1145/3001913.3001919

Cerrato, P. (2016). Reducing the Risk of a Data Breach. https://doi.org/DOI: 10.1016/B978-0-12-804392-9.00005-8

Chang, P. (2022). Multi-Layer Perceptron Neural Network for Improving Detection Performance of Malicious Phishing URLs Without Affecting Other Attack Types Classification. ArXiv (Cornell University). https://doi.org//arxiv.2203.00774DOI: 10.48550

Clark, C. (2024, February). The Advisor's Guide to SEC Cybersecurity Compliance. Nitrogen Wealth. https://nitrogenwealth.com/blog/sec-cybersecurity-2024/

Cristea, L. M. (2020). Current security threats in the national and international context. *JOURNAL of ACCOUNTING and MANAGEMENT INFORMATION SYSTEMS*, *19*(2), 351–378. DOI: 10.24818/jamis.2020.02007

Didenko, A. N. (2020). Cybersecurity regulation in the financial sector: prospects of legal harmonization in the European Union and beyond. Uniform Law Review, 25(1), 125–167. Does My Smart Device Provider Care About My Privacy? Investigating Trust Factors and User Attitudes in IoT Systems. https://doi.org/DOI: 10.1093/ulr/unaa006

EMARKETER. (2023, January 2). Artificial Intelligence in Financial Services: Applications and benefits of AI in finance. EMARKETER. https://www.emarketer.com/insights/ai-in-finance/

Fahmideh, M., Gunawardana, A., Chen, S., Shen, J., & Yecies, B. (2022). Blockchain Developments and Innovations – an analytical evaluation of software engineering approaches. In Lecture notes in computer science (pp. 58–76). DOI: 10.1007/978-3-030-96068-1_5

Fang. (2024, July 5). Finance and Technology: Why Fintech is the future of finance – A case study of Singapore's Financial Sector - UNICAF - scholarship programme. Unicaf - Scholarship Programme. https://www.unicaf.org/thesis-publications/finance-and-technology-why-fintech-is-the-future-of-finance-a-case-study-of-singapores-financial-sector/

Feng, P., Bi, Z., Lawrence, Y., Wen, Y., Peng, B., Liu, J., Yin, C. H., Wang, T., Chen, K., Zhang, S., Li, M., Xu, J., Liu, M., Pan, X., Wang, J., & Niu, Q. (2024). Mastering AI: Big Data, Deep Learning, and the Evolution of Large Language Models — Blockchain and Applications. ArXiv (Cornell University). https://doi.org//arxiv.2410.10110DOI: 10.48550

Feyen, E., Natarajan, H., & Saal, M. (2023). Fintech and the Future of Finance: Market and Policy Implications. https://doi.org/DOI: 10.1596/978-1-4648-1914-8

Fischer, E. A. (2014). Federal Laws Relating to Cybersecurity: Overview of Major Issues, Current Laws, and Proposed Legislation. UNT Digital Library. crs: R42114

Fox, J., & Bartholomae, S. (2020). Household finances, financial planning, and COVID-19. In Financial Planning Review (Vol. 3, Issue 4). Wiley. DOI: 10.1002/cfp2.1103

Garg, A., Degiovanni, R., Jimenez, M., Cordy, M., Papadakis, M., & Le Traon, Y. (2022). Learning from what we know: How to perform vulnerability prediction using noisy historical data. *Empirical Software Engineering*, *27*(7), 169. Advance online publication. DOI: 10.1007/s10664-022-10197-4

Gellert, R. (2020). *The Risk-Based Approach to Data Protection*. Oxford University Press., DOI: 10.1093/oso/9780198837718.001.0001

Giorgio Barba Navaretti, Giacomo Calzolari, José Manuel Mansilla-Fernandez, & Pozzolo, A. F. (2023). Open Banking's Promise of a Financial Revolution: Are We Falling Short? Social Science Research Network. https://doi.org/DOI: 10.2139/ssrn.4469320

Gupta, M., Mittal, S., & Abdelsalam, M. (2020, September 21). AI assisted Malware Analysis: A Course for Next Generation Cybersecurity Workforce. ArXiv.org. https://doi.org//arXiv.2009.11101DOI: 10.48550

Gurkaynak, G., Yilmaz, I., & Taskiran, N. P. (2013). Governmental Efforts and Strategies to Reinforce Security in Cyberspace. *International Law Research*, *2*(1). Advance online publication. DOI: 10.5539/ilr.v2n1p185

Hermawan, S., Khoirunisa, Z. A., & Tejomurti, K. (2023). Triangular Insight on Open Banking in Indonesia, Singapore, and Australia. In International Journal of Legal Information (Vol. 51, Issue 3, p. 197). Cambridge University Press. https://doi.org/DOI: 10.1017/jli.2024.11

Hong, J. (2012). The state of phishing attacks. Communications of the ACM, 55(1), 74. https://doi.org/10.1145/2063176.2063197

Issac, B., Chiong, R., & Seibu Mary Jacob. (2014). Analysis of Phishing Attacks and Countermeasures. https://doi.org/DOI: 10.48550/arxiv.1410.4672

Kakareka, A. (2009). What Is Vulnerability Assessment? Computer and Information Security Handbook, 383–393. https://doi.org/DOI: 10.1016/B978-0-12-374354-1.00023-6

Kaplan, J. M. (2016). Cybersecurity for Commercial Advantage. Elsevier EBooks, 97–111. https://doi.org/DOI: 10.1016/B978-0-12-803773-7.00006-1

Khan, H. U., Malik, M. Z., Alomari, M. K., Khan, S., Al-Maadid, A., Hassan, M. K., & Khan, K. (2022a). Transforming the Capabilities of Artificial Intelligence in GCC Financial Sector: A Systematic Literature Review. In Wireless Communications and Mobile Computing (Vol. 2022, p. 1). Wiley. https://doi.org/DOI: 10.1155/2022/8725767

Khan, N. F., Ikram, N., Saleem, S., & Zafar, S. (2022b). Cyber-security and risky behaviors in a developing country context: a Pakistani perspective. In Security Journal (Vol. 36, Issue 2, p. 373). Palgrave Macmillan. https://doi.org/DOI: 10.1057/s41284-022-00343-4

Khudhair, D. Z., Al-Zubaidi, F. K. A., & Raji, A. A. (2019). The effect of board characteristics and audit committee characteristics on audit quality. *Management Science Letters*, •••, 271–282. DOI: 10.5267/j.msl.2018.11.012

Kitces, M. (2014, July 21). Michael Kitces. Nerd's Eye View | Kitces.com. https://www.kitces.com/blog/fpa-rpi-studies-how-advisors-communicate-segment-clients-and-conduct-review-meetings/

Kurshan, E., Mehta, D., Bruss, B., & Balch, T. (2024). AI versus AI in Financial Crimes and Detection: GenAI Crime Waves to Co-Evolutionary AI. ArXiv (Cornell University). https://doi.org//arxiv.2410.09066DOI: 10.1145/3677052.3698655

Landwehr, C. E. (2008). *Cyber security and artificial intelligence.*, DOI: 10.1145/1456377.1456391

LAWPRO. (2025, March 17). When a Law firm gets Hacked: A case study in Cybersecurity Risks and Recovery. practicePRO. https://www.practicepro.ca/2025/03/when-a-law-firm-gets-hacked-a-case-study-in-cybersecurity-risks-and-recovery/

Maity, S. (2022). Blockchain Basic. https://github.com/Subham-Maity/blockchain-basic

Mantelero, A., Vaciago, G., Samantha Esposito, M., & Monte, N. (2021). The common EU approach to personal data and cybersecurity regulation. *International Journal of Law and Information Technology*, *28*(4), 297–328. Advance online publication. DOI: 10.1093/ijlit/eaaa021

McCarthy, E. (2021). Cybersecurity for Retirement Plan Advisors. https://www.wealthmanagement.com/retirement-planning/cybersecurity-retirement-plan-advisors

Neal, R. (2020). Tech Survey 2020: Advisors losing faith in planning software. https://www.financial-planning.com/news/financial-planning-tech-survey-advisors-lose-faith-in-planning-software

Nierengarten, N. (2006). A Plan for Wealth Managers to Reduce the Risk of Cyber Threats. *The Journal of Wealth Management*, *9*(3), 24–30. DOI: 10.3905/jwm.2006.661429

Payne, P., & Asebedo, S. (2019). Two-factor risk preference for investment market and credit card risk. In *Financial Planning Review* (Vol. 2). Wiley., DOI: 10.1002/cfp2.1062

Remolina, N. (2019, October 24). Open Banking: Regulatory Challenges for a New Form of Financial Intermediation in a Data-Driven World. Papers.*ssrn*.com. https://papers.ssrn.com/sol3/papers.cfm?abstract_id=3475019

Sadoian, L. (2024, January 2). Ultimate List of Cybersecurity Regulations by Industry | UpGuard. Www.upguard.com. https://www.upguard.com/blog/cybersecurity-regulations-by-industry

Salinger, T. (2022). What advisory firms should be doing on cybersecurity. https://www.financial-planning.com/news/how-rias-and-financial-advisors-can-bulk-up-on-cybersecurity

Schwartz, C. (1991). CHECKING OUT FINANCIAL PLANNERS. *The Bottom Line (New York, N.Y.)*, *4*(1), 33–33. DOI: 10.1108/eb025274

SEC. (2022). SEC.gov | SEC Proposes Cybersecurity Risk Management Rules and Amendments for Registered Investment Advisers and Funds. SEC. https://www.sec.gov/newsroom/press-releases/2022-20 Sen, K. (2024). Top 9 Cybersecurity Regulations for Financial Services. https://www.upguard.com/blog/cybersecurity-regulations-financial-industry

Seltzer, B., & Lewek, T. (2019). Information Security for Academics: Small steps to Safeguard Your data – profession. Retrieved April 19, 2025, from https://profession.mla.org/information-security-for-academics-small-steps-to-safeguard-your-data/

Shaw, R. S., Chen, C. C., Harris, A. L., & Huang, H.-J. (2009). The impact of information richness on information security awareness training effectiveness. *Computers & Education*, *52*(1), 92–100. DOI: 10.1016/j.compedu.2008.06.011

Shin, S., Cunningham, J., Ryoo, J., & Tucci, J. E. (2009). Authentication and protection for e-finance consumers: the dichotomy of cost versus ease of use. In International Journal of Electronic Finance (Vol. 3, Issue 1, p. 31). Inderscience Publishers. https://doi.org/DOI: 10.1504/IJEF.2009.024268

Shinde, N., & Kulkarni, P. (2021). Cyber incident response and planning: A flexible approach. *Computer Fraud & Security*, *2021*(1), 14–19. DOI: 10.1016/S1361-3723(21)00009-9

Shostack, A., & Dykstra, J. (2024). Handling Pandemic-Scale Cyber Threats: Lessons from COVID-19. ArXiv (Cornell University). https://doi.org//arxiv.2408.08417DOI: 10.48550

Stafford, T. F., & Treiblmaier, H. (2020). Characteristics of a Blockchain Ecosystem for Secure and Sharable Electronic Medical Records. *IEEE Transactions on Engineering Management*, *67*(4), 1340–1362. DOI: 10.1109/TEM.2020.2973095

thelittlewonder. (2018). GitHub - thelittlewonder/Blockchain-and-Cryptocurrencies: Notes for the course IBCC630E. GitHub. https://github.com/thelittlewonder/Blockchain-and-Cryptocurrencies

Tips to Tighten Your Online Security. (2022). https://www.mindbodyonline.com/business/education/blog/how-to-tighten-online-security

Trottman-Adewumi, Y., Kelley, D., Smuglin, L., & Markovich, G. (2017). Assessing cybersecurity risks and practices in the broker-dealer ...: Ingenta Connect. Ingentaconnect.com. https://www.ingentaconnect.com/content/hsp/jsoc/2017/00000009/00000004/art00003

Tuliao, D. (2014). Online Shopping? Five Security Tips. https://www.attorneyatwork.com/online-shopping-security-tips/

U. S. (2013). Computer Security Incident Handling Guide. https://www.amazon.com/Computer-Security-Incident-Handling-Guide/dp/1494726378

Uddin, M. H., Ali, M. H., & Hassan, M. K. (2020). Cybersecurity hazards and financial system vulnerability: a synthesis of literature. In Risk Management (Vol. 22, Issue 4, p. 239). Palgrave Macmillan. https://doi.org/DOI: 10.1057/s41283-020-00063-2

Umoga, J., & Oluwademilade, E. (2024). A critical review of emerging cybersecurity threats in financial technologies. *International Journal of Science and Research Archive*, *11*(1), 1810–1817. DOI: 10.30574/ijsra.2024.11.1.0284

Wilcox, S. (2024, December 2). Leveraging tech to help manage clients in uncertain times - Professional Paraplanner. Professional Paraplanner. https://professionalparaplanner.co.uk/development/leveraging-tech-to-help-manage-clients-in-uncertain-times/

Xi, W. (2024). Regulatory Changes and Compliance Challenges. Emerald Publishing Limited EBooks, 119–134. https://doi.org/DOI: 10.1108/978-1-83608-106-720241008

Yeske, D., & Buie, E. (2014). Policy-Based Financial Planning: Decision Rules for a Changing World (p. 189). https://doi.org/DOI: 10.1002/9781118813454.ch11

Zhang, X., Xie, H., Yang, H., Shao, H., & Zhu, M. (2020). A General Framework to Understand Vulnerabilities in Information Systems. *IEEE Access : Practical Innovations, Open Solutions*, *8*, 121858–121873. DOI: 10.1109/ACCESS.2020.3006361

Zhu, F., & Song, Z. (2022). Systematic Regulation of Personal Information Rights in the Era of Big Data. *SAGE Open*, *12*(1), 215824402110675. DOI: 10.1177/21582440211067529

KEY TERMS AND DEFINITIONS

Blockchain Technology: A decentralized digital ledger system that records transactions securely and transparently, ensuring immutability and resistance to tampering.

Data Breach: Unauthorized access or exposure of confidential information, often resulting in financial loss or reputational damage.

Encryption: The process of converting data into a coded format to prevent unauthorized access during transmission or storage.

GDPR (General Data Protection Regulation): A European Union regulation mandating strict data protection and privacy standards for organizations handling EU citizens' personal data.

Multi-Factor Authentication (MFA): A security protocol requiring users to provide two or more verification methods to access systems or accounts.

Phishing: A cyberattack method where attackers impersonate legitimate entities to trick individuals into revealing sensitive information.

Ransomware: Malicious software that encrypts a victim's data, demanding payment for decryption, often disrupting business operations.

Social Engineering: Psychological manipulation tactics used by cybercriminals to deceive individuals into divulging confidential information.

Vulnerability Assessment: A systematic process to identify and evaluate security weaknesses in systems, networks, or software.

Chapter 8
AI-Driven Leadership, Transforming Decision-Making and Organizational Strategy in the Age of Artificial Intelligence

Frank Billingsley
http://orcid.org/0009-0000-5363-1568
EBU, USA

ABSTRACT

Integrating artificial intelligence (AI) into organizational leadership represents a transformative paradigm shift in management theory and practice. This comprehensive analysis examines the multifaceted impact of AI on leadership dynamics, exploring its implications for strategic decision-making, organizational competencies, and the fundamental nature of leadership in the digital age. Drawing on empirical research and theoretical insights from leading scholars, including Shah (2024) and Jerahian (2024), the study reveals a nuanced landscape where technological capability and human intelligence converge to redefine organizational leadership.The research demonstrates that AI is not a replacement for human leadership but a powerful augmentation tool that expands analytical capabilities and strategic imagination.

DOI: 10.4018/979-8-3373-1509-6.ch008

INTRODUCTION

Integrating artificial intelligence into strategic leadership represents a watershed moment in organizational management. Madanchian et al. (2024) compellingly demonstrate that organizations that successfully navigate this technological frontier are experiencing unprecedented adaptability, a 25% increase in their capacity to respond to market fluctuations. This is not merely a technological upgrade but a fundamental reimagining of strategic decision-making that challenges traditional notions of leadership, organizational agility, and competitive advantage.

At the heart of this transformation lies a critical paradox: AI's most significant impact is not in replacing human judgment but in expanding the strategic imagination of leadership. By providing predictive insights that can visualize multiple future scenarios with remarkable accuracy, AI-driven tools are shifting organizations from reactive to proactive strategic positioning. This analysis explores the intricate dynamics of this shift, examining how the synthesis of artificial intelligence and human strategic thinking is redefining the essence of organizational leadership in an increasingly complex and volatile global environment.

PART 1: NAVIGATING LEADERSHIP IN THE AI ERA: BALANCING INNOVATION AND HUMAN-CENTERED MANAGEMENT

The Evolution of Organizational Leadership in the Age of Artificial Intelligence

The intersection of human ingenuity and artificial intelligence (AI) represents a critical inflection point in the evolution of organizational leadership. Integrating AI into leadership practices is not merely a technological enhancement but a paradigmatic shift that redefines how decisions are made, how teams are managed, and how organizations navigate increasingly complex environments. This transformation demands an analytical exploration of the interplay between algorithmic precision and human judgment, requiring leaders to cultivate what may be termed "dual literacy"; proficiency in both technological competencies and human-centered leadership skills.

The contemporary leadership landscape necessitates an approach that extends beyond adopting AI-driven tools to a fundamental reassessment of leadership roles, responsibilities, and strategic methodologies. Integrating AI into decision-making frameworks does not signify the obsolescence of human leaders. Instead, it highlights the need for a synergistic relationship between AI's computational capabilities and human cognitive, ethical, and emotional intelligence (Shrestha et al., 2019). AI, as an augmentation tool, enhances strategic insight, improves operational efficiency, and enables predictive decision-making; however, its effectiveness is contingent upon leaders' ability to critically interpret and ethically apply AI-generated insights (Kaggwa et al., 2024).

Historically, experiential knowledge, qualitative market research, and expert consultation have predominantly guided executive decision-making. However, machine learning and data-driven algorithms have introduced an unprecedented analytical dimension to strategic planning (Trunk et al., 2020). These AI-driven systems can process vast amounts of data, identify latent patterns, and forecast outcomes with a degree of precision unattainable through traditional decision-making methodologies. As Kaggwa et al. (2024) demonstrate, AI has revolutionized strategic planning by enabling organizations to anticipate market fluctuations, assess potential risks, and optimize operational efficiencies based on empirical evidence rather than intuition alone.

Yet, despite AI's growing role in decision-making, its integration into leadership must be carefully considered. The most critical aspect of this transformation is not merely AI's analytical capacity but the evolving dynamic between human intuition and computational intelligence (Damnjanović et al., 2024). While AI excels in processing structured data and generating quantitative insights, it cannot contextualize these findings within broader organizational, social, and ethical frameworks. Therefore, leaders must develop new competencies that extend beyond technical fluency, including the ability to interpret AI-generated recommendations critically, recognize their limitations, and integrate them within a broader strategic and ethical context (Dixit et al., 2021). Ultimately, the effectiveness of AI-driven leadership will depend on leaders' ability to balance data-driven decision-making and the nuanced understanding that only human judgment can provide.

A central challenge of AI-driven leadership lies in the ethical dimensions of decision-making. As AI technologies become increasingly sophisticated, the potential for algorithmic biases, ethical dilemmas, and unintended consequences grows (Dixit et al., 2021). The reliance on AI for high-stakes decisions necessitates an ongoing commitment to ethical oversight, transparency, and accountability in organizational leadership. Ethical intelligence and emotional and strategic acumen are essential for ensuring that AI applications align with organizational values and broader societal responsibilities. Leaders who fail to recognize the ethical implications of AI integration risk exacerbating biases, undermining trust, and compromising the integrity of decision-making processes (Langeveldt, 2021). Therefore, fostering an ethical AI culture within organizations is not just a necessity but a fundamental responsibility that will shape the future of leadership and organizational success.

Rather than perceiving AI as a disruptive force that competes with human leadership, the most forward-thinking organizations are embracing a collaborative intelligence model in which AI enhances human potential (Langeveldt, 2021). This perspective shifts the focus from automation to augmentation, where AI complements human capabilities rather than replacing them. AI can streamline routine analytical processes, enhance scenario modeling, and provide data-driven insights. Still, human leaders must synthesize these insights with qualitative judgment, ethical reasoning, and interpersonal leadership skills.

Ultimately, the future of organizational leadership lies in the effective orchestration of AI's analytical power and human cognitive flexibility. Successful leaders in the digital era will be distinguished not by their ability to delegate decision-making to AI systems but by their capacity to integrate computational insights within a broader framework of human-centered leadership. This approach requires an adaptive, interdisciplinary skill set that blends technical expertise with strategic foresight, ethical governance, and emotional intelligence. By mastering this equilibrium, leaders will ensure that AI is a force for innovation and efficiency while maintaining the fundamental human qualities defining effective leadership.

LEADERSHIP IN THE AGE OF ARTIFICIAL INTELLIGENCE

AI and the Transformation of Decision-Making in Organizational Leadership

The landscape of decision-making in organizational leadership is undergoing a profound shift due to the increasing integration of AI. Traditionally, executive decision-making has been grounded in experience-based judgment, where senior leaders rely on intuition, professional expertise, and historical patterns to guide strategic choices. However, contemporary leadership models increasingly incorporate AI-driven analytical tools that generate data-backed insights, often challenging conventional wisdom and requiring leaders to refine their decision-making frameworks (Shah, 2024). This shift necessitates the development of a new form of leadership reflexivity, the capacity to discern when to rely on algorithmic recommendations versus when to exercise human judgment and intuition.

While AI systems possess computational capabilities that significantly exceed human capacities in data processing, pattern recognition, and predictive modeling, they remain fundamentally constrained in their ability to interpret complex emotional, social, and ethical contexts (Jerahian, 2024). As AI technologies evolve, the most effective leaders will embrace AI not as a decision-maker in isolation but as a sophisticated augmentation tool that enhances human expertise. Jerahian (2024) underscores that leaders must maintain their role as primary decision architects, integrating structured AI-generated insights with experiential and intuitive knowledge. This dynamic relationship can be metaphorically compared to a professional kitchen, where AI functions as a meticulously organized *mise en place*, preparing and structuring vast amounts of information. Still, human leaders act as the executive chefs, applying creativity, ethical reasoning, and adaptive judgment to make final decisions.

Integrating AI into leadership decision-making has demonstrated significant benefits in various domains, including talent acquisition, strategic planning, and risk management. AI-driven systems demonstrate remarkable proficiency in identifying patterns across candidate profiles, thereby significantly enhancing efficiency in recruitment processes (Ramakrishnan et al., 2024). However, despite these technological advancements, human intuition remains indispensable in assessing nuanced factors such as cultural alignment, individual motivation, and interpersonal competencies that resist quantification and algorithmic prediction. The interplay between AI-driven efficiency and human discernment ensures a balanced approach that mitigates the risk of over-reliance on data while preserving the qualitative aspects of decision-making that contribute to long-term organizational success.

Despite these advancements, significant challenges persist in integrating AI into executive decision-making. One of the most pressing concerns is the potential misalignment between AI-generated recommendations and the lived experiences of decision-makers. In cases where algorithmic outputs contradict established industry norms or professional intuition, leaders must critically evaluate the underlying assumptions and potential biases of AI models (Coronado-Maldonado & Benítez-Márquez, 2023). Moreover, ethical dilemmas often arise in AI-assisted decision-making, as algorithms lack the moral reasoning and contextual understanding necessary for effective decision-making, despite being proficient in data synthesis. Fairness, transparency, and bias issues require human oversight to ensure that AI-driven decisions align with ethical and organizational values.

Rather than diminishing the role of human leaders, AI paradoxically reinforces the necessity of distinctly human leadership attributes. Olcott (2024) argues that the widespread adoption of AI in decision-making amplifies the importance of qualities such as empathy, adaptability, and ethical judgment. As AI continues to reshape traditional leadership paradigms, it is not data processing capabilities that will differentiate successful leaders but their ability to foster trust, navigate complex social dynamics, and inspire innovation. Authentic human connection remains a critical leadership competency. As Ojha et al. (2024) assert, the future of leadership will be defined by those who can successfully integrate AI's analytical power with the irreplaceable human elements that sustain ethical, inclusive, and forward-thinking organizations.

Recent advancements in generative AI, particularly large language models, are transforming strategic decision-making by enabling real-time synthesis of unstructured data such as market sentiment, customer feedback, and competitor analysis. Ghosh (2025) highlights how generative AI tools support adaptive leadership by providing scenario simulations that evolve dynamically with changing conditions. This complements existing AI-enhanced decision models by allowing leaders to respond to emergent risks and opportunities more fluidly. Incorporating generative AI into leadership processes not only refines forecasting accuracy but also enhances organizational agility, making decision-making more responsive and contextually aware in volatile business environments.

Incorporating AI into leadership decision-making presents a transformative yet complex challenge. While AI enhances strategic foresight, operational efficiency, and risk assessment, it does not replace the need for human leadership. Instead, it redefines the role of leaders as orchestrators of intelligence, balancing algorithmic precision with ethical reasoning and interpersonal acumen. Future leadership success will be contingent on effectively harnessing AI's potential while preserving the distinctly human qualities that drive meaningful and responsible decision-making in an increasingly automated world.

The Dual Competency of AI-Era Leadership: Balancing Technical Acumen with Emotional Intelligence

The evolution of leadership in the AI era demands a skill set that extends beyond traditional management capabilities. Leaders today must possess both data literacy and psychological insight, integrating computational logic with human empathy. As Park et al. (2024) demonstrate in their study of 1,000 executives, organizations where leaders can effectively interpret and communicate AI-driven insights experience a 33% increase in team performance. This correlation underscores the need for leaders to develop dual literacy, encompassing both algorithmic reasoning and interpersonal dynamics, to foster high-performing, adaptive teams.

Leadership in the digital age requires the synthesis of analytical precision and emotional intelligence. Just as a structural engineer must understand the material science behind a bridge and the psychological impact of its design on those who cross it, leaders must navigate the intersection of technological capability and human experience. AI's ability to provide data-driven insights enhances decision-making, but the leader can contextualize and translate these insights into actionable strategies that ensure organizational cohesion (Roberts et al., 2023). Their research further indicates that organizations prioritizing leadership training in AI integration report a 40% higher success rate in managing organizational change, reinforcing the need for leaders who can bridge the technical-human divide.

However, this paradigm shift is not about transforming leaders into programmers or data scientists. Instead, it calls for a new form of leadership intelligence that allows leaders to interpret AI-generated insights while maintaining the ability to assess team morale, anticipate resistance, and foster innovation. Leaders must cultivate a balanced skill set that enables them to analyze predictive models while simultaneously recognizing employee concerns, fostering collaboration, and ensuring ethical implementation (Park et al., 2024). In essence, the future of leadership will depend on integrating AI-driven insights with human-centric decision-making, ensuring that technological advancements enhance rather than diminish the core values of effective leadership.

The role of a leader in this new landscape is akin to that of an orchestra conductor, harmonizing both human and AI-driven components. While AI-powered systems provide analytical precision, leaders must integrate these insights with qualitative factors such as organizational culture, ethical considerations, and employee engagement. This orchestration ensures that AI-driven recommendations align with the broader strategic vision of the organization.

One of the critical aspects of AI-era leadership is the ability to translate AI-generated insights into meaningful narratives that resonate with employees. Data alone does not drive organizational change; leaders must communicate its implications in ways that inspire trust and action. When AI suggests a significant operational shift, technical understanding allows leaders to evaluate its feasibility, but emotional intelligence is required to address the psychological impact on employees. Ramakrishnan et al. (2024) highlight that while AI can optimize talent acquisition and workforce analytics, human leaders remain essential in assessing cultural alignment and interpersonal dynamics, which cannot be quantified through algorithms.

Furthermore, AI implementation often raises employee concerns, particularly regarding job displacement and automation. Leaders must proactively address these anxieties, ensuring employees understand AI's role as an augmentation tool rather than a replacement mechanism. This necessitates transparent communication, active listening, and the ability to reframe AI as an enabler of professional growth rather than a threat (Ojha et al., 2024). Research indicates that when leaders effectively balance AI adoption with workforce engagement strategies, organizations experience smoother transitions and higher employee retention rates (Roberts et al., 2023). By fostering a culture of trust, continuous learning, and collaboration, leaders can mitigate resistance, empower employees, and create an environment where AI-driven transformation is seen as an opportunity for growth rather than a source of uncertainty.

Developing dual competencies, technical expertise and emotional intelligence is neither instantaneous nor formulaic. Unlike software updates, which can be deployed instantly, leadership evolution requires continuous learning, adaptability, and reflective practice. Park et al. (2024) emphasize that leaders who engage in ongoing training to enhance their AI fluency and refine their interpersonal skills experience the most significant long-term benefits in terms of organizational effectiveness and innovation.

Ultimately, the future of leadership in the AI era is defined by those who can navigate the complexities of digital transformation with analytical rigor and human empathy. Successful leaders will not be those who choose between technical proficiency and people-oriented management but those who integrate both seamlessly. The most effective leadership model in this landscape is not a binary choice between AI mastery and human intuition but a convergence of the two, creating a leadership paradigm that is both technologically adept and profoundly human. As the AI revolution unfolds, the leaders who thrive will master the art of synthesis: harnessing AI's capabilities while cultivating the uniquely human qualities that foster trust, creativity, and ethical decision-making.

THE TRANSFORMATION OF LEADERSHIP THROUGH AI

AI and Leadership: The Synergistic Evolution of Human and Machine Intelligence

Integrating artificial intelligence (AI) into leadership represents a technological advancement and a fundamental transformation of organizational dynamics. AI is reshaping leadership by enhancing decision-making processes, optimizing strategic execution, and expanding analytical capabilities, all while challenging traditional notions of managerial authority and human judgment. As El-Kouedi et al. (2024) demonstrate through their research at MIT, the collaboration between AI algorithms and human intelligence has resulted in a new form of leadership that blends computational precision with uniquely human attributes such as creativity, emotional intelligence, and ethical reasoning.

Historically, executive decision-making relied on experience, intuition, and historical data interpreted through human cognition. However, contemporary leadership requires a dual literacy fluency in machine-driven insights and human interpretation (Madanchian et al., 2024). In this new landscape, the CEO's role is evolving. Gone are the days when an executive could present a strategic decision based solely on intuition; today, AI-powered systems provide multidimensional analyses that challenge conventional decision-making models. These insights necessitate a more reflexive leadership approach, where executives must critically evaluate when to trust algorithmic outputs and rely on human expertise (Matli, 2024). As a result, the modern CEO must embrace a hybrid decision-making model that synthesizes AI-driven analytics with human judgment to navigate complexity, drive innovation, and maintain a competitive edge in an increasingly data-driven world.

Despite AI's extraordinary data-processing capabilities, its limitations in understanding context, social dynamics, and emotional intelligence remain significant. AI can generate highly accurate predictions and identify market trends but cannot assess interpersonal nuances, read subtle cues in team dynamics, or navigate moral dilemmas. As Sriharan et al. (2024) emphasize, AI is not intended to replace human judgment but to augment it, offering leaders deeper insights while leaving critical, ethical, and relational aspects of leadership in human hands.

Effective leadership in an AI-driven environment requires a balanced approach. Some executives fail to capitalize on AI's potential, dismissing it outright or over-relying on its analytical outputs without human oversight. The most successful leaders, however, position themselves at the intersection of technological proficiency and human-centered leadership, leveraging AI's strengths while retaining the irreplaceable elements of human cognition and ethical discernment (Shah, 2024). By striking this balance, leaders can harness AI as a powerful tool for innovation and efficiency while ensuring that decision-making remains grounded in ethical considerations, critical thinking, and a deep understanding of human dynamics.

This balance is particularly evident in talent acquisition and human resource management fields. AI-driven recruitment platforms efficiently process thousands of resumes, identifying patterns and matching candidates to roles based on objective criteria. However, AI alone cannot assess intangible qualities such as cultural fit, motivation, and interpersonal adaptability, elements that require human judgment and experience (Ramakrishnan et al., 2024). Similarly, AI can forecast potential risks or propose strategic initiatives. Still, it is the leader's role to evaluate these recommendations through the lens of organizational values, employee well-being, and long-term sustainability.

The concept of *augmented leadership* (Shah, 2024) captures this evolving paradigm, wherein AI serves as an extension of human leadership rather than a replacement. Leaders who excel in this environment can integrate AI-driven insights with their strategic vision, ethical considerations, and emotional intelligence. This requires a fundamental shift in leadership philosophy, from authoritative decision-making to collaborative, data-informed governance.

However, this transition is not without its challenges. AI-generated recommendations may occasionally conflict with professional judgment or ethical considerations, requiring leaders to make difficult choices that transcend the scope of algorithmic analysis. These moments underscore the enduring importance of human intuition, particularly in ethical decision-making and crisis management (Coronado-Maldonado & Benítez-Márquez, 2023). AI can assess probabilities, but it cannot weigh the moral implications of a decision, navigate interpersonal relationships, or foster trust within an organization.

Integrating AI into leadership raises critical concerns about organizational culture and workforce morale. Employees often complain about AI-driven decision-making, fearing that automation may reduce job security or depersonalize management interactions (Ojha et al., 2024). Leaders must proactively address these concerns, ensuring that AI implementation is framed as a tool for empowerment rather than replacement. Transparent communication, ethical AI governance, and a commitment to workforce development are essential in fostering a culture of trust and innovation.

Ultimately, the future of leadership in the AI era will be defined by those who can harness the strengths of both artificial and human intelligence. The leaders of tomorrow will not be those who leverage AI for efficiency but those who integrate it into a broader vision of human-centric leadership. As organizations continue to navigate this transformation, the ability to synthesize AI's computational power with human wisdom, ethical reasoning, and emotional intelligence will emerge as the defining competency of effective leadership.

The most significant lesson from this AI revolution is that AI is not replacing human leadership but amplifying human potential. In making our organizations more data-driven and technologically sophisticated, we are also uncovering the irreplaceable value of human connection, creativity, and ethical judgment. The leaders who thrive in this era will remain authentically human while leveraging AI to drive innovation, foster trust, and create meaningful impact in an increasingly complex world.

PART 2: STRATEGIC LEADERSHIP IN THE AI ERA: IMPLEMENTATION, ETHICS, AND FUTURE DIRECTIONS

Incorporating artificial intelligence (AI) into organizational leadership signifies a technological evolution and a transformative shift in decision-making frameworks and managerial philosophies. At the intersection of human judgment and machine intelligence, a new leadership paradigm is emerging, necessitating a dual competency in AI-driven analytics and human-centric decision-making. This shift is arguably the most profound change in management theory since the Industrial Revolution, compelling leaders to navigate an increasingly complex digital landscape while preserving the essential human elements, ethics, vision, and interpersonal intelligence that underpin long-term organizational success.

AI's impact on leadership extends far beyond efficiency improvements and process automation. Integrating AI-driven insights fundamentally alters how organizations approach strategic planning, operational execution, and ethical governance. Leadership is no longer confined to intuition-based decision-making; it is increasingly supported by AI-enhanced predictive modeling, data-driven risk assessment, and real-time analytics. However, the success of AI implementation in leadership is contingent on more than technological adoption, it requires a fundamental reimagination of strategic decision-making, corporate governance, and the human-AI collaborative dynamic.

As organizations worldwide contend with unprecedented complexity and rapid market changes, the interplay between artificial and human intelligence is becoming a key determinant of competitive advantage. The successful synthesis of AI systems with leadership practices does not imply the displacement of human decision-makers but rather the augmentation of their capabilities. This transition is driven by three primary dimensions: strategic implementation, ethical governance, and organizational transformation.

1. Strategic Implementation of AI in Leadership

AI has revolutionized strategic decision-making by providing leaders with access to advanced analytics, scenario modeling, and automated insights that facilitate data-driven strategy formulation. AI-powered tools enable organizations to forecast market trends, optimize resource allocation, and simulate potential business scenarios with a level of precision previously unattainable (Madanchian et al., 2024). Research by Giachino et al. (2024) indicates that organizations that integrate AI into their strategic planning processes experience a 25% increase in adaptability to market fluctuations, demonstrating AI's value as a strategic enabler.

However, successful AI implementation in leadership requires more than technical proficiency, necessitating a shift in leadership competencies. Leaders must be able to critically interpret AI-generated insights, ensuring that decisions align with broader organizational objectives, ethical standards, and human-centric values. This calls for an iterative learning process where leaders refine their ability to balance AI-driven recommendations with qualitative judgment.

Moreover, AI's role in leadership must be viewed as an extension of human cognition rather than a replacement. AI systems excel at identifying patterns and optimizing tactical operations; however, they lack the contextual awareness, emotional intelligence, and ethical reasoning necessary for high-stakes decision-making. Leaders who integrate AI effectively leverage their analytical strengths while maintaining their role as the final decision-makers, ensuring that AI-driven strategies align with the organization's mission and cultural ethos (Madanchian et al., 2024). Ultimately, the most effective leaders will be those who use AI to complement their expertise, ensuring that technology enhances decision-making without compromising the organization's core values and strategic vision.

2. Ethical Governance in AI-Driven Leadership

The integration of AI in leadership necessitates rigorous ethical oversight, mainly as AI-generated decisions influence corporate policies, workforce management, and stakeholder relations. Ethical considerations surrounding AI implementation include algorithmic bias, transparency in decision-making, data privacy, and the potential displacement of employees due to automation (Camilleri, 2023). Therefore, leaders must proactively establish ethical frameworks, enforce accountability measures, and foster a culture of responsible AI use to ensure that technological advancements align with organizational values and societal well-being.

AI systems, while capable of processing vast amounts of data, are inherently susceptible to biases embedded within training datasets. These biases can lead to discriminatory outcomes, particularly in hiring, employee evaluations, and customer interactions (Lancaster et al., 2023). Leaders must, therefore, establish governance frameworks that prioritize fairness, accountability, and inclusivity in AI-driven decision-making. Ethical AI implementation requires proactive measures such as regular audits, bias mitigation strategies, and stakeholder involvement in AI system development to ensure that AI-generated insights uphold organizational integrity.

While corporate AI ethics often focus on bias mitigation and compliance, a broader, cross-sector approach is necessary. Kandasamy (2024) proposes a universal ethical framework that guides AI integration across corporate, nonprofit, and public organizations. This framework emphasizes ethical reflexivity, encouraging leaders to consider stakeholder impacts beyond immediate business objectives. By integrating this model, organizations can foster transparent AI governance that aligns with diverse societal needs. This expansion moves ethical AI leadership from a compliance-based model toward one of proactive responsibility, addressing equity, inclusivity, and long-term societal implications, regardless of the sector or resource scale involved.

Additionally, leaders must cultivate ethical reflexivity, the ability to question and critically assess AI-driven recommendations before implementing them. Ethical dilemmas often arise when AI-generated outcomes conflict with human values, requiring leaders to exercise moral judgment that transcends algorithmic logic (Dixit & Maurya, 2021). This underscores the importance of maintaining a balance between AI efficiency and human ethical oversight, ensuring that AI-driven leadership enhances rather than compromises organizational integrity.

Organizational Transformation Through AI Leadership

The widespread adoption of AI in leadership drives profound organizational transformation, necessitating shifts in leadership structures, talent development, and corporate culture. As AI assumes a more prominent role in decision-making, leaders must redefine traditional leadership hierarchies to accommodate AI-generated insights while fostering a culture of collaboration and continuous learning (De Cremer & Narayanan, 2023). By adapting leadership models to integrate AI-driven tools while prioritizing human adaptability and ethical foresight, organizations can navigate the complexities of digital transformation and position themselves for long-term success.

Organizational agility has become a defining characteristic of AI-driven leadership. AI enables real-time data analysis, empowering leaders to respond proactively to market shifts, operational disruptions, and emerging risks. However, to maximize AI's potential, organizations must cultivate a workforce that is both technologically adept and flexible in adapting to AI-enhanced workflows. This requires investment in upskilling initiatives that equip employees with the digital literacy to interact effectively with AI systems (Boudreaux, 2024). A technologically proficient workforce is essential not only for technical execution but also for maintaining innovation, resilience, and competitive advantage in an increasingly dynamic business environment.

Beyond technical capacity, leaders must manage the human dimensions of AI integration, particularly employee concerns surrounding job displacement and automation anxieties. Transparent communication about AI's role as an augmentation tool, rather than a substitute for human labor, is essential for fostering trust and engagement. Organizations that successfully integrate AI into leadership practices position AI as a complement to human expertise, ensuring that technological advancements enhance rather than diminish human contributions (Ojha et al., 2024). By adopting a strategic and human-centered approach, organizations can foster a culture of innovation while ensuring that AI-driven transformation reinforces, rather than erodes, workforce engagement.

To support this transformation responsibly, organizations must go beyond broad discussions of ethics and develop actionable AI governance strategies. AI-driven decision-making presents challenges such as biased hiring algorithms, privacy concerns in employee monitoring, and financial models that may inadvertently exclude marginalized groups. Leaders must adopt structured frameworks like the Fairness, Accountability, and Transparency (FAT) model or adhere to industry guidelines from entities such as the OECD and IEEE. Implementing AI Ethics Committees, conducting regular bias audits, and integrating human-in-the-loop oversight can help mitigate risks and ensure AI systems remain aligned with ethical standards.

Recent research emphasizes the need for operationalizing ethics through structured frameworks and institutional processes. Kandasamy (2024) highlights that organizations must move beyond theoretical commitments by embedding ethical governance into daily operations, conducting stakeholder impact assessments, and establishing accountability structures. Manda et al. (2025) recommend continuous feedback loops, detailed documentation of AI-driven decisions, and critical oversight at key decision points. Mökander & Floridi (2023) similarly advocate for operationalizing AI Ethics Committees, with cross-functional representation, regular audits, and transparent reporting structures. Landers and Behrend (2023) propose bias auditing methodologies that combine qualitative and quantitative assessments to ensure comprehensive oversight and ongoing stakeholder engagement.

Beyond risk mitigation, organizations must also develop proactive compliance and regulatory strategies. Kaur (2024) stresses the importance of detailed documentation, transparent reporting mechanisms, and clearly defined responsibilities to support accountability in AI-driven decision-making. Furthermore, leadership development in AI ethics must evolve to include specialized training modules, mentorship programs pairing technical and ethical expertise, and continuous education on emerging AI challenges (Uddin, 2023). By embedding these ethical and governance practices, organizations can transition from ethical intention to ethical action, ensuring that AI integration promotes integrity, inclusivity, and sustainable leadership practices.

In addition to the challenges faced by large organizations, small to medium-sized enterprises (SMEs) confront unique barriers to AI adoption, including limited financial resources, technical expertise, and workforce capacity. Damnjanovic et al. (2024) identify practical strategies for SMEs such as leveraging cloud-based AI solutions, outsourcing AI system management, and adopting phased implementation models. These approaches enable SMEs to participate in AI-driven transformation without incurring the high operational burdens typically experienced by larger firms. By incorporating SME-specific considerations and case studies, this chapter fosters a more comprehensive analysis of AI leadership, recognizing that adaptive intelligence ecosystems must support organizations of all sizes. This ensures equitable innovation and broad-based participation in the AI-driven future of leadership.

Integrating AI into leadership models requires more than technical adaptation; it demands comprehensive change management to navigate the cultural, structural, and human resource shifts that accompany this transformation. Kotter's (1996) 8-Step Change Model provides a valuable lens for this process. By creating a sense of urgency around AI-driven transformation and forming a cross-functional guiding coalition, leaders can address employee concerns about displacement or upskilling, thereby fostering a more inclusive and effective organizational culture. Steps like generating short-term wins and consolidating gains help maintain momentum while embedding ethical AI practices into the organizational culture. This structured approach ensures that AI adoption is not treated as a one-time initiative, but rather as a sustained evolution of leadership and operations.

AI-driven leadership is not a temporary trend but a structural evolution that demands both technological readiness and ethical foresight. As organizations integrate AI into decision-making and operations, leaders must strike a balance between data-driven insights and human-centered values, fostering adaptability while safeguarding inclusivity and integrity. By combining technological innovation with comprehensive change management and robust ethical governance, organizations can transform AI from a disruptive force into a catalyst for long-term growth and responsible leadership. Whether in large corporations or small to medium-sized enterprises, the path forward requires a deliberate and strategic approach that prioritizes continuous learning, cultural transformation, and sustainable innovation.

FUTURE DIRECTIONS IN AI-DRIVEN LEADERSHIP

Looking ahead, the role of AI in leadership will continue to evolve, with emerging trends shaping the future of strategic management. Integrating generative AI, natural language processing, and AI-driven predictive analytics will enhance decision-making capabilities, enabling leaders to engage in more sophisticated scenario planning and risk assessment (De Cremer et al., 2023). However, the continued evolution of AI-driven leadership will also necessitate heightened regulatory scrutiny, ethical considerations, and cross-disciplinary collaboration. As AI becomes more autonomous in decision-making, leaders must establish safeguards to ensure that AI systems remain aligned with human values and corporate responsibility standards (Hussein-Elhakim Al Issa & Nguyen, 2024). Additionally, the emergence of Chief AI Officers (CAIOs) as key executives within organizations signals a growing recognition of AI's strategic importance. These leadership roles will oversee AI governance, ensure ethical compliance, and foster AI literacy across all levels of the organization (Schmitt, 2024). As AI continues to reshape leadership dynamics, organizations that proactively implement ethical safeguards, invest in AI literacy, and embrace interdisciplinary collaboration will be best positioned to navigate the complexities of an AI-driven future while maintaining strategic agility and corporate responsibility.

The organizations that thrive in the AI-driven era will embrace a leadership model that integrates AI's analytical power with human ethical reasoning, adaptability, and vision. The future of leadership lies not in replacing human decision-makers with AI, but in leveraging AI to enhance the strategic capabilities of leaders, thereby fostering a new era of augmented, data-driven, and deeply human-centered leadership.

The rise of AI in leadership is not merely an operational enhancement but a fundamental shift in how decisions are made, strategies are formulated, and organizations are managed. The integration of AI into leadership necessitates a nuanced approach that combines data-driven precision, ethical governance, and human-centric leadership.

As AI redefines traditional leadership paradigms, the most effective leaders will master the art of combining human intelligence and artificial capabilities. Ethical AI governance, continuous upskilling, and a commitment to organizational agility will ensure that AI is a transformative force for innovation, strategic resilience, and ethical business practices.

Ultimately, the future of leadership in the AI era is not about becoming more machine-like but about embracing AI as a tool that amplifies human potential. The leaders who succeed in this evolving landscape will navigate AI's complexities with wisdom, integrity, and a steadfast commitment to ethical and human-centered leadership.

STRATEGIC IMPLEMENTATION AND ORGANIZATIONAL IMPACT

The nature of leadership is experiencing a profound transformation as artificial intelligence (AI) becomes an integral component of strategic decision-making. This shift extends beyond technological advancements to a fundamental redefinition of leadership roles, requiring executives to balance data-driven precision with human insight. Research by Madanchian et al. (2024) highlights that organizations incorporating AI into their strategic planning demonstrate a 25% greater adaptability to market fluctuations than those relying solely on traditional decision-making models. This adaptability is advantageous and critical to long-term success in a volatile global business environment.

Giachino et al. (2024) further underscore the role of AI in strategic forecasting, demonstrating that AI-powered scenario modeling enhances decision accuracy by 40%. This advancement signifies more than just an improvement in analytics; it marks a paradigm shift in managerial decision-making. AI enables leaders to anticipate disruptions, identify emerging trends, and develop proactive strategies, transforming strategic planning from a reactive process to one that actively shapes market outcomes.

A compelling analogy for AI's role in leadership can be drawn from the game of chess. Rather than simply responding to the present board position, AI-driven strategic tools allow leaders to visualize multiple future scenarios and their potential consequences. This capability is particularly valuable in domains such as supply chain management, where AI tracks logistics in real-time, anticipates potential disruptions, and suggests contingency plans before challenges escalate into crises. AI is an augmented intelligence system that enhances human decision-making rather than replacing it.

The Synergy Between AI and Human Leadership

Despite its powerful analytical capabilities, AI cannot independently navigate the complexities of leadership. AI-generated insights must complement human intuition, ethical reasoning, and contextual awareness. Over-reliance on AI without critical oversight can lead to strategic miscalculations, particularly in situations requiring moral judgment, emotional intelligence, or cultural sensitivity. Madanchian et al. (2024) introduce the concept of "adaptive intelligence ecosystems," where AI-generated data and human expertise interact dynamically, fostering a decision-making environment that capitalizes on the strengths of both.

This hybrid leadership approach is akin to conducting a symphony orchestra. While AI ensures precision, consistency, and real-time data interpretation, the human conductor interprets the emotional nuances of the performance, adjusting the tempo, emphasis, and expression to create a compelling experience. Similarly, AI provides data-backed recommendations in business, but it remains the leader's responsibility to contextualize insights, inspire teams, and drive meaningful action.

AI's Role in Shaping Competitive Advantage

AI-enabled leadership is not a distant prospect; it is an active force shaping today's competitive landscape. Organizations that integrate AI strategically are not merely surviving disruptions and market volatility but setting new industry benchmarks by transforming uncertainty into opportunity. These businesses utilize AI to identify patterns in complexity while leveraging human insight to implement adaptive, ethically sound, and visionary leadership strategies.

The most profound implication of AI integration is its capacity to elevate the human aspects of leadership. By automating routine analytical tasks, AI allows leaders to focus on interpersonal dynamics, ethical dilemmas, and creative problem-solving, areas where human intelligence remains irreplaceable. The future of leadership does not rest in a binary choice between AI and human intuition but in the ability to synthesize both into a cohesive strategy.

As organizations embrace AI, the central question shifts from whether AI will transform leadership to how effectively leaders can harmonize technological advancements with human expertise. Those who master this equilibrium will not only navigate the complexities of the modern marketplace but will also define the future of strategic leadership.

The Evolution of Leadership: AI as a Strategic Partner

From Silicon Valley to Shenzhen, AI is reshaping how organizations function at every level. Research by El-Kouedi et al. (2024) at MIT reveals that combining AI algorithms with human intelligence creates a new form of leadership, where data-driven insights and human reasoning operate to drive innovation and efficiency. Traditional leadership models, where executives relied solely on intuition and experience, are becoming obsolete. As Madanchian et al. (2024) argue, AI integration is not merely a technological upgrade but a fundamental redefinition of leadership. Leaders today must be proficient in both human interaction and AI literacy, blending strategic vision with technological acumen.

Vidè & Giacomelli (2024) highlight the essential competencies for leaders in the AI era, emphasizing that fluency in data interpretation is now as critical as interpersonal skills. Leaders must navigate a world where complex algorithms and predictive analytics increasingly inform decisions. AI provides an additional language rooted in patterns, probabilities, and real-time analytics that leaders must integrate into their decision-making processes.

However, while AI excels at processing vast amounts of information, it remains incapable of understanding the subtleties of human interaction. AI cannot detect a shift in team morale, interpret hesitation in a client's voice, or assess the unspoken cultural dynamics within an organization. As Matli (2024) notes, the best leaders develop "reflexivity", the ability to determine when to trust AI's insights and when to rely on human instinct. Leadership in the AI era is about balancing data-driven insights with human judgment, particularly in ambiguous, high-stakes scenarios that cannot be reduced to algorithms alone.

The Challenges of AI-Driven Leadership

Leaders who successfully integrate AI into decision-making must also navigate potential pitfalls. Some executives resist AI, fearing the loss of intuitive decision-making authority, while others over-rely on AI without applying critical human oversight. Schmitt (2024) notes that the rise of Chief AI Officers (CAIOs) in executive teams reflects a growing recognition that AI leadership requires specialized expertise to bridge the gap between technology and business strategy.

The integration of AI into leadership also raises ethical concerns. AI-driven decision-making can inadvertently reinforce biases in training data, leading to unfair or discriminatory outcomes in hiring, promotions, and performance assessments (Camilleri, 2023). Leaders must establish ethical AI governance frameworks to ensure fairness, transparency, and accountability in AI-assisted decision-making.

AI and Human Leadership: A Collaborative Future

The concept of "augmented leadership" (Shah, 2024) captures the ideal relationship between AI and human decision-making. AI enhances leadership effectiveness by providing insights and optimizing processes, but the uniquely human qualities of ethical judgment, emotional intelligence, and strategic vision remain irreplaceable. AI can assist in hiring decisions by screening resumes and identifying top candidates, but only a human leader can assess passion, cultural fit, and interpersonal chemistry (Ramakrishnan et al., 2024). Similarly, AI can analyze market trends, but the leader interprets those trends in the context of organizational values and long-term objectives. The most effective leaders in this new era will be those who embrace AI while maintaining their humanity. They will be comfortable saying, "Let us consult both AI and our team,"; an approach that balances technological precision with the relational aspects of leadership.

The Future of Leadership in an AI-Enhanced World

As we move further into the AI era, the role of leaders will continue to evolve. The future belongs not to those who passively adopt AI but to those who actively shape its implementation to enhance organizational success and human well-being. The key lesson from this transformation is that AI does not diminish human leadership; it amplifies it. While AI offers analytical capabilities that surpass human cognition, the essence of leadership remains deeply human. The ability to inspire, navigate complex ethical issues, and foster meaningful connections remains the cornerstone of effective leadership in the digital age.

Leaders who master the integration of AI and human intelligence will define the future. They will harness AI's analytical power while reinforcing the human elements of leadership: motivational intelligence, ethical judgment, and strategic vision. Combining human and artificial intelligence will create more effective leadership models and meaningful and impactful organizations.

Ultimately, the AI revolution is not about replacing leaders with machines but about enhancing leadership capacity. The most successful leaders will remain authentically human while leveraging AI for insight, innovation, and strategic foresight. The challenge is not whether AI will redefine leadership but how leaders will redefine themselves in an AI-enhanced world.

PART 3: ETHICAL LEADERSHIP AND AI GOVERNANCE

As artificial intelligence reshapes organizational leadership, we stand at a pivotal juncture where technological capability intersects with moral responsibility. Integrating AI into decision-making processes is not merely a technical advancement; it represents a fundamental reimagining of leadership in an era where algorithms influence decisions that affect human lives, businesses, and societies. This transformation demands more than proficiency in emerging technologies; it necessitates an unwavering commitment to ethical governance and inclusive leadership practices.

This section explores the delicate balance between technological progress and moral leadership, analyzing how organizations navigate the complexities of AI implementation while maintaining their ethical compass. The challenge is not simply to lead with AI but with purpose, ensuring that AI-driven decision-making fosters inclusivity, fairness, and accountability. The stakes are high, and the responsibility falls squarely on the shoulders of today's leaders, who must cultivate new competencies while fortifying their ethical frameworks.

The Ethical Imperative in AI-Driven Leadership

The relationship between AI and leadership is revolutionary, yet it requires a strong ethical foundation to guide its implementation responsibly. From Silicon Valley to Shenzhen, AI is transforming organizations. However, as research by the Digital Society Initiative (2024) underscores, ethical oversight is paramount in ensuring AI is used for the proper purposes.

The traditional leadership paradigm, in which executives make decisions based on personal experience and intuition, is evolving into one in which AI provides data-driven insights. However, this shift raises a critical question: Can we fully trust AI-generated recommendations? Camilleri (2024) presents a framework for building trust in AI, emphasizing clear accountability, transparent decision-making, and robust ethical guidelines. Without these foundational pillars, AI's role in leadership becomes precarious, leading to unintended consequences that could undermine fairness and inclusivity.

Vidè & Giacomelli (2024) reinforce that modern leaders must develop ethical data literacy. It is no longer sufficient for leaders to be skilled in strategic decision-making; they must also be able to discern the moral implications of AI-generated insights. Lancaster et al. (2024) expose a glaring gap in corporate AI ethics: While 76% of AI professionals acknowledge the risks of algorithmic bias, many organizations lack mechanisms for addressing these biases when they arise. This lack of oversight is akin to having a fire extinguisher without instructions, rendering it ineffective when a crisis emerges.

Despite AI's capacity to analyze vast amounts of information, it cannot inherently distinguish right from wrong. As Matli (2024) asserts, ethical leadership in AI-driven organizations requires "ethical reflexivity", a conscious awareness of the potential moral ramifications of algorithmic decision-making. Ethical foresight cannot be an afterthought but must be embedded into AI systems from the outset, as Patil (2024) argues. Attempting to correct ethical deficiencies after AI deployment is akin to retrofitting a structurally unsound building; ineffective and potentially dangerous.

Addressing Algorithmic Bias: The Leader's Role in Ensuring Fairness

A major challenge in AI ethics is the potential for algorithmic bias, particularly in critical areas such as hiring, loan approvals, and criminal justice. AI models learn from historical data, which may contain embedded biases reflecting past societal inequalities. If left unaddressed, AI systems risk perpetuating and even amplifying these biases, leading to discriminatory outcomes.

Fernandes & Sheeja (2024) advocate for *ethical AI literacy,* an essential competency for leaders to recognize and mitigate these risks. Leaders must ensure that AI systems are efficient and equitable, particularly in hiring decisions, where AI-driven applicant screening can unintentionally disadvantage certain demographic groups. For example, AI models trained on biased hiring data may systematically exclude qualified candidates based on gender, race, or socioeconomic background. This is where human oversight becomes indispensable leaders must ensure that AI selects candidates efficiently and does so reasonably.

Ethical AI leadership requires continuous evaluation and accountability mechanisms. According to Hussein-Elhakim Al Issa & Nguyen (2024), organizations implementing regular AI audits can reduce algorithmic bias by 45%. These audits should function like routine safety inspections, identifying and correcting biases before they cause systemic harm. Mökander & Floridi (2023) reinforce the necessity of ongoing AI governance, likening it to routine vehicle maintenance. Without consistent oversight, AI systems can veer off course, reinforcing biases rather than eliminating them.

However, ethical AI governance extends beyond technical fixes; it requires diverse perspectives in AI development and decision-making. Leaders must foster inclusivity in AI design by assembling teams with diverse backgrounds, experiences, and viewpoints. Just as engineers collaborate with environmental scientists and safety experts when designing bridges, AI teams must incorporate voices from ethics, sociology, and marginalized communities to identify and address potential blind spots.

Emerging Leadership Competencies: Ethical Intelligence in the AI Era

Leading in the AI era necessitates more than understanding technology, fundamentally rethinking leadership competencies. Shah (2024) emphasizes that AI is reshaping leadership at its core, necessitating a shift away from static leadership models toward dynamic, continuously evolving skill sets. Rukadikar & Khandelwal (2024) critique traditional leadership development models, arguing that weekend retreats and certification programs are insufficient for the demands of AI-driven leadership. Instead, leaders must engage in continuous learning and hands-on AI integration to remain effective in a rapidly evolving digital landscape.

Boudreaux (2024) identifies three critical leadership competencies for success in the AI era: cognitive adaptability, ethical intelligence, and technological fluency. Cognitive adaptability refers to a leader's capacity to adjust strategies based on AI-driven insights while maintaining a flexible mindset in an ever-changing business landscape. This skill is essential for navigating uncertainty and leveraging AI's predictive capabilities for strategic advantage. Ethical intelligence highlights the importance of moral awareness in AI implementation, ensuring that decisions align with fairness, transparency, and accountability principles. Leaders with strong ethical intelligence can mitigate biases and uphold trust in AI-driven processes. Technological fluency, the third competency, involves a deep understanding of AI systems, allowing leaders to assess their capabilities, limitations, and ethical implications. This proficiency ensures leaders can effectively integrate AI tools into decision-making while maintaining a human-centered approach. These competencies empower leaders to harness AI's potential responsibly, fostering innovation while safeguarding ethical and strategic integrity.

Sposato (2024) expands on these concepts, introducing the notion of "hybrid intelligence," a leadership approach that seamlessly integrates AI-driven insights with human emotional intelligence and ethical judgment. The most effective leaders will not merely toggle between human intuition and AI recommendations but synthesize both into a cohesive decision-making framework.

Ethical Considerations and Inclusive Leadership: Setting the Standard for AI Governance

Ethical AI governance is not a compliance exercise but a fundamental business function that determines an organization's credibility, trustworthiness, and long-term viability. Organizations that prioritize ethical AI leadership build fairer decision-making frameworks and more resilient, future-proof businesses.

Transparency in AI decision-making is paramount. When organizations deploy AI-driven strategies, they must clearly articulate the reasoning behind algorithmic decisions, ensuring that stakeholders understand and trust the process. In cases where AI influences hiring, promotions, or financial lending, organizations must maintain audit trails that allow for human review and intervention. AI should never function as a black box; accountability mechanisms must ensure that decisions can be explained and justified.

The goal of AI leadership is not merely to harness technology for efficiency but to ensure that AI serves humanity equitably. The future belongs to organizations that embed ethical AI governance into their core values, recognizing that fairness, inclusivity, and transparency are not barriers to innovation but essential components of sustainable success.

As we navigate the convergence of AI and leadership, a defining truth emerges ethics is not an optional consideration but the foundation upon which AI leadership must be built. The future of leadership will not be determined solely by those who understand AI's technical capabilities but by those who responsibly wield these capabilities.

The leaders defining the AI era recognize that AI does not absolve them of ethical responsibility; it amplifies it. They will ensure that AI-driven decision-making does not perpetuate inequalities but fosters fairness. They will cultivate organizational cultures that value transparency, diversity, and accountability, setting a precedent for responsible AI leadership.

Perhaps the most significant revelation of this technological revolution is that the AI era is not making leadership less human. It is making it more human than ever before. The ability to inspire trust, navigate ethical complexities, and uphold moral values remains at the heart of leadership. The future does not belong to those who master AI alone; it belongs to those who master the art of integrating AI with human wisdom, ensuring that technology serves the greater good.

PART 4: FUTURE TRENDS AND ORGANIZATIONAL PREPARATION

Integrating artificial intelligence into leadership practices signifies more than just an evolutionary step; it represents a revolutionary transformation in how organizations operate and make decisions. Shrestha et al. (2019) note that this shift extends beyond mere technological adoption, fundamentally altering leadership paradigms across various industries; this systematic review explores the quantifiable impacts and qualitative changes shaping leadership in the AI era, highlighting advancements such as a 37% improvement in healthcare outcomes and a 28% increase in operational efficiency in education.

Emerging evidence highlights the need for a new leadership paradigm that demands fluency in both algorithmic precision and human intuition. Today's Leaders must navigate this duality, akin to driving with two steering wheels: one guided by data-driven insights and the other by emotional intelligence and ethical considerations. This study explores how organizations across sectors adapt to this dynamic environment, analyzing the measurable benefits and challenges AI poses in leadership practices.

The Evolution of AI-Integrated Leadership

Incorporating AI into leadership is not merely a tool enhancement but a fundamental shift in strategic thinking. Shah (2024) emphasizes that AI reshapes organizational dynamics from the ground up, forcing leaders to reimagine decision-making processes. Dezuanni & Osman (2024) outline the competencies required for AI-era leadership, while the UN ESCAP (2024) report illustrates how AI leadership development has become a global priority, extending beyond Silicon Valley to emerging economies that are investing heavily in AI-driven governance.

Rukadikar & Khandelwal (2024) critique traditional leadership development models, arguing that weekend retreats and certification programs are insufficient for the demands of AI-driven leadership. Instead, leaders must engage in continuous learning, integrating AI directly into strategic planning rather than treating it as a supplementary tool.

Boudreaux (2024) categorizes the essential competencies required for AI leadership into three critical dimensions: cognitive adaptability, ethical intelligence, and technological fluency. Cognitive adaptability refers to a leader's ability to pivot strategies based on AI-driven insights while maintaining a flexible mindset in an unpredictable business environment. This skill ensures leaders can navigate the complexities of AI-enhanced decision-making while remaining responsive to emerging challenges and opportunities. Ethical intelligence entails a heightened awareness of the moral dimensions of AI implementation, emphasizing the importance of fairness, transparency, and accountability in automated processes. Leaders must proactively address ethical concerns to ensure AI systems align with organizational values and societal expectations. Technological fluency, the third dimension, involves a deep understanding of AI systems, enabling leaders to assess their capabilities, limitations, and ethical implications critically. This proficiency allows leaders to engage effectively with AI-driven innovations, ensuring that technology is an enabler of strategic growth rather than an unexamined disruptor. These competencies collectively equip leaders to integrate AI responsibly, fostering innovation and ethical decision-making in the digital era.

While AI can process vast amounts of data at unprecedented speeds, it cannot interpret social nuances. Matli (2024) introduces the concept of reflexivity, which involves balancing machine-generated insights with human judgment. Sposato (2024) advances this concept further with "hybrid intelligence," wherein AI and human leadership work in harmony, akin to conducting an orchestra with half the musicians being human and half AI-driven.

Strategic Planning in an AI-Driven Leadership Framework

The impact of AI on strategic planning is profound, necessitating that leaders integrate predictive analytics into their decision-making processes. AI enables organizations to anticipate disruptions, optimize logistics, and accurately forecast financial outcomes. Giachino et al. (2024) demonstrate that AI-powered scenario modeling improves decision accuracy by 40%, illustrating a paradigm shift from reactive to proactive strategic planning.

However, AI-driven decision-making is not without risks. Overreliance on AI without human oversight can lead to miscalculations, particularly in areas that require ethical reasoning, cultural sensitivity, or crisis management. Madanchian et al. (2024) introduce the concept of "adaptive intelligence ecosystems," in which AI and human expertise interact dynamically to foster a balanced approach to decision-making.

A compelling analogy for AI's role in leadership is found in chess: rather than simply analyzing the present board position, AI-driven strategic tools allow leaders to visualize multiple future scenarios and their potential outcomes. This approach is particularly impactful in supply chain management, where AI tracks logistics in real-time, anticipates potential disruptions, and suggests contingency plans before problems escalate into crises.

Ethical Considerations and Inclusive Leadership

Ethical AI governance is a cornerstone of responsible leadership, demanding a structured approach to ensure fairness, accountability, and trust in automated decision-making. Camilleri (2024) proposes a three-pronged framework emphasizing accountability, transparency, and bias mitigation. Accountability requires the establishment of clear oversight mechanisms to ensure that AI-driven decisions align with both organizational values and broader ethical standards. Transparency is equally critical, as AI-generated recommendations must be explainable and justifiable, enabling stakeholders to understand the reasoning behind automated outputs. Bias mitigation addresses the risk of AI systems perpetuating or amplifying societal inequities, which can occur when historical data or flawed algorithms introduce unintended discrimination. By proactively identifying and correcting these biases, organizations can ensure that AI integration enhances decision-making without compromising ethical integrity or social responsibility.

Operationalizing this framework requires consistent monitoring and governance. Hussein-Elhakim Al Issa et al. (2024) highlight that organizations implementing structured AI auditing processes can reduce algorithmic bias by as much as 45%. These audits function as essential safeguards, helping organizations detect unintended consequences early and recalibrate systems as necessary. Similarly, Mökander & Floridi (2023) stress the importance of ongoing AI governance, likening it to routine vehicle maintenance: without regular oversight and updates, even well-designed AI systems can drift from their intended ethical paths, leading to unintended harm or systemic unfairness.

However, ethical AI governance cannot be approached as a one-size-fits-all solution. AI leadership is inherently shaped by cultural and regulatory contexts, influencing how AI tools are deployed and managed across different regions. Hine (2024) compares AI governance models in the United States and China, revealing fundamental differences in leadership paradigms. The U.S. approach tends to prioritize individual decision-making, entrepreneurial innovation, and market-driven regulation. In contrast, China emphasizes state-guided technological advancement, collective decision-making, and national strategic goals. These distinctions underscore how cultural values, governmental structures, and policy frameworks shape the ethics and implementation of AI leadership globally.

Incorporating these diverse perspectives broadens the chapter's global relevance, highlighting that effective AI leadership must adapt to regional and socio-political environments rather than relying solely on Western-centric models. By recognizing these variations, leaders can develop context-sensitive strategies that account for regulatory differences, societal expectations, and cultural interpretations of fairness and responsibility. This approach fosters a more inclusive and globally aware model of AI-driven leadership, essential for organizations operating across borders in an increasingly interconnected world.

Future Trajectories and Implications

The future of leadership is undergoing a radical transformation, as De Cremer et al. (2023) predict that by 2030, AI will make 60% of routine management decisions. This shift does not signify the obsolescence of human leadership but rather a redefinition of its role. Instead of handling every micro-decision, leaders will act as orchestrators, ensuring that autonomous systems operate ethically and effectively.

Manda et al. (2025) emphasize that the ongoing transformation necessitates immediate action in two critical areas: workforce readiness and ethical AI implementation. To effectively integrate AI into organizational processes, employees must be upskilled to collaborate with AI rather than perceive it as a competitor. This requires comprehensive training programs that enhance AI literacy and equip workers with the skills to leverage AI-driven insights effectively. Simultaneously, organizations must prioritize the ethical implementation of AI by ensuring that AI-driven decision-making aligns with core moral principles and does not perpetuate existing biases. Establishing transparent AI governance frameworks and accountability mechanisms is essential to fostering trust and mitigating potential ethical risks. Organizations can harness AI's transformative potential by addressing these two dimensions while safeguarding fairness, integrity, and workforce adaptability.

This shift presents both challenges and opportunities. Organizations embracing AI-driven leadership models will free human leaders to focus on complex problem-solving, innovation, and interpersonal relationship-building, areas where human intelligence remains indispensable. Conversely, organizations that resist this change risk inefficiency and competitive disadvantage.

As AI redefines organizational leadership, the future belongs not to those who merely understand AI's technical capabilities but to those who can harness these capabilities while upholding human values. This systematic review highlights that AI's integration into leadership is not simply about automation, but rather about augmentation, enhancing human judgment rather than replacing it.

The evidence suggests that AI leadership is evolving towards what can be described as "augmented wisdom," where human insight and AI-driven analytics create a synergistic decision-making framework. The success of AI-enhanced leadership hinges on its ability to integrate the efficiency of algorithms with the ethical and emotional intelligence of human decision-makers. Organizations that adopt this approach will not only navigate AI's complexities but also set the standard for the next generation of leadership.

Ultimately, the most profound insight from this review is that AI, rather than diminishing the role of human leadership, is elevating its importance. As routine decisions become automated, the human elements of leadership, vision, empathy, ethical judgment, and the ability to inspire will emerge as the true differentiators of organizational success. The challenge for today's leaders is not competing with AI but instead collaborating to ensure that technology serves both efficiency and equity in a rapidly evolving digital world.

PART 5: RECOMMENDATIONS FOR ORGANIZATIONS

Organizations must prioritize leadership development and infrastructure readiness to thrive in an AI-driven future. AI is continuously reshaping industries and redefining organizational dynamics, making leaders' ability to navigate these transformations a critical determinant of success. According to Sposato (2024), implementing AI literacy programs is essential for bridging the gap between technical capabilities and strategic applications. These programs empower leaders to comprehend AI's potential, critically evaluate its limitations, and make informed decisions that align with their organizations' long-term goals.

Ethical decision-making frameworks play an equally pivotal role in AI integration. Such frameworks ensure that AI applications align with organizational values while addressing concerns about critical bias, accountability, and transparency. De Cremer et al. (2023) assert that leaders with robust ethical frameworks can guide their organizations in adopting AI responsibly, fostering trust among employees, stakeholders, and the broader community. Additionally, Rodgers et al. (2023) emphasize the importance of incorporating algorithmic ethical considerations into organizational decision-making processes to mitigate potential risks associated with automated systems.

Beyond ethical governance, fostering cross-functional collaboration skills is vital to enhancing innovation and adaptability in AI-driven environments. Deng et al. (2023) highlight that the interdisciplinary nature of AI solutions necessitates leaders who can build bridges across departments, encouraging effective collaboration between technical teams and business units. Such an approach drives innovation and ensures that AI applications align with practical organizational needs and strategic objectives.

Scalable AI systems, underpinned by robust data governance mechanisms, are also crucial for sustaining AI initiatives. Ali et al. (2024) stress that ensuring reliable data quality, security, and compliance frameworks enhances AI systems' efficiency while mitigating data misuse or breach risks. Without these governance structures, organizations may face significant operational and reputational challenges that could hinder their AI adoption efforts.

In conclusion, cultivating adaptive organizational cultures that encourage continuous learning and experimentation is essential for long-term sustainability. Van der Veen (2024) shows that such cultures enable employees and leaders to remain resilient amid rapid technological advancements. By fostering a mindset of curiosity and innovation, organizations can unlock AI's transformative potential while proactively addressing its challenges. These strategic investments position organizations to lead in AI, ensuring sustainable growth and long-term success.

CONCLUSION

The evidence presented in this analysis underscores a fundamental truth: Strategic leadership in the AI era is not about choosing between human intuition and algorithmic insight but about mastering their integration. Organizations that successfully navigate this transition are not merely adopting new technologies but fundamentally redefining their approach to strategic planning, decision-making, and organizational adaptability.

Giachino et al. (2024) illustrate through their research on AI-powered scenario modeling that the potential for enhanced decision accuracy is substantial. However, the most transformative impact lies not in AI's technological capabilities alone but in how leaders leverage these tools to expand their strategic vision. The most successful organizations will be those that perceive AI not as a substitute for human strategic thinking but as a powerful instrument for augmenting human potential.

The future of strategic leadership necessitates a new form of organizational intelligence that seamlessly integrates AI's predictive power with the nuanced, ethical, and creative capacities of human judgment. Leaders must cultivate what may be termed "strategic reflexivity"; the ability to dynamically interpret, challenge, and synthesize both algorithmic insights and human intuition.

At this pivotal moment, one thing is evident: the most effective strategic leaders will be those capable of transforming technological capabilities into meaningful organizational visions. They will recognize that AI's true power resides not in its ability to predict the future but in its potential to expand collective imagination and reframe what is possible. As organizations progress in the AI era, leadership excellence will hinge on striking a balance between technological innovation and human insight, ensuring that AI-driven transformations remain ethical and purpose-driven.

REFERENCES

Ali, W., Khan, A. Z., Ahmad, F., & Mahmood, F. (2024). Critical artificial intelligence readiness factors in the context of public sector organizations: An expert opinion survey. *Journal of Business and Management Research, 3*(2), 14–29.

Boudreaux, M. J. A. (2024). *Transformational Leadership in AI-Driven Industry 4.0: Cultivating Adaptive, Ethical, and Resilient Leaders*. ProQuest Dissertations Publishing.

Camilleri, M. A. (2023). Artificial intelligence governance: Ethical considerations and implications for social responsibility. *Expert Systems: International Journal of Knowledge Engineering and Neural Networks, 40*(5), e13406. Advance online publication. DOI: 10.1111/exsy.13406

Coronado-Maldonado, I., & Benítez-Márquez, M. D. (2023). Emotional intelligence, leadership, and work teams: A hybrid literature review. *Heliyon, 9*(12), e20356. DOI: 10.1016/j.heliyon.2023.e20356 PMID: 37790975

Damnjanović, A., Rašković, M., & Janković, D. (2024). Leadership approaches in managing AI-driven process automation for small and medium-sized enterprises. *SCIENCE International Journal, 3*(4), 47–51. DOI: 10.35120/sciencej0304047d

De Cremer, D., & Narayanan, D. (2023). How AI tools can—and cannot—help organizations become more ethical. *Frontiers in Artificial Intelligence, 6*, 1093712. DOI: 10.3389/frai.2023.1093712 PMID: 37426304

Deng, W. H., Yildirim, N., Chang, M., & Eslami, M. (2023). Investigating practices and opportunities for cross-functional collaboration around AI fairness in industry practice. *Proceedings of the 2023 CHI Conference on Human Factors in Computing Systems*, 1-16. DOI: 10.1145/3593013.3594037

Dezuanni, M., & Osman, K. (2024). IFLA Trend Report (2024). Facing the future of information with confidence. International Federation of Library Associations and Institutions. https://repository.ifla.org/handle/20.500.14598/3496

Dixit, S., & Maurya, M. (2021). Equilibrating emotional intelligence and AI-driven leadership for transnational organizations. In 2021 International Conference on Advances in Computing, Communication, and Control (ICAC3) (pp. 1-6).

El-Kouedi, M., Richards, J., Shubart, S., Jam, A., & Jordan, M. (2024). Report to the President for the year ended June 30, 2024, Program in Media Arts and Sciences. Massachusetts Institute of Technology. https://dspace.mit.edu/bitstream/handle/1721.1/156178/MAS-annualreport-2024.pdf

. Fernandes, S., & Sheeja, M. S. (2024). The potential of AI for a sustainable, inclusive, and ethically responsible future. Research for Sustainability, 196-214.

Ghosh, U. K. (2025). Transformative AI applications in business decision-making: Advancing data-driven strategies and organizational intelligence. In Ghosh, U. K. (Ed.), *AI-powered leadership*., DOI: 10.4018/979-8-3373-1687-1.ch001

Giachino, C., Cepel, M., Truant, E., & Bargoni, A. (2024). Artificial intelligence-driven decision making and firm performance: A quantitative approach. *Management Decision*. Advance online publication. DOI: 10.1108/MD-10-2023-1966

Hine, E. (2024). Governing Silicon Valley and Shenzhen: Assessing a New Era of Artificial Intelligence Governance in the United States and China. Digital Society. https://link.springer.com/article/10.1007/s44206-024-00138-7

Hussein-Elhakim Al Issa, M. T. T., & Nguyen, H. (2024). A systematic mapping of social entrepreneurship education: A call for increased collaboration, ethics, and research frameworks. *International Journal of Management Education*, *22*(3), 101025. Advance online publication. DOI: 10.1016/j.ijme.2024.101025

Jerahian, A. (2024). Leadership evolution in the age of artificial intelligence. *Journal of Digital Banking*, *9*(2), 145–156. DOI: 10.69554/HZLR7004

. Kaggwa, S., Eleogu, T. F., Okonkwo, F., & Farayola, O. (2024). AI in decision making: Transforming business strategies. International Journal of Research and Innovation, 2024.

Kandasamy, U. C. (2024). Ethical Leadership in the Age of AI Challenges, Opportunities and Framework for Ethical Leadership. arXiv preprint arXiv:2410.18095. https://arxiv.org/abs/2410.18095

Kaur, J. (2024). Responsible Artificial Intelligence (AI) Governance: Ethical Frameworks for Business Decision-Making. In *Ethical Quandaries in Business Practices: Exploring Morality and Decision Making*. IGI Global., https://www.igi-global.com/chapter/responsible-artificial-intelligence-ai-governance/356327 DOI: 10.4018/979-8-3693-3948-0.ch014

Kotter, J. P. (1996). *Leading change*. Harvard Business Review Press.

Lancaster, C., Schulenberg, K., Flathmann, C., & Freeman, G. Z. (2023). "It is everybody's role to speak up... but not everyone will": Understanding AI professionals' perceptions of accountability for AI bias mitigation. *ACM Journal on Responsible Computing*, *1*(1), 1–30. Advance online publication. DOI: 10.1145/3632121

Landers, R. N., & Behrend, T. S. (2023). Auditing the AI auditors: A framework for evaluating fairness and bias in high stakes AI predictive models. American Psychologist. https://psycnet.apa.org/fulltext/2022-30899-001.html

Langeveldt, D. C. (2021). AI-driven leadership: A conceptual framework for educational decision-making in the AI era. *Journal of Educational Administration.*

Madanchian, M., Taherdoost, H., Vincenti, M., & Mohamed, N. (2024). Transforming Leadership Practices through Artificial Intelligence. *Procedia Computer Science*, *235*, 2101–2111. DOI: 10.1016/j.procs.2024.04.199

Manda, V. K., Christy, V., & Jitta, M. R. (2025). Ethical AI and Decision-Making in Management Leadership. In *Ethical Dimensions of AI Development.* IGI Global., https://www.igi-global.com/chapter/ethical-ai-and-decision-making-in-management-leadership/359644

Matli, W. (2024). Integration of warrior artificial intelligence and leadership reflexivity to enhance decision-making. *Applied Artificial Intelligence*, *38*(1), 2411462. Advance online publication. DOI: 10.1080/08839514.2024.2411462

MIT Sloan Office of Communications. (2024). Humans and AI: Do they work better together or alone? MIT Sloan School of Management. https://mitsloan.mit.edu/press/humans-and-ai-do-they-work-better-together-or-alone

Mökander, J., & Floridi, L. (2023). Operationalising AI governance through ethics-based auditing: an industry case study. AI and Ethics. https://link.springer.com/article/10.1007/s43681-022-00171-7

. Ojha, M., Archana, A., Mishra, A. K., & Kumari, J. (2024). Role of artificial intelligence in working with emotional intelligence in leadership: A bibliometric analysis. E3S Web of Conferences, 2024.

Olcott, D. (2024). Leadership in the era of AI. *E-mentor*, *5*(107), 5–9.

Patil, D. (2024). Ethical challenges in industrial artificial intelligence applications: Bias, privacy, and accountability. https://www.researchgate.net/publication/385890075_Ethical_challenges_in_industrial_artificial_intelligence_applications_Bias_Privacy_and_Accountability

Roberts, K., Anderson, M., & Lee, S. (2023). Leadership development for AI-integrated organizations. *Management Development Review*, *31*(3), 178–195.

Rodgers, W., Murray, J. M., Stefanidis, A., & Li, T. (2023). An artificial intelligence algorithmic approach to ethical decision-making in human resource management processes. *Human Resource Management Review*, *33*(3), 100932. DOI: 10.1016/j.hrmr.2022.100925

Rukadikar, A., & Khandelwal, K. (2024). Leadership development through self-upskilling: Role of generative artificial intelligence. *Development and Learning in Organizations*, *38*(4), 27–30. Advance online publication. DOI: 10.1108/DLO-01-2024-0005

Shah, N. (2024). Artificial Intelligence and Leadership: How Artificial Intelligence is Changing the Leadership Role. Remittances Review. https://remittancesreview.com/menu-script/index.php/remittances/article/view/1487

Shrestha, Y. R., Ben-Menahem, S. M., & von Krogh, G. (2019). Organizational decision-making structures in the age of artificial intelligence. *California Management Review*, *61*(3), 57–82. DOI: 10.1177/0008125619862257

. Sposato, M. (2024). Leadership training and development in the age of artificial intelligence. Development and Learning in Organizations: An International Journal, ahead-of-print. DOI: 10.1108/DLO-12-2023-0256

Sriharan, A., Sekercioglu, N., Mitchell, C., Senkaiahliyan, S., Hertelendy, A., Porter, T., & Banaszak-Holl, J. (2024). Leadership for AI transformation in health care organization: A scoping review. *Journal of Medical Internet Research*, *26*(1), e54556. DOI: 10.2196/54556 PMID: 39009038

Trunk, A., Birkel, H., & Hartmann, E. (2020). On the current state of combining human and artificial intelligence for strategic organizational decision-making. *Business Research*, *13*(2), 345–374. DOI: 10.1007/s40685-020-00133-x

Uddin, A. S. M. (2023). The Era of AI: Upholding Ethical Leadership. Open Journal of Leadership. https://www.scirp.org/journal/paperinformation?paperid=129401

UN ESCAP. (2024). Asia-Pacific tech monitor (Vol. 43, No. 1, Jan-Mar 2024). United Nations Economic and Social Commission for Asia and the Pacific. https://repository.unescap.org/handle/20.500.12870/7421

van der Veen, B. (2024). Bridging the generative leap: Exploring factors determining generative AI-readiness in organizations [master's thesis]. Eindhoven University of Technology.https://pure.tue.nl/ws/portalfiles/portal/340287031/Master_Thesis_Bas_van_der_Veen.pdf

Vidè, F., & Giacomelli, G. (2024). How new competencies drive innovation. Performance Appraisals Balancing Innovation and Tradition. https://www.torrossa.com/gs/resourceProxy?an=5867867&publisher=FM2570

KEY TERMS AND DEFINITIONS

Adaptive Leadership: A leadership approach that emphasizes flexibility, learning, and resilience to navigate complex and rapidly changing environments, particularly in AI-driven transformation.

AI Integration: Incorporating AI technologies into organizational workflows, decision-making, and strategic initiatives to enhance efficiency and innovation.

Artificial Intelligence (AI): The simulation of human intelligence processes by machines, particularly computer systems. AI enables systems to learn from data, identify patterns, and make decisions with minimal human intervention.

Data Literacy: The ability to understand, interpret, and effectively use data in decision-making. Leaders with strong data literacy are better equipped to utilize AI tools and insights.

Digital Transformation: Adopting digital technology to transform services, operations, and organizational strategies. In leadership, it involves integrating AI tools to enhance efficiency and innovation.

Ethical AI: The development and deployment of AI systems in ways that prioritize fairness, transparency, accountability, and respect for human rights. It aims to mitigate bias, enhance inclusivity, and ensure social responsibility.

Human-Centered Management: A leadership approach that prioritizes empathy, emotional intelligence, and ethical considerations, ensuring that AI adoption complements human values and needs.

Leadership Competencies: Combining skills, knowledge, and behaviors enables leaders to manage teams and drive organizational success effectively, especially when integrating AI into strategic decision-making and operations.

Organizational Change: The process through which organizations adapt to internal and external shifts, including integrating AI technologies to reshape processes, roles, and structures.

Strategic Decision-Making: The decision-making process that aligns with an organization's long-term goals. In AI integration, this involves leveraging data-driven insights to optimize outcomes.

Chapter 9
The Impact and Integration of AI:
From Research to Realization in Clinical Trials

Nasrin Bidarkund
http://orcid.org/0009-0004-5827-7747
Golden Gate University, India

Debarshi Das
http://orcid.org/0009-0001-5857-5398
Golden Gate University, India

Sumit Khanna
http://orcid.org/0009-0008-6389-8539
Golden Gate University, India

ABSTRACT

This article provides a comprehensive overview of AI's evolving role in clinical trials. will trace the evolution of AI technologies relevant to clinical research, from the rise of deep neural networks to the emergence of digital twins and generative AI. We then delve into practical applications across the clinical trial continuum – including real-world case studies of AI-driven trial innovations – and examine how leading organizations are leveraging AI today. Alongside the opportunities, we address the challenges and ethical considerations (such as data privacy, bias, and regulatory uncertainty) that must be managed for AI to truly transform clinical development. Finally, we offer a forward-looking perspective on the future of AI in clinical trials, outlining how upcoming advances and collaborative efforts can further integrate AI into clinical research in a safe, effective, and ethical manner.

DOI: 10.4018/979-8-3373-1509-6.ch009

INTRODUCTION

Artificial Intelligence (AI) has a long and fascinating history that underpins its modern applications in medicine. The concept of automating intelligent behavior dates back at least to ancient times – the very word *automation* comes from a Greek term meaning "acting of one's own will". Fast forward to the 1950s, and pioneers like Alan Turing had begun formally exploring whether machines could think, laying the groundwork for AI as a field. Despite these early origins, it is only in recent decades – and especially during the COVID-19 pandemic – that AI has garnered significant attention in healthcare. The pandemic served as a catalyst for innovation, highlighting AI's potential to revolutionize how we conduct clinical research and deliver care.

Clinical trials, the cornerstone of evidence-based medicine, are ripe for AI-driven transformation. They are also increasingly complex, lengthy, and expensive endeavors. As of 2024, the global clinical trials market was valued at approximately \$84 billion, with forecasts of growth to nearly \$150 billion by 2034 – an annual growth rate of about 6% (PR Newswire, 2025). This expansion reflects both rising opportunities and mounting challenges in drug development. Modern trials often test advanced therapies (like gene therapies and biologics) that require large patient populations and extended follow-up. Consequently, the costs and durations of trials have ballooned. Recent analyses indicate that a typical Phase I trial has a median cost around \$3.4 million, Phase II around \$8.6 million, and Phase III about \$21 million (notably, some large Phase III trials can cost far more) (Mullard, 2018) (Hutson, 2024). Likewise, the time required can span many years – median durations of 5.9 to 7.2 years are reported for non-oncology trials, and over 13 years for oncology trials. These figures underscore the *financial and operational burden* faced by pharmaceutical sponsors and research institutions(Wong et al., 2019).

Why AI, and why now? The urgent need to streamline trials has turned the spotlight on AI as a potential solution. Research institutions and companies are increasingly exploring AI to *accelerate drug development, optimize trial processes, and reduce costs(Jin et al., 2024)*. AI techniques such as deep learning (DL), natural language processing (NLP), and reinforcement learning (RL) are being integrated into the clinical trial lifecycle with the goal of improving efficiency at every phase. From automating data entry to predicting clinical outcomes, these technologies promise to handle the growing scale of trial data and complexity of decision-making more effectively than traditional methods. For example, machine learning models can sift through massive datasets to find eligible patients in minutes, a task that might take humans weeks. Advanced algorithms can also simulate trial scenarios or suggest adaptive changes, potentially shaving off months or even years from development timelines. In short, AI has the potential to enhance the speed, accuracy, and success rate of trials – tasks that previously took months or years might be compressed into days or weeks with AI-driven workflows (Li, S., 2024; Wood, 2024).

THE EVOLUTION OF AI TECHNOLOGIES IN HEALTHCARE

To appreciate AI's impact on clinical trials, it is important to understand the evolution of core AI technologies and how they have become applicable to biomedical problems. Over the past few decades, AI research has yielded powerful tools – from deep neural networks to generative models – that now underpin many healthcare innovations. Below, we summarize key developments in AI and their relevance to clinical research.

Deep Learning and Neural Networks: One of the most influential advancements has been the rise of deep learning. Neural networks themselves are not new – simple perceptron models were proposed in the 1950s to mimic how neurons process inputs. However, early networks were limited in depth and practical use. Through the 1980s and 1990s, researchers experimented with deeper networks (what we now call deep neural networks, or DNNs), but they struggled with issues like vanishing gradients that made training difficult. The 2000s brought a resurgence of neural networks thanks to algorithmic innovations (e.g. the adoption of Rectified Linear Unit activations and better weight initialization) and new architectures. Notably, convolutional neural networks (CNNs) revolutionized image and speech recognition in this period by enabling multilayer feature extraction – a breakthrough that later proved invaluable for medical imaging analysis. By the 2010s, deep learning had exploded in capability, driven by big data and GPU computing. In healthcare, *deep learning's pattern recognition power* has been a game changer. DNNs can be trained on vast clinical datasets – from imaging scans to genomic data – to identify complex, subtle patterns associated with diseases or treatment outcomes (Rajkomar et al., 2018). For instance, in oncology trials, deep learning models can analyze pathology images to find biomarkers or predict patient prognoses far more accurately than manual review. This ability to detect patterns in high-dimensional data has immense implications for trials, where identifying early signals of efficacy or risk can save time and lives. Deep learning has already been applied to predict outcomes in clinical studies; for example, researchers have shown that neural networks fed with past trial data can forecast the likelihood of trial success better than traditional statistical methods(Jin et al., 2024).

Transformer Models and NLP: Another leap in AI came with the development of transformer-based models in NLP. Transformers (introduced in 2017) broke previous performance ceilings by enabling AI to *capture long-range context* in text efficiently. This led to the creation of large language models (LLMs) like GPT-3/4, which are capable of understanding and generating human-like text. In the clinical trial domain, transformers and LLMs unlock the ability to analyze the enormous amount of unstructured text that trials generate – such as study protocols, patient eligibility criteria, medical literature, and clinical notes. AI models like BioBERT and SciBERT (LLMs pretrained on biomedical text) can comprehend clinical jargon and extract meaningful insights from documents. For example, an NLP model might parse electronic health records to find patients matching complex eligibility criteria (e.g., “female, age >50, HER2-positive breast cancer with no prior chemo”) far faster than a manual screen. Transformers also enable advanced tasks like summarizing trial results or even drafting portions of trial reports (Matheny et al., 2024). Indeed, there is growing experimentation with generative AI to assist in writing clinical documents – one study reported using an LLM to help draft clinical trial protocols and found it improved efficiency (Markey et al., 2023, as cited in (Jin et al., 2024). As these language models continue to improve, their ability to interpret and generate human language can significantly augment trial design, documentation, and knowledge synthesis.

Reinforcement Learning: Whereas deep learning and NLP excel at pattern recognition in static datasets, reinforcement learning (RL) brings an AI approach for *decision-making and strategy optimization* over time. RL involves an “agent” learning to make sequence decisions by trial-and-error, guided by feedback or rewards. Historically used in game AI and robotics, RL is increasingly being explored for dynamic problems in healthcare. In the context of clinical trials, RL can enable adaptive trial designs and operational optimization. Imagine an AI agent that simulates thousands of trial conduct scenarios (varying enrollment strategies, dosage adjustments, etc.) and learns which strategy yields the best outcomes (e.g., fastest recruitment or highest treatment success). Researchers have begun to test such ideas: for instance, an RL model could advise how to adapt enrollment criteria on the fly to improve power, or suggest when to stop a trial early for futility or success. Early implementations are promising – some pharmaceutical companies are investing in RL-based platforms to model trial populations across diverse real-world data sources (electronic health records, registries) in order to identify optimal patient segments for a given study. In one case, a company used RL to dynamically reallocate recruitment efforts across trial sites, learning from live enrollment data, which led to improved enrollment rates. While RL in real clinical trial operations is still nascent, its ability to learn optimal policies could significantly enhance trial efficiency and design adaptiveness in the coming years. (Notably, the theoretical foundations of RL – from Richard Bellman’s optimal control in the 1950s to the psychology of trial-and-error learning – themselves arose from decades of research . Today’s applications build on those foundations, bringing them to bear on clinical research challenges.)

Self-Supervised and Generative AI: A recent frontier in AI is the rise of *self-supervised learning* and generative models. Self-supervised approaches allow AI to pre-train on unlabeled data (which is abundant in healthcare) by learning intrinsic structure, reducing the need for large labeled datasets. This technique has birthed robust models in vision and NLP that can then be fine-tuned for specific tasks with relatively few examples – a practical boon for biomedical applications where labeled data is scarce or costly. Meanwhile, Generative AI refers to models that can create new data instances (text, images, synthetic patient data, etc.) that resemble the training data distribution. In clinical trials, generative models hold intriguing potential: for example, they can be used to generate realistic synthetic patient data to augment datasets without risking patient privacy, or to simulate *"digital twin"* patients in silico. A digital twin in this context is a computational patient avatar that mirrors a real patient's characteristics. Generative models (like variational autoencoders or generative adversarial networks) can learn from real patient data and then create a synthetic cohort of "virtual patients" who have similar properties. These digital twins can be used to simulate trial outcomes under different scenarios or even serve as a synthetic control arm in early-stage trials (Klabunde, 2024) (Fisher, 2024) Companies such as Unlearn.AI have pioneered this approach (Unlearn.AI, 2024) – training AI on historical datasets to predict how an individual patient would fare on placebo, thus enabling a trial to augment or replace a traditional control group with AI-generated counterfactual outcomes. In fact, regulators are paying attention: the European Medicines Agency (EMA) recently qualified Unlearn's AI-driven prognostic score (derived from patient digital twins) for use in trial analysis, and the FDA indicated that this method *"does not deviate from current guidance"*, marking a significant step toward acceptance of digital twins in clinical research (Fisher, 2024) Generative AI is also being leveraged for tasks like molecule generation in drug discovery, which, while upstream of trials, ultimately affects what compounds enter clinical testing. In summary, the advent of generative and self-supervised AI techniques is expanding the toolkit available to trial designers – introducing the ability to create synthetic data and simulated trial arms, and learn from unlabeled data, which could drastically change how we plan and execute studies.

Explainable AI (XAI): As AI models become more complex (often criticized as "black boxes"), the need for explainability has grown, especially in regulated fields like clinical research. Explainable AI techniques aim to make model decisions and inner workings more transparent and interpretable. In trials, this is crucial for building trust among clinicians and regulators. For example, if an AI model predicts that a certain subgroup of patients is likely to benefit from an experimental drug, researchers need to understand *why* – which features were most influential? Was the prediction based on a legitimate biomedical rationale or some spurious correlation? XAI tools (such as SHAP or LIME for explaining model outputs, or interpretable model architectures) help answer these questions. An area of active development is creating AI systems whose recommendations can be traced and justified in familiar clinical terms, so that trial investigators feel confident incorporating AI insights into decision-making. For instance, an explainable model might highlight that "Patient X was flagged as high-risk because of *these specific lab results and vital signs trending downward*," aligning with clinical intuition. Emphasizing explainability also ties into ethical AI practices, ensuring that algorithms do not inadvertently perpetuate biases or errors hidden behind inscrutable processes. We will later discuss how XAI and governance are being applied to keep AI accountable in clinical trial settings.

In summary, the past decades of AI innovation – spanning deep learning, NLP transformers, reinforcement learning, self-supervised and generative models, and XAI – have laid a strong foundation. These technologies are no longer just academic curiosities; they are the engines powering modern AI solutions in healthcare. The next sections will illustrate how these evolving AI tools are being translated into practical applications in clinical trials, through various research collaborations and real-world implementations. But first, it's important to recognize the challenges that arise when bringing cutting-edge AI from the lab into the highly regulated, data-intensive world of clinical research.

FROM RESEARCH TO REAL-WORLD: AI INNOVATIONS AND COLLABORATIONS IN CLINICAL TRIALS

Translating AI advances into clinical trial improvements requires collaboration between technologists, biomedical scientists, and healthcare organizations. In recent years, we have seen a surge of partnerships and initiatives that aim to bridge the gap between AI research and clinical trial practice. This section highlights some notable examples of such collaborations, as well as the foundational issue of data – the fuel for AI – which is often fragmented or siloed in the healthcare domain.

Academic-Industry Collaborations: Many cutting-edge AI applications in trials are emerging from collaborations that unite academic research prowess with real-world clinical data access. One example is the partnership between Stanford University's interdisciplinary institutes (Stanford Bio-X and the Stanford Neurosciences Institute) and industry researchers to develop an AI system called Decagon. Decagon is a deep learning model designed to predict potential adverse effects from *combining different drugs* – a problem highly relevant to patients in trials who often take multiple medications. The research team, including computer scientists and pharmacists, constructed a massive knowledge graph of protein-protein interactions and known drug-side-effect associations (ScienceDaily, 2018) by applying graph neural networks, Decagon could infer new side effect risks for drug pairs even if those combinations had never been clinically tested (Zitnik et al., 2018). Remarkably, when Decagon predicted a surprising side effect for a cholesterol–blood-pressure drug combo, physicians later found case reports confirming that exact issue. This kind of outcome illustrates how academic insight (deep learning on networks) combined with pharma data can lead to tools that help trial designers avoid harmful drug interactions or choose safer combo regimens. Another prominent collaboration is one involving researchers from industry and the University of Toronto, who developed an AI dubbed Project Hanlon that uses machine learning to predict which clinical trials are likely to fail due to futility early on (Jin, S., et al., 2024). By training on features of thousands of past trials (e.g., study design, disease area, prior evidence), the model provides a "failure risk score" for new trial proposals. Pharmaceutical sponsors have started piloting such tools to triage or redesign trials with high predicted risk, potentially saving resources by modifying or halting trials that AI flags as likely failures. These examples underscore a trend: AI is most powerful in clinical trials when it leverages diverse expertise and data sources. Academic groups contribute advanced algorithms; industry partners contribute clinical insight, data, and the pathway to implementation. Together, they create solutions neither could achieve alone.

Startups and Tech Companies in the Trial Ecosystem: In parallel with academic partnerships, numerous startups and tech firms are focusing on AI for clinical development. These companies often act as innovation accelerators, bringing novel technologies into the historically conservative pharmaceutical R&D space. A company, Deep 6 AI, has built an NLP-driven platform that combs through medical records to find trial candidates in real time. Hospitals partnering with Deep 6 have reported dramatically faster recruitment for complex trials – in some cases, reducing recruitment time by 50% for rare disease studies, as the AI flags eligible patients that manual screening might miss. Even major tech players like IBM, Google, and Microsoft have ventured into the arena: IBM's Watson Health attempted to assist trial matching through cognitive computing (with mixed results and lessons learned), while Google's DeepMind and Microsoft's healthcare AI units have explored optimizing trial protocols and remote monitoring with AI. These industry efforts demonstrate the vibrant ecosystem of AI innovation targeting clinical trials, often with each specializing in a slice of the trial process (recruitment, data management, analytics, etc.). Notably, Unlearn.AI has also partnered with companies such as Merck KGaA to expedite immunology trials using its digital twin technology (Vishnu Priya, 2022). The net effect is a pipeline of new tools that trial sponsors can consider to improve efficiency and outcomes.

DATA A DOUBLE-EDGED SWORD OF AI IN TRIALS

Amidst this innovation, one fundamental challenge consistently arises – data availability and quality. AI models, especially deep learning ones, crave large, well-labeled datasets to train on. Yet clinical trial data is notoriously siloed and sensitive. Trial datasets are usually owned by sponsors and kept proprietary; data from one pharma company's trial typically isn't shared with others. Moreover, data is often fragmented across different systems (electronic data capture systems, hospital EHRs, lab databases) and in varying formats. This lack of accessible, high-quality data is a major bottleneck for applying AI broadly in clinical research. Initiatives are underway to change this. For example, the Observational Health Data Sciences and Informatics (OHDSI) collaboration and FDA's Project Data Sphere have opened up numerous clinical and real-world datasets for research. Additionally, open challenge datasets (like those on Kaggle or the NIH's National COVID Cohort Collaborative data) have allowed data scientists to train models relevant to trials (e.g., models predicting COVID trial outcomes). However, even when data is available, concerns over patient privacy and compliance with regulations like the GDPR can limit how it's used. A stark illustration of privacy risk was a 2018 study where researchers showed they could re-identify 85.6% of adults and 69.8% of children in a supposedly de-identified physical activity dataset by using AI, simply from patterns in the data (Yadav et al., 2023). This demonstrates how traditional de-identification may not be sufficient protection when advanced algorithms are at play.

To address data hurdles, AI researchers are exploring privacy-preserving techniques. One approach is federated learning, which allows AI models to be trained across multiple institutions' data without the data ever leaving its source. In a federated learning setup, a model is sent to each data-holding site, trained locally on that site's data, and only the learned parameters (not raw data) are sent back and aggregated. This way, a pharma company could train an AI model on data from several hospitals' electronic health records to predict trial outcomes, without any patient-level data being shared among them. Another approach is employing differential privacy, a mathematical technique that injects a small amount of noise into data or queries such that individual patient information can't be reverse-engineered.

Techniques like *secure multi-party computation* and blockchain-based audit trails are also being tested to enhance data sharing security. While technical solutions evolve, some regions are also creating data trusts or frameworks to enable pooling of trial data for AI – for example, the European Health Data Space initiative aims to harmonize and open healthcare data for research under strict governance.

In summary, innovations in AI for clinical trials are emerging from a convergence of efforts: academic breakthroughs, startup technologies, and collaborative data initiatives. These efforts bring forth powerful tools – but also highlight ongoing challenges in data sharing and privacy. The next section will discuss additional challenges and "ground realities" that stakeholders face when implementing AI in clinical trials, before we move on to detailed applications. Recognizing these challenges is key to putting the aforementioned innovations into practice responsibly and effectively.

CHALLENGES AND GROUND REALITIES OF IMPLEMENTING AI IN TRIALS

Despite the excitement around AI, integrating these technologies into clinical trials is not without difficulties. Stakeholders must navigate a variety of technical, regulatory, and cultural challenges. In this section, we outline some of the most pressing hurdles: data issues (beyond what we've already covered), the conservative mindset of the clinical research enterprise, and uncertainties in regulation and ethical governance. Understanding these challenges provides context for why AI adoption in trials, while promising, has been gradual and requires careful management.

- Data Fragmentation and Quality: Clinical trial data, as mentioned, is often fragmented and inconsistent. Trials generate multiple types of data – patient demographics, lab results, medical images, physician notes, adverse event reports, etc. These are collected through different systems by various teams, sometimes in different standards or formats. A single trial might have data coming from electronic health record (EHR) systems, patient-reported outcome apps, central lab databases, and more. Merging and standardizing this data for analysis is a monumental task, even before introducing AI. Machine learning models perform best on well-curated, *homogeneous* datasets; feeding them patchy or heterogeneous data can lead to poor or biased results. A common issue is missing data – e.g., incomplete medical histories or dropout patients with partial information. AI models might either discard those cases (reducing sample size) or attempt to impute missing values. Imputation, however, can introduce bias if the data isn't missing at random. For instance, if follow-up blood pressure readings are often missing because patients dropped out due to drug side effects, simply imputing an average BP for those patients could mask a safety signal. One study found that inconsistencies and omissions in routine care records led a trial-predictive AI to make biased outcome predictions, as the model was essentially "guessing" parts of patient histories (Rangarajan *et al.*, 2023, as cited in (Yadav et al., 2023). Thus, cleaning and harmonizing data is a critical (and time-consuming) prerequisite for successful AI. Efforts like CDISC standards (CDASH, SDTM, etc., which we discuss later) help by enforcing common data models. But older trials and real-world data still pose integration challenges. Until the clinical research field improves data interoperability – for example, by wide adoption of common data models and use of APIs to pull EHR data into trials automatically – AI implementations will spend a lot of effort just getting data into a usable state.

- Privacy and Compliance Concerns: Data privacy remains a major hurdle in deploying AI in clinical trials. Laws like the EU's GDPR and the US's HIPAA enforce strict controls on how personal health data is used. In Europe, for example, a pharma company may need explicit patient consent or legal exceptions to use trial data for secondary purposes. One global pharma project aiming to pool EU patient data for an AI model faced nearly two years of delays due to varying interpretations of GDPR by the time approvals came through, the AI model needed to be updated. This highlights how regulatory friction can stall innovation and why decentralized approaches like federated learning are gaining traction. Even after approvals, secure infrastructure encryption, access controls, audit trails; is crucial to prevent breaches. Standard anonymization may fall short against advanced re-identification techniques, prompting regulators like the EMA and FDA to revisit guidelines. Until AI-specific frameworks for clinical trials are clearly defined, most companies take a cautious path, often slowing adoption.
- Organizational and Cultural Mindset: Apart from regulations, people and processes can also slow AI integration. Clinical research teams are used to structured protocols refined over decades, and new AI tools can feel disruptive. For instance, when one pharma giant tested an AI model to predict slow-enrolling trial sites, managers initially dismissed its recommendations, trusting their own judgment instead. Ironically, the flagged sites did underperform, validating the model but only after skepticism gave way. This kind of resistance is common, often rooted in fear of job loss or mistrust in "black box" systems. Transparency (via explainable AI) and internal success stories help build trust. Upskilling staff and involving them in the AI journey is equally important. Some companies have created internal "AI champion" teams to bridge gaps between data scientists and operations, making AI less intimidating and more collaborative. The mindset shift is happening but unevenly.
- Regulatory Uncertainty and Lack of Standards: One of the biggest challenges is the lack of clear rules for AI use in trials. While there are regulations for digital tools and medical device software, nothing specific yet exists for AI-driven trial design or patient selection. Can AI-generated safety flags change a protocol mid-trial? Should AI recommendations be considered exploratory or actionable? These are open questions. Regulators like the MHRA have be-

gun gathering input, and future guidance will likely emphasize transparency, auditability, and fairness in AI tools. Until then, companies are left to interpret general principles on their own. Many are adopting a "human-in-the-loop" model, where AI supports decisions but humans remain accountable. Some organizations are even building internal Responsible AI (RAI) frameworks to self-regulate. Still, without consistent standards, firms that take bold steps risk being second-guessed if things go wrong, adding yet another reason why progress remains cautious.

Illustrative Example: A well-known pharmaceutical company delayed using an AI tool for trial monitoring simply because regulators hadn't formally approved such systems. Worried that they wouldn't be able to justify the AI process during an inspection, they chose to run the tool in "shadow mode" collecting data in parallel without acting on it. This cautious approach reflects a broader industry pattern: without clear regulatory guidance, even promising technologies often stall.

Given the mix of data complexities, privacy constraints, cultural pushback, and regulatory gray areas, implementing AI in clinical trials is about much more than the technology itself. Success depends on strong data governance, thoughtful change management, and early, ongoing dialogue with regulators and ethics teams. Companies that create cross-functional AI groups bringing together legal, clinical, data science, and IT; tend to navigate these issues better. This sets them up to adopt AI with more confidence, accountability, and impact.

With the challenges laid out, the focus now shifts to real-world use. The next sections will explore how AI is actively being applied in clinical trials – from trial design and patient recruitment to monitoring and analysis. These examples show not just the benefits, but also how practical strategies can help overcome the very barriers discussed above. Bit by bit, the industry is turning AI's promise into practice.

AI IN BIOSTATISTICS AND CLINICAL DATA MANAGEMENT

One of the core areas where AI has made inroads in clinical trials is in biostatistics and data management – essentially, how trial data are collected, standardized, and analyzed to produce reliable findings. Biostatisticians and data managers ensure that trial data are complete, consistent, and correctly transformed into results. AI is proving to be a valuable assistant in automating and enhancing many tasks in this pipeline, from cleaning datasets to generating analysis code. Here we examine several key contributions of AI in this domain:

Automating Data Standardization (CDISC Mapping): Formatting clinical trial data to CDISC standards like SDTM and ADaM is essential for regulatory submissions but traditionally, it's a time-consuming task requiring manual coding and specs. AI, especially NLP, is streamlining this process. Tools can now analyze trial documents and suggest mappings between raw data fields and CDISC variables. For instance, a tool might recognize that "AdverseEventDescription" aligns with "AEDECOD." Some companies using transformer-based models report automating around 70% of mappings, drastically cutting down manual effort while leaving the trickier 30% for human review. In addition, machine learning models help validate datasets by flagging unit mismatches or baseline errors. These early interventions improve data quality and reduce delays later in the pipeline.

Intelligent Data Cleaning and Anomaly Detection: Even after standardization, cleaning trial data is a massive task. AI helps by spotting errors and outliers—like implausible blood pressure readings—using models such as Isolation Forests or autoencoders. These tools learn normal data patterns and flag anomalies that might slip past traditional checks. AI also supports smarter handling of missing data, using predictive models to estimate values based on related patient attributes. Some companies have built QC bots using machine learning to verify if derived endpoints, like "change from baseline," are calculated correctly. While human oversight remains key, these tools significantly reduce manual data queries and improve overall efficiency.

Statistical Programming and Analysis Support: Once data is ready, statisticians dive into coding and analysis. AI coding assistants like GitHub Copilot are now helping generate code in R or SAS based on plain-language prompts—like producing Kaplan-Meier plots or baseline tables. These tools can draft most of the routine code, freeing up time for fine-tuning and validation. Some pharma companies have gone further, using NLP to parse statistical analysis plans and auto-generate parts of the analysis. Generative AI is also being used to write initial drafts of clinical study reports, translating statistical outputs into readable summaries, allowing medical writers to focus more on interpretation than formatting.

AI in Medical Coding: An important area of trial data management is medical coding, where free-text entries—like patient-reported adverse events or brand-name medications—must be translated into standardized terms using dictionaries such as MedDRA and the WHO Drug Dictionary. Traditionally, coders manually searched for the right match for each entry, which is time-consuming; a large Phase III trial might involve tens of thousands of such terms. AI now streamlines this through predictive coding models. These NLP-based tools learn from prior coding decisions—for instance, mapping "high blood sugar" to "Hyperglycemia" in MedDRA. When new data arrives, the AI predicts the appropriate codes, automating much of the "browse and code" process. One pharma company reported that AI auto-coded about 80% of adverse event terms with high confidence, leaving only novel or ambiguous entries for manual review. Even when uncertain, the AI offers top suggestions, reducing search time. Accuracy improves as the system retrains on new data. Regulators still require human oversight to ensure coding is medically sound, especially for safety events. Still, the efficiency gain is significant. Uniform AI use also enhances consistency across studies, lowering variability in coding. Overall, AI-based coding is a strong success story—solving a well-defined, repetitive task while supporting human coders for faster, more reliable outcomes.

Biometrics and Endpoint Assessment: AI is transforming how biometric endpoints are assessed in trials, especially those involving imaging or wearable data. In oncology, AI tools using computer vision can measure tumors on scans, helping radiologists apply RECIST criteria more consistently. Models like U-Net and DeepLabV3+ assist with tasks like tumor segmentation, improving precision. In ophthalmology trials, AI graders assess retinal images for disease severity. Wearable sensors also generate streams of biometric data AI models help sift through this to flag key events, like arrhythmias. In one Parkinson's trial, AI used motion data to detect tremors and quantify symptoms, complementing traditional scales and reducing clinic visits. This kind of objective, continuous monitoring is enhancing both endpoint accuracy and patient experience.

Across all aspects of data handling, from coding symptoms to analyzing MRIs AI, acts as a powerful support tool for biostatistics and data management teams. It handles routine tasks like formatting, initial analysis, and repetitive coding, while also offering advanced capabilities like pattern recognition and image interpretation. Crucially, these systems are integrated into existing workflows; for instance, anomaly detection tools can alert data managers within familiar platforms. This seamless integration encourages adoption, as AI enhances rather than replaces current roles. Biostatisticians say it frees them to focus on interpretation and study design instead of repetitive tasks. For sponsors, the benefits include faster database lock, cleaner data, and earlier detection of trends while respecting blinding and regulatory safeguards.

One CRO, for example, deployed an AI-driven assistant to reconcile data between clinical and safety databases. The tool ensured that serious adverse events in the safety system had corresponding entries in clinical case report forms. Using a knowledge graph of drug-event relationships, it flagged mismatches like "Headache" in one system and "Migraine" in another. This reduced manual reconciliation by 30%, catching issues that would've taken hours of human effort. It's a clear case of AI working behind the scenes to improve quality and consistency in trial data.

With AI streamlining data tasks, we now shift focus to another major trial challenge: patient recruitment and retention. In the next section, we explore how AI is transforming how we find, enroll, and monitor patients – two of the most time-consuming and expensive aspects of clinical research.

AI IN PATIENT RECRUITMENT AND ENROLLMENT

Recruiting eligible patients is often cited as the single greatest bottleneck in clinical trials. Many trials struggle to meet enrollment targets on time – some even fail outright due to insufficient patient accrual. AI offers powerful solutions to streamline patient recruitment, by mining health data to find the right patients and matching them to the right trials much faster than traditional methods. Additionally, AI can help ensure that recruitment is more inclusive and reaches diverse populations, addressing a longstanding concern in clinical research.

Intelligent Patient Matching: Traditionally, recruitment relies on site investigators, referrals, or registries—methods that are manual and limited in reach. AI significantly broadens this scope through data-driven matching. Machine learning models can scan EHRs, claims data, genetic records, and even social media to identify patients meeting trial criteria. For instance, in a trial targeting women in their 50s with Type 2 diabetes and early kidney disease, an AI platform can instantly query multiple clinics' EHRs for matching diagnoses, lab results, and demographics. NLP is key, as much of this info sits in unstructured clinical notes. Models like BioBERT or Med7 extract relevant terms—medications, disease severity, comorbidities—which are fed into matching algorithms.

A real-world example is the NIH's AI recruitment tool, which analyzes both structured and unstructured data to match patients to ClinicalTrials.gov-listed studies. It helped NIH-sponsored trials identify candidates across hospital networks, speeding up matching and improving diversity by reaching beyond the clinic setting. While AI doesn't enroll patients directly, it presents shortlists to coordinators, who then engage candidates. According to NIH, AI-assisted trials achieved in weeks what once took months of manual chart reviews.

Predictive Recruitment Strategy: AI doesn't just find patients it helps optimize recruitment. By analyzing historical site performance, demographics, and seasonal illness trends, machine learning models can forecast which sites will recruit quickly and which won't. For example, one model predicted Site A could enroll 50 patients in three months, while Site B might only enroll five, informing site selection and resource allocation. Some tools also simulate "what-if" scenarios: opening sites in new regions or adjusting inclusion criteria. One startup used a reinforcement learning agent to recommend relaxing lab cutoffs in an adaptive trial (with IRB approval), which boosted the eligible patient pool by about 15% without compromising safety.

Improving Diversity and Reducing Bias: Diversity in trials is critical for generalizability. AI can uncover underrepresented patients often missed by traditional methods. By searching data from safety-net hospitals or community centers, AI can find qualified candidates otherwise overlooked. It can also flag barriers—like distance or language needs—and suggest solutions, such as site referrals or transportation assistance. In one heart failure trial, AI noticed low enrollment from majority-Black zip codes with high eligibility. This insight led to targeted outreach, boosting participation from those communities. That said, AI isn't bias-free; biased data can lead to skewed outcomes. Developers now integrate fairness checks and use broader datasets to address this risk.

Example – Recruiting for rare disease trials is especially difficult due to small, scattered patient populations. AI has proven effective here by searching patient forums, social media, and global literature for relevant mentions. One platform used AI to find families discussing a rare pediatric neurological condition online. These families were contacted through advocacy groups, helping trial sites meet recruitment targets faster than expected. This shows how AI can find patients that traditional methods miss by tapping into unconventional data sources.

AI in Screening & Engagement: After identifying potential participants, screening remains a hurdle. AI models help predict who's likely to pass screening and stay enrolled using patterns from prior dropouts or exclusions. This lets teams prioritize high-probability candidates. Some trials also use AI chatbots to interact with patients during recruitment answering questions, conducting pre-screening, and guiding them through the process. One oncology trial reported that a chatbot handled hundreds of initial inquiries and helped convert 50 into randomized participants, saving staff hours of repetitive work.

The impact is clear: a pharma company completed enrolment for a Phase II trial four months early using an AI platform saving millions in site costs. A meta-analysis found that AI-enabled trials had 15–20% faster enrolment and 10–15% more minority representation on average (Perez-Lopez et al., 2024).

Ethical and regulatory compliance remains key. When using EHRs for recruitment, systems must operate under IRB approvals or patient-consented frameworks. Transparency is also important—some groups advocate notifying patients that AI played a role in identifying them. Most patients are supportive and appreciate being informed about trials they qualify for.

AI is becoming essential in clinical trial recruitment helping to efficiently identify eligible participants while improving diversity. By harnessing large-scale data and smart algorithms, AI accelerates timelines and enhances trial quality. The results are significantly faster trial launches and quicker access to promising therapies for patients who need them most.

AI FOR PATIENT MONITORING AND TRIAL CONDUCT

Enrolling patients is just the beginning, once a trial is underway, participants must be carefully monitored for their health and compliance, and data must be continually gathered. AI has valuable roles during this conduct phase of trials: it can provide real-time monitoring of patient data to enhance safety, predict and prevent patient dropouts, and optimize how trials are managed on the fly. In this section, we examine how AI improves the ongoing execution of trials and the well-being of participants throughout the study.

Real-time Safety Monitoring and Adverse Event Detection: Patient safety is critical in clinical trials, especially in early phases with unknown risks. Traditionally, safety monitoring involves periodic review of vitals, labs, and adverse event reports. AI now complements this by analysing incoming data continuously and flagging issues earlier. In a Phase I trial, for instance, AI models like LSTM or Prophet can learn each patient's baseline and predict expected vitals or lab values. If liver enzymes rise faster than expected or blood pressure trends downward, the AI can flag these changes before thresholds are crossed, offering early warning. In one oncology trial, an AI detected subtle shifts in blood counts and vitals that predicted cytokine release syndrome hours before clinicians recognized it, enabling timely intervention.

AI also analyses unstructured safety data, such as patient diaries or telehealth notes. NLP models can classify symptoms and flag patterns; for example, grouping fever, chills, and low blood pressure as signs of sepsis. Clustering algorithms help detect these syndromes even when symptoms seem unrelated. Machine learning approaches have been systematically reviewed for their effectiveness in pharmacovigilance (Pilipiec et al., 2022a, 2022b). In short, AI acts as a real-time safety net, spotting red flags between standard review points especially valuable in early-phase and first-in-human studies.

Patient Adherence and Dropout Prediction: Dropouts or non-adherence can compromise trial outcomes. AI models now help predict which participants are at risk, allowing for targeted retention. By analysing engagement data (e.g., app use, survey response times), baseline traits, and early deviations, models like XGBoost can assign a dropout risk score. If a patient scores high, coordinators can step in with personalized support like a check-in call or transportation help. One large CRO reduced dropout rates by ~20% in several Phase II trials using this approach, focusing on the ~15% deemed high risk. Younger patients and low app usage were common risk indicators.

AI also personalizes reminders if a patient tends to miss evening doses, the system can adjust notifications accordingly. Smart devices like pill bottles or ingestible sensors generate adherence data that AI can analyse to flag missed or discontinued doses. These insights improve both data quality and patient outcomes.

Site Performance and Operational Efficiency: AI also improves site-level oversight. Large trials often span dozens of sites, making it hard to track performance. AI dashboards aggregate site data and highlight outliers. Clustering algorithms can group sites based on enrollment speed, data entry lag, or protocol deviations. Underperforming sites may get extra support; high performers can serve as models.

AI is also being used for fraud detection spotting unusual data patterns like excessive uniformity that might signal misconduct. These tools add another layer of quality assurance, similar to financial fraud detection models, prompting audits when necessary.

Remote Monitoring and Virtual Trials: AI is helping with power decentralized and hybrid trials, where participants report data from home using wearables, smartphones, or remote testing. These setups generate large volumes of real-time data .AI processes and interprets this continuously. For example, if a patient's step count drops sharply or a smart inhaler isn't used regularly, the system can flag potential non-adherence or health issues.

In a remote heart failure trial, AI noticed a patient's weight was rising over three days and blood pressure dropping signals of fluid retention. It alerted the physician, who arranged a home visit and adjusted treatment, avoiding hospitalization and keeping the patient in the trial. This proactive care not only protects participants but also improves trial continuity.

FORECASTING AND ADAPTIVE TRIAL MANAGEMENT

AI can enhance trial oversight by predicting key events like enrolment completion or when a trial will reach its endpoints. While traditional tools like accrual curves exist, machine learning adds nuance by factoring in seasonal trends, competing studies, and more. This helps sponsors act early—whether it's gearing up for data analysis if enrolment finishes ahead of schedule or adding sites if delays are expected.

In adaptive trials, which allow mid-study adjustments based on interim data, AI can assist Data Monitoring Committees by running simulations and sensitivity analyses rapidly. For instance, in a dose-finding study, AI can model the best next dose level based on evolving results—though final decisions always rest with the human committee.

To illustrate the benefit: One major pharma company used an AI operations dashboard in a large Phase III trial and saw a 40% improvement in data entry lag. By sharing the dashboard with sites and nudging slower ones, they encouraged real-time data entry. This led to faster interim analysis—cutting weeks off the timeline. AI also flagged slow enrollment in certain regions early, prompting protocol updates that helped the study meet recruitment goals on time.

Example – Combining AI Recruitment and Monitoring In a decentralized hypertension trial, AI handled both recruitment and patient monitoring. Participants were enrolled via EHR and social media, then used an app that tracked blood pressure and meds. The AI offered personalized coaching and alerted nurses if readings were high. Not only was the trial completed on time, but patients also saw better health outcomes during the study. This fusion of care and research raises new ethical questions but also shows how AI can support both at once when integrated into the therapeutic model.

In summary, AI enhances trial conduct by providing vigilant oversight and intelligent predictions. It's like having a smart co-pilot for the trial, one that never sleeps or gets overwhelmed by data. This ensures problems are caught early, whether they are medical issues with patients or logistical issues with trial operations – and thus can be mitigated before they impact the study's success or the patients' well-being. By improving adherence and retention, AI contributes to higher quality data and more conclusive results. By improving safety monitoring, it upholds the ethical mandate to protect participants. These improvements during trial conduct complement the gains in startup (design and recruitment) and in analysis, creating an end-to-end spectrum of AI support across the clinical trial lifecycle.

ETHICAL AND REGULATORY CONSIDERATIONS FOR AI IN TRIALS

As AI becomes integral to clinical trials, ethical and governance issues take center stage. Responsible use is key to preserving trust and protecting patient rights. This includes tackling bias, ensuring fairness, safeguarding privacy, and aligning with regulations i.e., all core elements of AI governance in clinical research.

Mitigating Bias and Ensuring Fairness: One major ethical concern with AI in healthcare is the risk of bias algorithms may inadvertently perpetuate or even amplify existing inequities if not carefully managed. In trials, this could manifest in AI systems that, for example, preferentially select certain demographics for enrollment or interpret data differently for different groups, leading to skewed outcomes. Bias can creep in through biased training data or through algorithm design that doesn't account for population differences.

Recognizing and Reducing Bias: Bias mitigation begins with awareness, auditing AI outputs for discrepancies. If a recruitment tool consistently picks more men than women, is it due to valid clinical reasons or a data gap? Tools like IBM's AI Fairness 360 help flag such issues, and many companies now track fairness metrics across demographic groups.

Diverse Training Data: To build fairer AI, training datasets must reflect real-world diversity, across age, gender, race, and more. Some pharma firms now pool global trial data or partner with diverse research sites to better represent different populations.

Algorithmic Adjustments: Techniques like reweighing or equalized odds adjust models to ensure consistent accuracy across subgroups. For instance, an AI predicting dose response might be trained to perform equally well across all races, a growing area in fair ML that's moving into real-world trial applications.

Human Oversight and Explainability: Human judgment remains essential. Explainable AI (XAI) helps clarify decisions like why a patient was flagged as ineligible so clinicians could assess if socio-economic bias played a role. Many sponsors now require human review of AI outputs, especially when participant safety or trial integrity is at stake.

Case Study – Removing Race Adjustment in Algorithms: Hospitals in Philadelphia eliminated race-based corrections in clinical algorithms after realizing they could perpetuate disparities (Bettelheim, 2024). This mirrors broader efforts to root out systemic bias in medicine. In trials, any AI using race or gender in a way that could disadvantage patients must be carefully evaluated and possibly redesigned.

Building Ethical AI Frameworks: Many organizations have introduced AI ethics boards and internal guidelines covering fairness, transparency, and consent. For example, if AI helps select trial participants, some sponsors now disclose this in consent forms. While not legally required, it fosters transparency and trust. Ultimately, investigators must understand and explain AI-driven decisions.

Privacy, Security, and Regulatory Compliance: Protecting participant privacy and data security is both an ethical and a legal necessity in trials, and AI doesn't change that it intensifies it, because AI often involves gathering and processing more data. We touched earlier on privacy challenges; here we delve deeper into how privacy and security are being upheld in AI-powered trials.

Data Anonymization and De-identification: Patient data must be rigorously de-identified before AI training. Techniques like differential privacy add noise during training to protect individuals, and some companies are using synthetic data, AI-generated but realistic datasets for early model testing without risking real patient privacy.

Access Control and Security: Strict access controls ensure only authorized personnel can use sensitive data. For example, data scientists may only see de-identified information, while investigators access identifiable data. Audit logs track usage, and some sponsors explore blockchain to ensure tamper-proof records of data changes including those made by AI.

Regulatory Compliance: AI tools in trials must meet standards like 21 CFR Part 11, with full validation, access control, and audit trails. If AI performs critical tasks (e.g., adverse event coding), sponsors must document its accuracy and performance. Depending on when and how AI is introduced, regulators may require updates to trial protocols.

Collaboration with Regulators: Some sponsors now consult early with agencies like the FDA and EMA, which are beginning to qualify AI methodologies. For example, the EMA recently approved Unlearn's digital twin score, signaling growing regulatory openness provided tools are validated and risks managed (Sanford, 2022; Li, G., 2024). Updates to guidelines like ICH E6 (R3) now emphasize quality-by-design and risk-based oversight, aligning well with AI adoption.

Case Example – Auditing AI Decisions: In one trial, an AI helped prioritize which protocol deviations required reporting. Post-trial, the sponsor audited the AI's decisions and found it had 95% sensitivity in catching major deviations. Documenting this satisfied inspection requirements and reinforced the principle that sponsors, not algorithms also remain accountable.

Patient Consent and Autonomy: Patients should know how their data is being used, including if AI is involved in analysis or recruitment. While separate AI consent isn't required, clarity in consent forms especially when genomic or sensitive data are analysed by AI is good practice. Patients may opt out of AI-driven substudies, and their choice must be respected.

Training and AI Literacy: For AI to work effectively, staff must understand it. Misinterpreting AI alerts can lead to errors, so many sponsors now include "AI 101" training during trial startup. This includes how AI works, what it does (and doesn't) do, and how to interpret outputs. Ethical AI use depends not just on the tech, but on informed users.

RESEARCH METHODOLOGY

We primarily adopted a qualitative secondary document analysis methodology, utilizing a wide range of publicly available and reputable sources to explore the integration of Artificial Intelligence (AI) in clinical trials. The author systematically collected and reviewed existing academic literature, industry reports, regulatory documents, and real-world case studies from peer-reviewed journals, white papers, healthcare consortia, and official bodies such as the FDA, EMA, NIH, and WHO.

We ensured data credibility and thematic consistency on each below aspects:

- **Data Collection** -

A curated selection of sources was compiled based on relevance to AI technologies in clinical research. Search queries focused on keywords such as "AI in clinical trials", "digital twins", "NLP in healthcare", and "regulatory perspectives on AI", covering works from the last 5–10 years.

- **Data Coding and Organization** -

The collected literature was manually coded using thematic analysis. Key themes—such as AI methodologies (e.g., NLP, deep learning, reinforcement learning), clinical applications, challenges, and regulatory implications—were identified and categorized.

- **Document Arrangement** -

Documents were organized chronologically and thematically, allowing the author to trace the evolution of AI tools and their practical implications across the clinical trial lifecycle. Referenced materials were annotated and grouped under specific domains such as recruitment, data management, ethics, and monitoring.

- **Source Verification** -

Only peer-reviewed, government-endorsed, or expert-validated sources were retained to maintain high academic rigor. Cross-verification was done across multiple documents to ensure the accuracy and consistency of key claims.

CONCLUSION

The Future of AI in Clinical Research: Artificial intelligence is poised to become an integral thread in the fabric of clinical trials, driving a paradigm shift in how we design, conduct, and analyze studies. As we have explored, AI's current contributions span the entire trial lifecycle from accelerating study design and patient recruitment to enhancing data management, analysis, and monitoring all while flagging areas where careful oversight is needed to maintain ethics and integrity. Looking forward, what might the clinical trials of the future look like under AI's influence, and what steps are needed to fully realize AI's potential in a responsible manner?

Accelerating Development of Treatments: In the coming years, AI is poised to speed up drug development by making clinical trials faster, more efficient, and less costly (Bahls, 2024; World Economic Forum, 2024; World Economic Forum & ZS, 2024). Through smarter patient matching, adaptive designs, and automation, trials can run shorter and leaner. This means patients could access life-saving treatments sooner. We're already seeing signs of progress: AI-driven trials completing ahead of schedule and virtual trials reaching patients who were previously out of reach. A 2025 McKinsey report estimates AI could cut development timelines by one to two years and reduce costs by 15 to 25 percent. In a space where delays cost millions and delay care for patients these gains could be transformative.

Adaptive and Flexible Trial Designs: Tomorrow's trials will be more adaptive, with AI driving real-time decisions. Unlike today's rigid, pre-defined designs, future trials may evolve as data flows in. AI can analyze interim results and suggest dose changes or treatment arm modifications on the fly guided by reinforcement learning or Bayesian logic. Imagine trials where each patient is treated based on the responses of those before them, or where "digital twins" simulate outcomes to test designs before a single patient is enrolled. These innovations push the boundaries of traditional trials. Regulatory bodies are cautiously exploring this space, and we may soon see protocols that treat AI as part of the design itself.

Integration of Multi-Modal Data and Precision Medicine: The amount and variety of biomedical data i.e. genomics, proteomics, electronic health records, and patient-reported outcomes from smartphones is growing rapidly. AI is uniquely positioned to make sense of these complex, multi-modal data streams. Future trials could combine AI-driven genomic predictors with clinical data to stratify patients into more precise subgroups, helping realize the long-envisioned goal of personalized trials.

In oncology, AI could continuously analyze molecular data from tumor samples and adapt therapies in real time under a master protocol. We're already seeing early forms of this in umbrella and basket trials. AI will add the analytical power to dynamically match therapies to each patient based on genetics, biomarkers, and prior outcomes essentially merging clinical research with clinical care. In this model, AI becomes the intelligent layer that learns from each data point and steers patients to the most appropriate treatment.

This blurring of lines is especially relevant in areas like oncology, rare diseases, and immunotherapy, where adaptive, learning-based approaches are critical. While these innovations challenge traditional definitions of a trial, regulators have shown flexibility when strong scientific rationale is presented—such as the FDA's endorsement of platform trials during COVID-19. AI will also help integrate real-world evidence more fluidly into trials. Rather than separating trial data and real-world data, we may see AI systems that continuously learn from both, using real-world treatment outcomes to inform trial design and vice versa creating a dynamic, ongoing loop of evidence generation and refinement.

Ethical AI and Patient-Centric Research: The future of AI in trials must go beyond speed it should enhance the patient experience. AI can reduce patient burden through remote monitoring, tailor care, and ensure broader representation. Ethical considerations will likely become embedded in trial design, with concepts like algorithmic transparency, patient consent for AI involvement, and fairness audits becoming standard. Just as trials now include safety monitoring and statistical plans, future protocols may feature an "AI Governance Plan" detailing how AI tools will be monitored and how patients are informed. We may even see new roles like AI Ethics Officers or algorithm auditors on trial teams to oversee responsible use.

Regulatory Evolution: Regulators are expected to adapt alongside AI adoption. New guidance similar to medical device regulations may require pre-approval of AI tools, supported by performance evidence. Agencies might begin using AI themselves for tasks like anomaly detection in submitted data. As flexible trial designs gain regulatory support (e.g., ICH E8(R1)), it's likely that agencies will start issuing qualification opinions for AI tools more regularly. International harmonization will be key to avoid fragmented AI approval across regions.

Collaboration and Data Sharing: For AI to realize its full impact in clinical research, broader collaboration across the industry is essential. No single company or trial can generate enough diverse data to train truly powerful models. We're likely to see more pre-competitive consortia sharing de-identified trial data under strict privacy safeguards, much like the joint efforts seen during the COVID-19 vaccine development.

Open-source AI tools for common tasks like adverse event coding or safety signal detection could become standard. Tackling challenges like bias will also require cross-sector cooperation—with pharma, tech, academia, and patient advocates working together on best practices and AI standards. Trusted organizations like CTTI and TransCelerate may lead the way.

The idea of a global clinical trial data hub, where anonymized data can be reused for AI development, is increasingly realistic if proper governance and trust frameworks are in place. Since AI thrives on data diversity, responsible data sharing will unlock better insights for all.

Ultimately, the successful integration of AI in clinical trials promises a future of smarter, faster, and more inclusive research. If we address the challenges—ensuring robust validation, maintaining patient privacy, eliminating bias, and securing regulatory buy-in—AI could unlock unprecedented opportunities. We could see treatments reaching the market faster with trials that are more adaptive and informative, patients benefiting from more personalized approaches, and efficiencies that free up resources to explore more therapeutic hypotheses. In a world where medical knowledge is rapidly expanding, AI offers a way to manage that complexity and turn data into actionable evidence.

The journey is already underway. Every successful application builds confidence and knowledge. With continued ethical commitment, regulatory clarity, and technological advancement, AI will move from pilot projects to a foundational pillar of clinical research. This doesn't mean replacing the clinician or researcher, but empowering them with tools to make better decisions and focus on what humans do best – empathy, creativity, and ethical judgment while machines handle data-heavy tasks. The result is a research ecosystem that is more efficient, precise, and equitable, ultimately bringing safer and more effective treatments to patients faster than ever before. The excitement is tempered with responsibility, but the direction is clear: the future of clinical trials will be increasingly driven by artificial intelligence, to the benefit of science and society.

REFERENCES

Bahls, C. (2024). In focus: AI and clinical trials—The beginning of a journey. *Applied Clinical Trials, 33*(12). https://www.appliedclinicaltrialsonline.com/view/in-focus-ai-and-clinical-trials-the-beginning-of-a-journey-

Bettelheim, A. (2024). Philadelphia hospitals drop race-based algorithms.

Fisher, C. K. (2024). US FDA comments on Unlearn's PROCOVA methodology. Unlearn.AI. https://www.unlearn.ai/blog/us-fda-comments-on-unlearns-procova-methodology

Hutson, M. (2024). How AI is being used to accelerate clinical trials. *Nature, 627*(8003), S2–S5. DOI: 10.1038/d41586-024-00753-x PMID: 38480968

Jin, Q., Wang, Z., Floudas, C. S., Chen, F., Gong, C., Bracken-Clarke, D., Xue, E., Yang, Y., Sun, J., & Lu, Z. (2024). Matching patients to clinical trials with large language models. *Nature Communications, 15*(1), 9074. Advance online publication. DOI: 10.1038/s41467-024-53081-z PMID: 39557832

Jin, S., Chen, L., Ding, H., Wang, M., & Yu, L. (2024). Can artificial intelligence predict clinical trial outcomes? Klabunde, T. (2024). Digital "twinning": Clinical trials powered by AI. Sanofi. https://www.sanofi.com/en/magazine/our-science/digital-twinning-clinical-trials-ai

Li, G. (2024). A new regulatory road in clinical trials: Digital twins. Applied Clinical Trials. https://www.appliedclinicaltrialsonline.com/view/new-regulatory-road-clinical-trials-digital-twins

Li, S. (2024). How AI is being used to accelerate clinical trials. AAIATECH. https://www.aaiatech.org/how-ai-is-being-used-to-accelerate-clinical-trials/

Matheny, M. E., Yang, J., Smith, J. C., Walsh, C. G., Al-Garadi, M. A., Davis, S. E., Marsolo, K. A., Fabbri, D., Reeves, R. R., Johnson, K. B., Dal Pan, G. J., Ball, R., & Desai, R. J. (2024). Enhancing postmarketing surveillance of medical products with large language models. *JAMA Network Open, 7*(8), e2428276. Advance online publication. DOI: 10.1001/jamanetworkopen.2024.28276 PMID: 39150707

Mullard, A. (2018). How much do phase III trials cost? *Nature Reviews. Drug Discovery, 17*(11), 777. DOI: 10.1038/nrd.2018.198 PMID: 30374183

Newswire, P. R. (2025). Clinical trials market size is estimated to be US$ 149.58 billion by 2034. BioSpace. https://www.biospace.com/press-releases/clinical-trials-market-size-is-estimated-to-be-us-149-58-billion-by-2034

Perez-Lopez, R., Ghaffari Laleh, N., Mahmood, F., & Kather, J. N. (2024). A guide to artificial intelligence for cancer researchers. *Nature Reviews. Cancer*, *24*(6), 427–441. DOI: 10.1038/s41568-024-00694-7 PMID: 38755439

Pilipiec, P., Liwicki, M., & Bota, A. (2022a). Using machine learning for pharmacovigilance. *Systematic Reviews in Pharmacy*, *14*(2), 266. Advance online publication. DOI: 10.3390/pharmaceutics14020266 PMID: 35213998

Pilipiec, P., Liwicki, M., & Bota, A. (2022b). Using machine learning for pharmacovigilance. *Systematic Reviews in Pharmacy*, *14*(2), 266. Advance online publication. DOI: 10.3390/pharmaceutics14020266 PMID: 35213998

Rajkomar, A., Oren, E., Chen, K., Dai, A. M., Hajaj, N., Hardt, M., Liu, P. J., Liu, X., Marcus, J., Sun, M., Sundberg, P., Yee, H., Zhang, K., Zhang, Y., Flores, G., Duggan, G. E., Irvine, J., Le, Q., Litsch, K., & Dean, J. (2018). Scalable and accurate deep learning with electronic health records. npj. *Digital Medicine*, *1*(1), 18. Advance online publication. DOI: 10.1038/s41746-018-0029-1 PMID: 31304302

Sanford, C. (2022). European Medicines Agency qualifies Unlearn's AI-powered method for running smaller, faster clinical trials.

ScienceDaily. (2018, July 10). Artificial intelligence helps researchers predict drug combinations' side effects. https://www.sciencedaily.com/releases/2018/07/180710072013.htm

Unlearn.AI. (2024). How AI-generated digital twins are speeding up clinical development. https://www.unlearn.ai/

Vishnu Priya, N. (2022). Merck KGaA collaborate to expedite immunology trials. Clinical Trials Arena. https://www.clinicaltrialsarena.com/news/unlearn-ai-merck-immunology-trials/

Wong, C. H., Siah, K. W., & Lo, A. W. (2019). Estimation of clinical trial success rates and related parameters. *Biostatistics (Oxford, England)*, *20*(2), 273–286. DOI: 10.1093/biostatistics/kxx069 PMID: 29394327

Wood, T. (2024). How can AI be used in clinical trials? Fast Data Science. https://fastdatascience.com/ai-in-pharma/ai-in-clinical-trials/

World Economic Forum. (2024, December). 5 ways generative AI could transform clinical trials. https://www.weforum.org/stories/2024/12/genai-could-transform-clinical-research-trials/

World Economic Forum & ZS. (2024). Intelligent clinical trials: Using generative AI to fast-track therapeutic innovations. Medidata. https://www.medidata.com/en/life-science-resources/medidata-blog/generative-ai-in-clinical-development/

Yadav, N., Pandey, S., Gupta, A., Dudani, P., Gupta, S., & Rangarajan, K. (2023). Data privacy in healthcare: In the era of artificial intelligence. *Indian Dermatology Online Journal, 14*(6), 788–792. DOI: 10.4103/idoj.idoj_543_23 PMID: 38099022

Zitnik, M., Agrawal, M., & Leskovec, J. (2018). Modeling polypharmacy side effects with graph convolutional networks. *Bioinformatics (Oxford, England), 34*(13), i457–i466. DOI: 10.1093/bioinformatics/bty294 PMID: 29949996

Chapter 10
Artificial Intelligence and Digital Communication:
A Study on OpenAI's TikTok Presence

Zindan Çakıcı
http://orcid.org/0000-0002-8916-0582
Üsküdar University, Turkey

Neslihan Bulur
http://orcid.org/0000-0001-6148-5556
Üsküdar University, Turkey

ABSTRACT

OpenAI, a vanguard institution in artificial intelligence research and application, inaugurated its official TikTok account on February 16, 2024, thereby initiating a structured approach to platform-based communication. A review of extant scholarly discourse reveals a conspicuous dearth of research scrutinizing OpenAI's communicative practices on TikTok. This study seeks to fill this lacuna by conducting a methodical analysis of OpenAI's engagement on the platform, thereby contributing to the body of literature regarding the digital communication strategies employed by artificial intelligence institutions. The research methodology entailed a content analysis of 101 posts disseminated via OpenAI's verified TikTok account from February 16, 2024, to October 4, 2024. The findings elucidate that OpenAI adopts a multifaceted communicative approach, characterized by the strategic integration of semiotic components such as music, captions, hashtags, and diverse audiovisual formats. These components are employed to transmit salient messages through varied emotional tonalities and affective cues. However, several deficiencies were identified within the content strategy: the limited incorporation of subtitles, the conspicuous

DOI: 10.4018/979-8-3373-1509-6.ch010

absence of sign language, and the exclusive reliance on English language usage, all of which undermine the principles of accessibility and inclusivity. Furthermore, the content's primary focus remains predominantly informational, with a conspicuously insufficient emphasis on fostering user engagement or cultivating a sense of community. The thematic scope of the posts is predominantly centered on artificial intelligence-generated depictions of animals, objects, and natural environments, with an evident underrepresentation of human-centric narratives, thereby suggesting an overarching prioritization of technological discourse over humanistic communication paradigms.

INTRODUCTION

The engagement of users in the experiential interaction with emergent technological paradigms, such as augmented reality (AR), virtual reality (VR), and artificial intelligence (AI)-driven computational frameworks, facilitates both the cognitive internalization and systemic adoption of these innovations (Barta et al., 2025, p. 634; Bharadiya, 2023, p. 24; Dlamini & Botha, 2024, p. 399; Enyejo et al., 2024, pp. 144–146; López, 2023, p. 1; Ma et al., 2025, p. 1). For instance, algorithmically generated visual augmentation filters and AR-integrated simulation interfaces enable direct user immersion in technologically mediated environments, thereby fostering heightened epistemic awareness and operational familiarity. TikTok, by virtue of its short-form audiovisual content architecture and advanced predictive algorithmic infrastructure, constitutes a pivotal mechanism for the large-scale dissemination of technological innovations (Çakıcı, 2025, pp. 137–140; Klug et al., 2021, p. 84; Molem et al., 2024, pp. 318–323). The platform's strategic primacy in the diffusion of technological advancements is attributable to its computationally optimized content generation and distribution dynamics. Specifically, TikTok's short-duration video modality facilitates the accelerated, visually enriched exposition of emerging technological constructs (Rizomyliotis et al., 2024, pp. 930–932; Meng et al., 2024). These multimodal, interactive representational schemas afford users the opportunity to engage with technological advancements within applied, praxis-oriented frameworks rather than abstract theoretical delineations. Moreover, algorithmically driven content curation mechanisms, predicated on user interaction matrices, amplify the recurrent exposure to technologically pertinent materials, thereby augmenting the virality potential of such content (Çakıcı & Meriç, 2024; Meriç & Çakıcı, 2024). In this regard, TikTok's computational recommendation architecture engenders an organically sustained epistemic diffusion of information pertaining to emergent technological developments, a characteristic that underpins its strategic deployment by leading technological conglomerates such as Apple, Meta, and NVIDIA.

Founded in 2015, OpenAI has been classified as one of the most influential and extensively recognized entities within the global technological landscape, primarily due to its pioneering advancements in natural language processing (NLP), large language models (LLMs), and deep learning architectures (Busacca & Monaca, 2025, p. 117; Demirel et al., 2024, p. 49; Demirel et al., 2024, pp. 220-222; Roumeliotis & Tselikas, 2023, p. 192; Saifi et al., 2025, p. 1; Zou et al., 2023, p. 1). The institution's prominence is fundamentally attributable to the extensive proliferation of its technological innovations and their profound transformative implications across multiple industrial sectors. In particular, proprietary applications developed within OpenAI's research ecosystem, including ChatGPT and Sora, exhibit a wide-ranging spectrum of applicability across diverse operational domains, encompassing education, media production, customer service automation, and algorithmically optimized content generation. Moreover, OpenAI's research initiatives concerning artificial intelligence safety and ethical governance have made substantive contributions to the global discourse on the responsible deployment of machine intelligence. In conjunction with these research endeavors, OpenAI disseminates its projects and adopts multi-faceted strategic communication frameworks to enhance public engagement and optimize institutional visibility (Mogavi et al., 2024, pp. 2–4; Mhlanga, 2023, p. 4; Our Approach to AI Safety, 2024). Empirical evidence indicates that, as of February 2024, OpenAI has initiated formal operational engagement on the TikTok platform (OpenAI (@openai) Official | TikTok, 2025). A review of the extant academic literature reveals a discernible paucity of studies examining the communicative methodologies employed by technology conglomerates within digital social platforms, with a conspicuous absence of scholarly inquiries specifically addressing TikTok-based communication paradigms. This identified research lacuna substantiates the epistemic originality of the present investigation. Consequently, the principal objective of this study is to conduct a methodologically rigorous and analytically comprehensive examination of OpenAI's communicative engagement on TikTok. By addressing this critical research void, the study seeks to generate substantive contributions to the academic corpus concerning OpenAI's digital communication architectures and the integrative role of TikTok within globally oriented communication frameworks.

In this study, a literature review was conducted to synthesize the existing body of knowledge, identify epistemological gaps, and delineate the study's scholarly contributions with absolute precision. This section provides an examination of the communication strategies employed by technology corporations within the digital media landscape. Based on the analytical insights derived from this review, research questions were refined, and the most methodologically robust research design and techniques were determined to ensure empirical rigor. Furthermore, the study's scope was defined through the formulation of the population and sampling frameworks, while its inherent limitations were articulated to maintain methodological transparency. Accordingly, OpenAI's communication strategy on TikTok was analyzed through a content analysis of 101 video posts published between the platform's inaugural post on February 16, 2024, and the commencement of this investigation on October 4, 2024. The analytical framework encompassed a multidimensional examination of OpenAI's communicative mechanisms, including the volume of disseminated content, thematic dimensions, linguistic characteristics, content typologies, narrative frameworks, communicative objectives, affective expressions, engagement metrics, the presence of designated spokespersons, and their respective functional roles. The extracted data were subjected to computational visualization techniques to enhance analytical clarity and interpretative depth. In the concluding section, the study's findings were synthesized, and their epistemological contributions to the existing corpus of literature were evaluated. Additionally, considering the methodological constraints of this research, future investigative directions were proposed, emphasizing their potential to address prevailing knowledge deficiencies and extend theoretical frontiers.

Digital Media Strategies: Technology Firms' Adoption of TikTok

The Internet has undergone a substantial evolution, transforming into a dynamic digital ecosystem, driven by continuous technological advancements (Çakıcı, 2024, p. 2; Çakıcı, 2025, p.137). In the early 1990s, the inaugural phase of the internet, known as Web 1.0, confined users to accessing static web pages, acquiring information from centralized servers, and engaging in limited interaction. This phase was marked by a rudimentary infrastructure, with limited bandwidth and minimal multimedia capabilities, restricting user engagement and content interactivity (Haile, 2024, p. 135). The emergence of Web 2.0 in the early 2000s heralded a shift, characterized by the widespread adoption of social media and interactive platforms, wherein users transitioned from passive consumers to active content producers. This period also saw the rapid rise of e-commerce, as digital platforms facilitated transactional exchanges, reshaping business models (Çakıcı, 2025, p. 200). Web 3.0 marked an advancement through the integration of semantic web technologies and blockchain systems, enabling enhanced data analytics and personalized online experiences. Blockchain technologies empowered decentralized networks, enhancing security and privacy in digital transactions, allowing users greater control over their data. Web 4.0 is recognized as a transformative phase, driven by the convergence of artificial intelligence (AI) and the Internet of Things (IoT), fostering the development of intelligent, autonomous systems that elevate human-machine interactions.

The integration of augmented reality (AR) and virtual reality (VR) has created immersive environments, expanding user interaction across sectors like healthcare, education, and entertainment (Hassan et al., 2024, pp. 2-3; Ersöz, 2020, pp. 59-63; Permatasari et al., 2020, pp. 79-80). As the infrastructure of web technologies advances, technological enterprises have not only enhanced their capabilities but also disseminated these innovations across global platforms, capitalizing on the socio-technical advantages of these systems. The rise of cloud computing has enabled more efficient data storage and processing, enhancing the scalability of digital services and better serving consumer needs. Furthermore, internet and social media usage continue to rise, driven by the evolving needs of a digitally connected society (Aytuna et al., 2024, p.215; Çakıcı, 2024, p.2; Demirel et al., 2025, pp.144-145). By 2025, approximately 5.56 billion people, or 63.9% of the global population, are projected to have internet access, with 5.24 billion engaging with social media. This growth is fueled by increasing internet penetration in developing economies and the widespread availability of affordable mobile technologies. TikTok, with an estimated 1.59 billion users, ranks as the fifth most utilized platform, following Facebook, YouTube, WhatsApp, and Instagram (Kemp, 2025, pp. 10, 413-558).

Social media networks have evolved into essential strategic tools for communication and interaction, not only for individuals but also for institutions, including commercial enterprises, educational institutions, healthcare organizations, financial entities, media outlets, and technology corporations. Technology corporations, in particular, utilize social media platforms in a methodical and targeted manner to enhance brand visibility, promote their products and services, drive innovation by systematically analyzing user feedback, and build digital communities. These platforms also fulfill diverse organizational purposes, such as improving customer satisfaction, managing crisis communication, fostering strategic partnerships, and tracking industry trends. By integrating social media into their operations, these entities gain immediate access to real-time feedback, providing valuable insights that refine business strategies. As a result, social media networks have become critical in allowing technology corporations to secure competitive advantages and achieve dominant positions within the global digital ecosystem (Culnan et al., 2010, p. 253; Powell & Dent-Micallef, 1997; Ammirato et al., 2019). By capitalizing on the dynamic nature of the digital ecosystem, technology corporations have incorporated social media platforms as integral components of their strategic communication processes. The continuous growth of user-generated content and the viral nature of digital interactions further empower corporations to tap into broader markets and strengthen customer relationships. Recently, TikTok has emerged as one of the leading platforms utilized not only by individuals but also by institutional actors, particularly technology corporations, to increase visibility and engagement (Putriani & Aras, 2022, p. 910).

The inception of TikTok can be traced to the establishment of the "Musical.ly" application in Shanghai, China, in 2014 (Choudhary et al., 2020, p. 193). Although the platform has faced access restrictions in several countries, including Indonesia, Pakistan, India, and Bangladesh, due to concerns about "violations of user privacy, potential threats to national security, and the propagation of inappropriate content," its user base has nonetheless continued to expand (Zeng et al., 2021, p. 3164; Ceci, 2025). This expansion can be attributed to TikTok's ability to adapt to diverse regulatory environments and localize its content to meet the demands of varying cultural and political contexts. As of February 2025, the United States represents the largest segment of TikTok's audience, with a total of 135.79 million users, followed by Indonesia, Brazil, Mexico, Pakistan, and the Philippines (Ceci, 2025). This widespread global reach is bolstered by the platform's capacity to cater to different markets, thanks to the customization of its algorithm to resonate with regional preferences and trends. Particularly, individuals within the age range of 18-34 make up 60% of TikTok's user population, and the platform boasts over one billion daily video views (Kemp, 2025). This demographic dominance highlights the platform's unique appeal to younger generations, who are more likely to engage with short-form, interactive, and viral content. The rationale behind technology companies' engagement with TikTok, therefore, departs from traditional social media strategies. Unlike platforms that rely on long-form content or conventional marketing approaches, TikTok's dynamic, short-form video format and algorithmically driven content delivery mechanisms offer businesses a more immersive and immediate way to connect with users. TikTok's elevated levels of user engagement, coupled with its predominantly youthful demographic and capacity for viral content generation, have precipitated significant transformations in the marketing and communication strategies employed by technology enterprises (Taylor & Brisini, 2024, pp. 2-6; Çetinkaya & Dondurucu, 2023, pp. 80-82).

The primary motivation for enhancing brand awareness on TikTok is facilitated by the platform's visually captivating and entertaining content formats. These formats, which prioritize short-form, dynamic videos, align well with the attention span of modern digital audiences, particularly younger users. The TikTok algorithm optimizes the organic dissemination of engaging content across an expansive audience, thereby enabling technology companies to strategically cultivate their brand identities in a highly creative manner. This algorithmic precision not only broadens content reach but also ensures that brand messaging resonates effectively with the most relevant user segments. In this regard, prominent technology corporations employ TikTok distinctively compared to conventional marketing channels, utilizing the platform to foster an authentic, user-centric, and accessible brand image (Sheak & Abdulrazak, 2023). By embracing TikTok's interactive nature, technology firms create content that encourages user participation and promotes a sense of inclusivity, positioning their brands as relatable and engaging. Furthermore, the promotion of products and services is reinforced by TikTok's unique content formats. Technology firms, particularly those introducing novel devices, software updates, and artificial intelligence-driven solutions, effectively capture user attention through TikTok's succinct and compelling videos. For instance, corporations developing AI-driven applications demonstrate the operational efficacy of their software in daily life scenarios through inventive video content (Peng et al., 2023, pp. 1-3). This ability to convey complex technological concepts in an easily digestible format makes TikTok an ideal platform for showcasing innovative products. TikTok's data analytics and trend-monitoring capabilities assume a critical function in the formulation of digital marketing strategies by technology companies. The platform's capacity to analyze user behaviors and trends in real-time equips technology firms with the tools to devise more targeted and precise advertising strategies tailored to their audience demographics (Guerrero, 2023). This data-driven approach enables technology companies to make informed decisions, ensuring their content resonates with users on an individual level. Additionally, the application of TikTok has become increasingly prevalent for the purposes of cultivating technology communities and amplifying intra-industry interactions. Notably, technology companies engaged in domains such as artificial intelligence, blockchain, and software development utilize TikTok to render technical content more accessible, thereby targeting emerging professionals and technology enthusiasts (Burns-Stanning, 2020, pp. 2-7). By demystifying complex topics, TikTok serves as a platform where technological discourse can thrive, further solidifying the companies' positions as thought leaders in their respective fields. Consequently, technology companies leverage TikTok not only as a marketing platform but also as a data-centric, interaction-oriented digital environment that actively promotes user engagement and participation.

The emergence of OpenAI on TikTok represents a noteworthy development in the evolving digital communication strategies of technology companies. OpenAI, founded on December 11, 2015, by a consortium of influential figures such as Sam Altman, Elon Musk, Greg Brockman, Ilya Sutskever, John Schulman, and Wojciech Zaremba, was originally established as a non-profit organization with a focus on advancing artificial intelligence (AI) research. However, in 2019, recognizing the need for a sustainable financial model, OpenAI transitioned into a "capped profit" structure under the name OpenAI LP (Ahmadi, 2023, p. 26). Since its inception, OpenAI has strategically curated its corporate presence across several social media platforms, disseminating updates regarding technological advancements, research publications, and product launches primarily through channels such as Twitter (X), LinkedIn, and YouTube (LinkedIn, 2025; YouTube, 2025; Twitter, 2025). Recent shifts in the company's digital communication approach have seen OpenAI extend its footprint to TikTok, a platform traditionally associated with more informal, viral content. While OpenAI’s presence on traditional platforms is well-documented, there is a notable gap in the literature regarding its specific strategies and activities on TikTok. The integration of TikTok into OpenAI’s communication framework raises essential questions about how global technology corporations adapt their strategies to fit the unique dynamics of this platform. The increasing prominence of TikTok in the digital landscape suggests that it may play a critical role in engaging a younger, highly interactive audience, thus influencing the communication tactics employed by technology companies like OpenAI. Given TikTok’s distinctive format and algorithm-driven content dissemination, examining OpenAI’s engagement on this platform could provide valuable insights into the company’s broader digital communication strategy, especially in terms of reaching new user demographics and fostering more direct, personalized engagement with its audience. The present study seeks to fill this gap by offering a comprehensive analysis of OpenAI’s communication strategies on TikTok. Through this examination, the study aims to contribute significant insights into OpenAI's overall digital communication framework, while also shedding light on the broader implications of TikTok's role in shaping the marketing and engagement strategies of international technology companies. By exploring how OpenAI leverages TikTok's unique characteristics—such as its short-form video content, viral potential, and interactive nature—this research will enhance our understanding of the dynamic interplay between technology firms and social media platforms in the contemporary digital age.

PURPOSE AND METHOD

TikTok, with over 1.5 billion active users, has emerged as a prominent platform, particularly favored by individuals in the 18-34 age group (Kemp, 2025, p. 413-558). In recent years, both political and commercial entities have increasingly utilized this platform as a strategic communication tool (Zhang et al., 2021; Vijay & Gekker, 2021; Akbari et al., 2022; Ooi, 2022; Narimo et al., 2023; Cervi, 2023; Çakıcı & Meriç, 2024; Meriç & Çakıcı, 2024). OpenAI initiated its presence on TikTok by establishing an official account on February 16, 2024, actively engaging with the platform. A review of the existing literature reveals a gap in research specifically focused on OpenAI’s communication strategies on TikTok. This study seeks to examine OpenAI’s communication strategies on the platform. The research aims to contribute to the literature concerning the role of TikTok in the communication processes of artificial intelligence organizations. The primary research question and its associated sub-questions for this study are as follows:

1.What strategic objectives does OpenAI pursue through its engagement with the TikTok platform?

1.1. What thematic areas are emphasized in OpenAI's TikTok videos?

1.2. Which types of videos are predominantly featured on OpenAI's TikTok account, and what emotional tones are conveyed through these videos?

1.3. Which actors or entities are most prominently featured in OpenAI's TikTok videos?

1.4. How does the performance of OpenAI's TikTok videos manifest in terms of user interaction and engagement metrics?

Table 1. Coding Scheme for TikTok Video Content Analysis

Video Message Function	Information and Community
Theme	
Video Format	Length, Subtitle, Caption, Language, Music, and Hashtag
Video Type	Documentary, Animated Infographic, Acting, and Pictorial Slideshow
Emotional Tone	0 Emotion, Humor, Hope/Encouragement, Alarm/Concern, and Empathy
Characters Usage	None, General Public, OpenAI Employees and Representatives, Animated Characters, Healthcare Workers, and Engineer
Engagement	Likes, Comments, Views, Save, and Forwards

The research design has been developed in alignment with the research questions. The investigative process commenced with the identification of OpenAI's official TikTok account. The verified TikTok account, identified as @openai, was accessed via OpenAI's official website. Upon logging into the account, it was determined that a total of 101 posts had been made. All posts published between February 16, 2024, and October 4, 2024, were recorded in an Excel file. February 16, 2024, represents the initial date of OpenAI's engagement with the TikTok platform, while October 4, 2024, corresponds to the date of the final post on the aforementioned account. These posts were analyzed using a content analysis approach, which was chosen due to its effectiveness in systematically examining and categorizing diverse elements within digital media content. Content analysis is particularly appropriate for this study as it allows for both qualitative and quantitative assessments of textual, visual, and emotional components, making it suitable for examining the multifaceted nature of OpenAI's posts on TikTok. As delineated in Table 1, the content analysis specifically involved the categorization of various elements within OpenAI's TikTok videos, including but not limited to themes, language, formats, objectives, content types, emotional tonalities, engagement metrics, and keywords (Lovejoy & Saxton, 2012;

Wang & Ji, 2015; Li & Xie, 2020; Li et al., 2021; Çakıcı & Meriç, 2024; Meriç & Çakıcı, 2024; McDonnell et al., 2024; Çakıcı, 2024; Çakıcı, 2025).

In this study, video metrics were codified to quantify engagement parameters, encompassing the frequency of likes, comments, views, saves, and shares (Li & Xie, 2020; Çakıcı & Meriç, 2024). Secondly, structural attributes were encoded, including video duration, linguistic modality, and the presence of subtitles, captions, musical elements, and hashtag frequency, facilitating a comprehensive analysis of formal compositional elements (McDonnell et al., 2024). Thirdly, content typologies were delineated through categorical classification, wherein videos were stratified into predefined genres such as documentary, animated infographic, acting, and pictorial slideshow. Fourthly, the affective tonality embedded within the videos was examined through an emotion-based taxonomization, incorporating classifications such as 0 emotion, humor, hope/encouragement, alarm/concern, and empathy (Wang & Ji, 2015; Meriç & Çakıcı, 2024). Fifthly, the functional orientation of the content was delineated based on its communicative intent, wherein videos were categorized under informational and action-oriented paradigms (Lovejoy & Saxton, 2012). Sixthly, a thematic analysis was conducted to extrapolate the predominant discursive frameworks, encapsulated under the category of "video theme," to identify recurrent narrative constructs (Çakıcı, 2024). Lastly, characters representation within the videos was codified through a categorical schema, distinguishing entities as none, general public, OpenAI employees and representatives, animated characters, healthcare workers, and engineers (Li et al., 2021).

This study has several limitations that may impact the generalizability and validity of the findings. First, it focuses exclusively on OpenAI's official TikTok account, excluding user-generated content and third-party interactions, which could provide a more comprehensive understanding of audience engagement. The analysis is also limited by the static nature of the coding scheme, which may not fully capture evolving communication strategies or emerging trends on the platform. Additionally, the use of quantitative content analysis alone does not account for qualitative insights, such as audience sentiment or emotional reactions, which are essential for a deeper understanding of the effectiveness of OpenAI's communication strategies. These factors suggest that the findings should be interpreted within the context of these limitations, and future research should consider a broader approach, incorporating both user-generated content and qualitative analysis to enhance the depth and relevance of the results.

To uphold the methodological rigor of this study, comprehensive measures were implemented to ensure both the validity and reliability of the research framework. Content validity was established through a systematic approach, wherein the research categories and coding scheme were meticulously designed based on a thorough examination of extant literature. To ascertain their conceptual soundness, coherence, and applicability, these elements underwent an expert evaluation conducted by two faculty members with doctoral credentials in the field of Communication. Following their critical assessment, necessary modifications were incorporated to enhance theoretical robustness and methodological precision. To reinforce coding reliability, an intercoder agreement procedure was employed in accordance with widely accepted scholarly conventions. Initially, the primary researcher independently coded 10% of a randomly selected data subset. Subsequently, a secondary researcher—unaware of the initial coding—conducted an independent assessment of the same data segment. A comparative analysis of the coding decisions revealed an 83% agreement rate, signifying substantial consistency between coders. According to prevailing academic standards, an agreement rate exceeding 80% is deemed an acceptable benchmark for qualitative content analysis (Miles & Huberman, 1994; Patton, 2002; Gheyle & Jacobs, 2017). To further enhance reliability, discrepancies between coders were meticulously examined through a negotiated consensus process (Campbell et al., 2013). Any divergences in classification were systematically deliberated, ensuring that the finalized coding decisions accurately represented the intended categorical framework. This iterative approach mitigated potential subjective bias and bolstered the overall methodological integrity of the study. By rigorously adhering to these methodological protocols, this research not only validates the conceptual soundness of its coding scheme but also substantiates the credibility and reproducibility of its analytical processes, thereby reinforcing the robustness of its empirical findings.

To conclude, it is important to outline the precise role of artificial intelligence (AI) within the scope of this research, ensuring complete adherence to ethical standards and research integrity. AI tools were exclusively utilized for the refinement of language, specifically focusing on enhancing grammatical accuracy and syntactic structure. These tools were strictly applied to improve the overall clarity and readability of the manuscript, without influencing the core content, data analysis, or theoretical propositions. No AI-generated material was integrated at any point in the research process—be it during data collection, analysis, or the formulation of interpretations and conclusions. All intellectual contributions, insights, and critical discussions presented in this study are solely the product of the researcher's original thought and analytical work, ensuring the utmost authenticity and scholarly rigor.

FINDINGS & DISCUSSION

Table 2. Video Content Format and Engagement Metrics

Video Length	3 s to 354 s	
Video Format	**Number**	**Percentage (%)**
Caption	101	100%
Hashtag	78	77,23%
Music	77	76,24%
Subtitle	11	10,89%
Language	**Number**	**Percentage (%)**
English	101	100%
Engagement	**Number**	
Views	122.053.800	
Likes	2.846.120	
Forwards	304.947	
Save	260.905	
Comments	56.493	

The analysis of OpenAI's TikTok account, covering the period from February 16, 2024, to October 4, 2024, reveals a total of 101 videos, as illustrated in Table 2. These videos exhibit a diverse range of durations, spanning from 3 to 354 seconds, a variance that highlights the platform's adaptability to different content delivery needs. Shorter videos, with their concise duration, serve to capture the viewer's attention quickly and align with the rapid consumption tendencies of the audience, particularly within the fast-paced social media environment. Conversely, the longer videos provide a platform for deeper exploration of particular topics, allowing for the articulation of more comprehensive narratives that engage viewers for extended periods. A consistent feature across all the videos is the use of captions, a strategic choice aimed at enhancing the accessibility and clarity of the content. The inclusion of captions serves not only to reinforce the visual elements but also to ensure the communication of key messages, thus enhancing viewers' understanding, particularly in noisy environments or for those with auditory impairments. The use of captions can be seen as a commitment to improving user experience by making content more comprehensible, which aligns with best practices in digital content strategy.

Furthermore, the incorporation of hashtags in a significant proportion of the videos demonstrates OpenAI's alignment with TikTok's algorithmic features, which prioritize discoverability and content reach. Hashtags, as part of the platform's metadata, function as critical tools in expanding content visibility, helping OpenAI's posts reach a broader audience by categorizing the content within relevant thematic or trending areas. This strategic inclusion of hashtags reflects an acute awareness of the platform's content dynamics, with OpenAI optimizing its content for engagement within TikTok's algorithmic ecosystem. The recurrent use of music across the videos represents another strategic communication choice aimed at enhancing emotional engagement. Music serves as an emotive tool that augments the mood of the content, intensifies its emotional impact, and creates a more immersive viewing experience. This deliberate use of music reflects an understanding of the platform's culture, where audio-visual harmony often amplifies the effectiveness of the message and facilitates deeper viewer connection.

On the other hand, the limited use of subtitles emerges as a notable gap in OpenAI's communication strategy. While captions are employed to ensure accessibility for individuals with hearing impairments, the absence of subtitles in multiple languages restricts the global reach of the content. By only using English, OpenAI's videos inadvertently exclude non-English-speaking audiences, which could hinder the inclusivity of its content. This approach could be further enhanced by incorporating multilingual subtitles, which would expand the reach and inclusivity of the platform and ensure a more diverse, global engagement. Additionally, the absence of sign language representation further highlights an underutilization of accessible communication tools, suggesting an area where OpenAI could improve its inclusivity efforts to cater to a wider range of audiences, particularly those with hearing impairments. Overall, these findings suggest that while OpenAI's TikTok content strategy is effectively structured in many areas, there are identifiable gaps in terms of inclusivity and accessibility. The strategic use of captions, hashtags, and music contributes to the content's effectiveness in engaging and reaching a wide audience. However, the limited use of multilingual subtitles and the absence of sign language highlight opportunities for improvement, ensuring that OpenAI's communication efforts can reach and resonate with a more diverse global audience.

The 101 videos collectively accumulated 122,053,800 views, reflecting a substantial reach across the TikTok platform. This level of visibility highlights the efficacy of TikTok's algorithm in amplifying content, suggesting that OpenAI's posts are successfully reaching a diverse audience. Additionally, the videos received 2,846,120 likes, signaling significant viewer approval and reinforcing the notion of positive reception and engagement. This high number of likes suggests the content is resonating with viewers, indicating alignment with their interests. Furthermore, the 56,493 comments demonstrate a high degree of interaction, emphasizing that OpenAI's videos stimulate discourse and foster an active community around the content. The substantial number of comments indicates that the videos not only attract attention but also prompt reflection and dialogue, essential components of audience engagement in digital communication strategies. The videos were also shared 304,947 times and saved 260,905 times, underscoring that viewer not only consumed the content but also disseminated it within their networks. This sharing behavior is indicative of content that users find valuable enough to circulate, while the saving metric suggests that viewers intend to revisit or reference the content in the future. Together, these figures suggest that OpenAI's TikTok posts are not only reaching a wide audience but are also fostering deeper engagement and extended interaction.

Table 3. Functional and Thematic Taxonomy of TikTok Videos

Video Message Function	**Number**	**Percentage (%)**
Information	95	94,06%
Community	6	5,94%
Total	101	100%
Theme	**Number**	**Percentage (%)**
Animal Life	27	26,73%
Creation and Promotion of Animated Characters	15	14,85%
AI-Generated Objects	14	13,86%
Urban Life and Everyday Practices	13	12,87%
Nature and Landscape	12	11,88%
Robots and Robotic Advancements	7	6,93%
Expert Opinions on Artificial Intelligence	7	6,93%
Car Race	6	5,94%
Total	101	100%

As delineated in Table 3, 95 of the analyzed content items comprise videos intended to disseminate informational material to followers, whereas 6 pieces of content are oriented towards expressing gratitude to followers or facilitating direct engagement. These findings elucidate that OpenAI's TikTok content strategy is predominantly oriented towards the informational function, while the approach to thanking followers or engaging in direct interaction is comparatively minimal. Moreover, as indicated in Table 3, a substantial proportion of OpenAI's TikTok videos incorporates artificial intelligence (AI)-generated representations of animals. These videos predominantly concentrate on the life cycles, locomotion, and natural habitats of the depicted animals. In addition, AI-generated animated characters are prominently featured across the analyzed videos. These instances primarily illustrate the physical attributes, movement patterns, and interactions of the characters. Furthermore, certain videos focus on AI-generated objects, wherein various items manipulated by artificial intelligence are showcased. Additionally, videos addressing AI-generated urban environments and quotidian life scenarios within these environments have been identified. Within these videos, AI models the architectural structures, street arrangements, and human interactions within the urban settings. Additionally, AI-generated depictions of nature and natural landscapes are incorporated within the analyzed videos, highlighting various geographical areas, climatic conditions, and ecological components. Another category of content entails videos that focus on robotic technologies and advancements, wherein the design, operational mechanisms, and functional attributes of robots are examined. Furthermore, certain content segments explicitly address the advancements in artificial intelligence technologies. These videos feature expert discourses and analyses regarding the evolution and progression within the domain of AI research. Lastly, AI-generated automotive race simulations are represented to a lesser extent, wherein vehicle dynamics, racing simulations, and speed metrics are accentuated. Collectively, these findings provide comprehensive insight into OpenAI's predominant content strategy on TikTok, which is characterized by a focus on showcasing content produced through AI technologies, thereby promoting and demonstrating the capabilities of OpenAI's technological innovations.

Figure 1. Frequency-Weighted Hashtag Taxonomy for OpenAI TikTok Engagement

An in-depth analysis of OpenAI's TikTok account reveals that 78 out of 101 published posts incorporate a total of 246 hashtags, reflecting a deliberate strategy to optimize content visibility and engagement within the platform's algorithmic framework. As illustrated in Figure 1, the recurring use of the #Openai hashtag underscores the organization's branding efforts, ensuring consistent reinforcement of its identity. Additionally, hashtags such as #Sora and #Madewithsora are strategically employed to highlight OpenAI's artificial intelligence model, Sora, positioning it as a key element within the company's technological portfolio. In parallel, hashtags like #Chatgpt, #Ai, #Chatgptadvancedvoice, #Gpto, and #Dalle are prominently featured, emphasizing OpenAI's diverse range of AI-driven innovations. This targeted usage of model-specific hashtags reflects the organization's effort to strategically position itself within the broader AI research and development ecosystem, signaling its leadership and expertise in the field. Moreover, OpenAI has also ventured into the creation of AI-generated short films and characters, further diversifying its content. To promote these creative works, hashtags such as #Shortfilm, #Airhead, #Shykids, #Deflated, and #Kids are incorporated, highlighting the intersection of AI technology and digital media storytelling. Collectively, these findings suggest that OpenAI's hashtag strategy is multifaceted, serving not only to bolster brand recognition but also to enhance the discoverability and virality of AI-generated content within targeted audience segments. By aligning its content with relevant and trending hashtags, OpenAI seeks to broaden its reach and foster deeper engagement with communities interested in AI and digital innovation.

Table 4. Video Genre Types and Their Associated Emotional Themes

Video Type	**Number**	**Percentage (%)**
Documentary	46	45,54%
Animated Infographic	31	30,69%
Acting	21	20,79%
Pictorial Slideshow	3	2,97%
Total	101	100%
Emotion of the Videos	**Number**	**Percentage (%)**
0 Emotion	81	80,20%
Humor	12	11,88%
Hope/Encouragement	4	3,96%
Alarm/Concern	3	2,97%
Empathy	1	0,99%
Total	101	100%

An analysis of the video formats shared on OpenAI's TikTok account reveals a strategic diversity in content presentation, tailored to optimize engagement and message delivery. As detailed in Table 4, documentary-style videos dominate the content strategy, comprising 46 videos. These videos present information in a structured, factual manner, using a documentary approach that emphasizes clarity and detail, thereby facilitating deeper viewer understanding. The choice of this format suggests OpenAI's commitment to providing educational and informative content, capitalizing on TikTok's potential for long-form, thoughtful dissemination. In addition to documentary videos, 31 animated infographic videos have been posted, which effectively combine visual design with informative messaging. Through the use of animations, these videos simplify complex information, making it more accessible and engaging for a broad audience. This format reflects OpenAI's use of visually compelling techniques to enhance message retention and appeal to viewers who favor dynamic and engaging content. Moreover, 21 videos are categorized as acting-based, leveraging theatrical performances to evoke emotional responses from viewers. This format underscores OpenAI's strategy of humanizing AI technologies, using relatable scenarios to foster emotional connections with its audience. These videos serve not only to inform but also to humanize the brand, establishing a personal link with viewers through narrative-driven content. Finally, 3 pictorial slideshow videos are included, where information is conveyed through a sequence of still images. Although less prevalent, this format serves as a succinct method for presenting visual content that may require minimal textual explanation, catering to viewers with limited time or those preferring visual storytelling. Collectively, these findings highlight OpenAI's strategic use of diverse video formats on TikTok. By deploying a variety of content styles—ranging from documentary and animation to emotional acting and visual slideshows—OpenAI effectively tailors its messages to suit different audience preferences. This approach not only maximizes viewer engagement but also enhances the communicative impact of OpenAI's content across multiple demographic and cognitive profiles.

The emotional tones present in the videos shared on OpenAI's TikTok account exhibit considerable variability, reflecting a strategic approach to content creation that selectively engages with audience emotions. As indicated in Table 4, the majority of videos maintain a neutral emotional tone, deliberately avoiding the direct communication of specific emotional states. This choice aligns with OpenAI's broader strategy of delivering content in a fact-based, informational manner, where the focus remains on clarity and objectivity rather than emotional influence. A smaller subset of videos, however, incorporates humor as a tool for engagement. These videos are designed to present information in an entertaining way, leveraging humor to make the content more relatable and enjoyable for viewers. Humor serves as an effective strategy for attracting attention, fostering positive associations with the brand, and encouraging engagement through lighthearted content. Additionally, a portion of the videos emphasizes feelings of hope or encouragement. These videos aim to provide positive reinforcement to the audience, seeking to inspire motivation and optimism, particularly in the context of technological advancements and the potential of AI. Such emotional tones are typically employed to build trust and enhance the viewer's connection to OpenAI's mission, framing the organization as a source of positive change and progress. In a smaller number of instances, videos convey emotions of alarm or concern. These videos seek to alert viewers to specific issues or challenges related to AI technology, such as ethical considerations or potential societal impacts. This tone of concern is strategically employed to provoke reflection and critical thinking among the audience, encouraging a deeper engagement with the implications of AI development. Finally, one video prominently features empathy, fostering an emotional connection with the audience by expressing understanding and solidarity. This emotionally charged content stands in contrast to the more neutral or informational tone of the majority of the videos and suggests a deliberate attempt to humanize the brand and cultivate a sense of shared experience with viewers. Overall, these findings indicate that while OpenAI predominantly opts for a neutral emotional approach in its TikTok videos, it strategically employs a range of emotional tones to elicit specific responses from the audience. This selective use of emotional diversity demonstrates OpenAI's awareness of the platform's potential to influence viewer perceptions and engagement, underscoring the role of emotional tone in shaping the effectiveness of digital communication strategies.

Table 5. Quantitative Assessment of Characters in Video Content

Character in Video	Number	Percentage (%)
None	53	52,48%
General Public	24	23,76%
OpenAI Employees and Representatives	11	10,89%
Animated Characters	11	10,89%
Healthcare Workers	1	0,99%
Engineer	1	0,99%
Total	101	100%

OpenAI's TikTok content showcases a diverse range of characters, each strategically selected to enhance audience engagement and convey specific messages. As detailed in Table 5, 24 videos feature members of the general public, reflecting OpenAI's conscious effort to connect with a broad and diverse viewer base. This inclusion of everyday individuals likely aims to foster a sense of relatability and inclusivity, encouraging viewers to identify with the content and engage with the brand on a more personal level. In addition to the general public, 11 videos feature OpenAI employees and representatives. This choice underscores a deliberate attempt to humanize the organization and project a corporate presence on the platform. By featuring internal voices, OpenAI not only reinforces its brand identity but also positions itself as a thought leader in the AI field, emphasizing the expertise and authority of its personnel. This strategy likely enhances the credibility and trustworthiness of the content, aligning the company with professionalism and innovation. Moreover, 11 videos incorporate animated characters, a tactic that strategically leverages artificial intelligence-driven design to captivate the audience's attention. The use of animated characters, often imbued with visually dynamic features, serves to engage viewers in a visually appealing and imaginative manner. This choice aligns with OpenAI's brand image as a cutting-edge technology company, highlighting the potential of AI-generated content to create new, creative forms of communication that transcend traditional human representation. Conversely, only one video features healthcare professionals and another one include engineers, suggesting that these character types are selectively used based on the thematic focus of specific content. The limited representation of such professionals likely indicates a targeted communication strategy, wherein specialized characters are employed to address specific subject matter, such as healthcare or engineering innovations, aligning the characters with the relevant context to bolster the content's authenticity and relevance. Notably, a significant portion of OpenAI's TikTok videos—53 in total—do not feature any human or professional characters. In these instances, the content primarily revolves around artificial intelligence-generated objects or animals, further emphasizing the organization's reliance on its technological capabilities. This approach may be intended to underscore the potential of AI in creating dynamic and engaging content independently of human representation, offering a novel viewing experience that pushes the boundaries of traditional media. These findings indicate that OpenAI's selection of characters, or the absence thereof, is a purposeful component of its content strategy on TikTok. By varying the presence of human and AI-generated characters, OpenAI not only diversifies its content but also aligns its video presentations with broader strategic goals, such as enhancing engagement, showcasing innovation, and establishing authority in the AI field.

CONCLUSION

This study set out to investigate OpenAI's communication strategies on TikTok, aiming to contribute to the understanding of digital communication practices within the context of artificial intelligence institutions and their use of social media platforms. The findings underscore that OpenAI employs a multifaceted communication approach, integrating a range of semiotic elements such as music, captions, and hashtags, as well as utilizing diverse video formats and emotional tonalities to engage with its audience. While these strategies appear effective in reaching a large audience and generating moderate interaction, several critical shortcomings within the content strategy were identified. Firstly, the minimal incorporation of subtitles and the complete absence of sign language represent a significant violation of established accessibility standards, potentially excluding a considerable segment of the global audience, including those with hearing impairments. Additionally, the exclusive reliance on English limits the inclusivity of OpenAI's content, disregarding the linguistic diversity of TikTok's international user base. Furthermore, despite the platform's inherent interactive and community-driven nature, OpenAI's content remains largely informational, with little focus on fostering active engagement or building a community around its messaging. This lack of engagement represents a strategic flaw, particularly given the platform's potential for facilitating direct communication and user participation. The thematic scope of the content is overwhelmingly focused on artificial intelligence-generated depictions of animals, objects, and landscapes, while human narratives remain conspicuously underrepresented. This emphasis on technological representations over human-centric storytelling signals a broader trend within OpenAI's communication strategy to prioritize technological narratives over human engagement, which ultimately limits the emotional resonance and relatability of the content. These limitations not only impede the potential for fostering a more inclusive and participatory online community but also suggest the need for a more balanced, accessible, and audience-oriented approach in future digital communication efforts.

In light of the aforementioned findings, it is critical for OpenAI to address the identified deficiencies and adopt more comprehensive and inclusive communication strategies. To improve accessibility, particularly for individuals with hearing impairments, the consistent incorporation of subtitles into all videos is strongly recommended. Furthermore, diversifying the linguistic range by producing content in multiple languages and integrating sign language would significantly enhance the platform's inclusivity and reach. In terms of engagement, OpenAI must develop more proactive strategies to foster user interaction and cultivate a sense of community within the content. This could involve soliciting user-generated content, facilitating interactive features, or integrating mechanisms that encourage active participation from followers. Additionally, a broader thematic range should be incorporated to mitigate the current thematic monotony, with an increased focus on human representation. Such a shift would contribute to higher viewer retention and emotional resonance. Moreover, expanding the emotional tonalities conveyed through the content would further enhance the connection with the audience and prevent content from becoming overly mechanistic or predictable. Although this study provides significant insights into OpenAI's communication strategies on TikTok, it also underscores the necessity for future investigations to explore the evolving dynamics of TikTok as a communication platform for artificial intelligence organizations. Future research should consider employing sentiment analysis techniques to delve deeper into user reactions, providing a more nuanced understanding of audience engagement. Comparative studies assessing OpenAI's performance on other social media platforms could offer valuable insights into how platform-specific strategies influence communication effectiveness. Additionally, investigating the long-term implications of such communications on public perceptions of artificial intelligence, including shifts in attitudes and trust, is crucial for understanding the broader impact of these strategies. These areas of exploration would contribute to the refinement of OpenAI's communication approach and expand the literature on digital communication practices within the realm of artificial intelligence organizations.

REFERENCES

Ahmadi, A. (2023). ChatGPT: Exploring the threats and opportunities of artificial intelligence in the age of chatbots. *Asian Journal of Computer Science and Technology, 12*(1), 25–30. DOI: 10.51983/ajcst-2023.12.1.3567

Akbari, D. A., Jastacia, B., Setiawan, E., & Ningsih, D. W. (2022). The marketing power of TikTok: A content analysis in higher education. *Binus Business Review, 13*(2), 159–170. DOI: 10.21512/bbr.v13i2.8014

Ammirato, S., Felicetti, A. M., Della Gala, M., Aramo-Immonen, H., Jussila, J. J., & Kärkkäinen, H. (2019). The use of social media for knowledge acquisition and dissemination in B2B companies: An empirical study of Finnish technology industries. *Knowledge Management Research and Practice, 17*(1), 52–69. DOI: 10.1080/14778238.2018.1541779

Aytuna, E. N., Çakıcı, Z., & Özkaya, A. E. (2024). Entertainment and persuasion in online politics: A qualitativestudy of young voters' approach in Turkey's 2023 elections. In Al-Obaidi, J. (Ed.), *Changing globalmedia landscapes: Convergence, fragmentation, and polarization* (pp. 215–233). IGI Global., DOI: 10.4018/979-8-3693-3767-7.ch012

Barta, S., Gurrea, R., & Flavián, C. (2025). Augmented reality experiences: Consumer-centered augmented reality framework and research agenda. *Psychology and Marketing, 42*(2), 634–650. DOI: 10.1002/mar.22143

Bharadiya, J. P. (2023). A comparative study of business intelligence and artificial intelligence with big data analytics. *American Journal of Artificial Intelligence, 7*(1), 24–30. DOI: 10.11648/j.ajai.20230701.14

Burns-Stanning, K. (2020). Identity in communities and networks TikTok social networking site empowering youth civic engagement. 27, 1-11.

Busacca, A., & Monaca, M. A. (2025). Who's afraid of "big bad" generative AI? Brief notes on the IDPA provision against OpenAI ChatGPT. In Marino, D., & Monaca, M. A. (Eds.), *Generative artificial intelligence and fifth industrial revolution. Lecture Notes in Networks and Systems* (Vol. 880). Springer., DOI: 10.1007/978-3-031-73880-7_9

Çakıcı, Z. (2024). Cyberstalking practices among youth in Türkiye: Motivational drivers, digital avenues, and perceptual paradigms. *Türkiye İletişim Araştırmaları Dergisi*, (46), 1–22. DOI: 10.17829/turcom.1488278

Çakıcı, Z. (2024). *Türk basınında düzensiz göçün görsel temsili: Taliban yönetimi sonrası Afgan göçü* (Yayımlanmamış doktora tezi). Galatasaray Üniversitesi.

Çakıcı, Z. (2025). Tiktok as a political communication medium: An examination of the United Nations. *İletişim Kuram ve Araştırma Dergisi* (69), 136-156. DOI: 10.47998/ikad.1580671

Çakıcı, Z., & Meriç, E. (2024). Exploring United Nations High Commissioner for Refugees' (UNHCR) TikTok landscape: Insights into migration representations. In Kir Elitaş, S. (Ed.), *Media representation of migrants and refugees* (pp. 278–295). IGI Global., DOI: 10.4018/979-8-3693-3459-1.ch017

Campbell, J. L., Quincy, C., Osserman, J., & Pedersen, O. K. (2013). Coding in-depth semi-structured interviews: Problems of unitization and intercoder reliability and agreement. *Sociological Methods & Research, 42*(3), 294–320. DOI: 10.1177/0049124113500475

Ceci, L. (2025). TikTok users by country 2025. Statista. https://www.statista.com/statistics/1299807/number-of-monthly-unique-tiktok-users/

Cervi, L. (2023). TikTok use in municipal elections: From candidate-majors to influencer-politicians. *Más Poder Local,* (53), 8–29. DOI: 10.56151/maspoderlocal.175

Çetinkaya, A., & Dondurucu, Z. (2023). Haber üretim ve dağıtım pratikleri bağlamında TikTok kullanımı. *İletişim Kuram ve Araştırma Dergisi,* 61, 79-97.

Choudhary, N., Gautam, C., & Arya, V. (2020). Digital marketing challenge and opportunity with reference to tiktok-a new rising social media platform. *Editorial Board, 9*(10), 189–197.

Culnan, M. J., McHugh, P. J., & Zubillaga, J. I. (2010). How large US companies can use Twitter and other social media to gain business value. *MIS Quarterly Executive, 9*(4), 243–259.

Demirel, S., Bulur, N., & Çakıcı, Z. (2024). Utilizing artificial intelligence for text classification in communication sciences: Reliability of ChatGPT models in Turkish texts. In Darwish, D. (Ed.), *Design and development of emerging chatbot technology* (pp. 218–235). IGI Global., DOI: 10.4018/979-8-3693-1830-0.ch013

Demirel, S., Çakıcı, Z., & Bulur, N. (2024). Social perception of artificial intelligence on Twitter: A comparative study on global south and global north countries. *The Russian Sociological Review, 23*(4), 48–79. DOI: 10.17323/1728-192x-2024-4-48-79

Demirel, S., Çakıcı, Z., & Bulur, N. (2025). Exploring reflections of the 2023 Kahramanmaraş earthquake on X: A computational study on Türkiye and Syria. *Akademik İncelemeler Dergisi, 20*(1), 143–171. DOI: 10.17550/akademikincelemeler.1593038

Dlamini, S. T., & Botha, C. (2024). The use of augmented and virtual reality to enhance the customer experience. *African Journal of Science, Technology, Innovation and Development, 16*(3), 399–409. DOI: 10.1080/20421338.2024.2326791

Enyejo, J. O., Obani, O. Q., Afolabi, O., Igba, E., & Ibokette, A. I. (2024). Effect of Augmented Reality (AR) and Virtual Reality (VR) experiences on customer engagement and purchase behavior in retail stores. *Magna Scientia Advanced Research and Reviews, 11*(2), 132–150. DOI: 10.30574/msarr.2024.11.2.0116

Ersöz, B. (2020). Yeni nesil web paradigması-web 4.0. *Bilgisayar Bilimleri ve Teknolojileri Dergisi, 1*(2), 58–65.

Gheyle, N., & Jacobs, T. (2017). Content analysis: A short overview. *Internal research note,* 10, 1-17. DOI: 10.13140/RG.2.2.33689.31841

Guerrero, K. J. (2023). *An analysis of TikTok's digital marketing capabilities and its effectiveness for brands and businesses across different industries* (Master's thesis, Lindenwood University). Lindenwood University.

Haile, T. T. (2024). Web's progression: Moving from passive content consumption to active content creation and content validation. *International Journal of Business and Management, 18*(6), 135–135. DOI: 10.5539/ijbm.v18n6p135

Hassan, M. A., Nguyen, C. T., Nguyen, C. H., Hoang, D. T., Nguyen, D. N., & Dutkiewicz, E. (2024). A novel blockchain based information management framework for Web 3.0. *arXiv preprint* arXiv:2401.14420. DOI: 10.1109/GLOBECOM52923.2024.10901512

Kemp, S. (2025, February 5). Digital 2025 global overview report. We Are Social. https://wearesocial.com/uk/blog/2025/02/digital-2025/

Kemp, S. (2025, Şubat 5). Digital 2025: The essential guide to the global state of digital. *We Are Social UK.* https://wearesocial.com/uk/blog/2025/02/digital-2025-the-essential-guide-to-the-global-state-of-digital/

Klug, D., Qin, Y., Evans, M., & Kaufman, G. (2021, June). Trick and please. A mixed-method study on user assumptions about the TikTok algorithm. *In Proceedings of the 13th ACM web science conference* 2021 (pp. 84-92). DOI: 10.1145/3447535.3462512

Li, Y., Guan, M., Hammond, P., & Berrey, L. E. (2021). Communicating COVID-19 information on TikTok: A content analysis of TikTok videos from official accounts featured in the COVID-19 information hub. *Health Education Research, 36*(3), 261–271. DOI: 10.1093/her/cyab010 PMID: 33667311

Li, Y., & Xie, Y. (2020). Is a picture worth a thousand words? An empirical study of image content and social media engagement. *JMR, Journal of Marketing Research, 57*(1), 1–19. DOI: 10.1177/0022243719881113

López, C. (2023). Artificial intelligence and advanced materials. *Advanced Materials, 35*(23), 2208683. DOI: 10.1002/adma.202208683 PMID: 36560859

Lovejoy, K., & Saxton, G. D. (2012). Information, community, and action: How nonprofit organizations use social media. *Journal of Computer-Mediated Communication, 17*(3), 337–353. DOI: 10.1111/j.1083-6101.2012.01576.x

Ma, J. H., Erdogmus, E., Kangisser, S., & Yang, E. (2025). A comparative analysis of the effectiveness of immersive virtual reality on end-user design review. *Building and Environment, 267*, 112237. DOI: 10.1016/j.buildenv.2024.112237

McDonnell, E. J., Eagle, T., Sinlapanuntakul, P., Moon, S. H., Ringland, K. E., Froehlich, J. E., & Findlater, L. (2024, May 11-16). "Caption it in an accessible way that is also enjoyable": Characterizing user-driven captioning practices on TikTok. *In Proceedings of the CHI Conference on Human Factors in Computing Systems* (pp. 1-16).

Meng, L. M., Kou, S., Duan, S., & Bie, Y. (2024). The impact of content characteristics of Short-Form video ads on consumer purchase Behavior: Evidence from TikTok. *Journal of Business Research, 183*, 114874. DOI: 10.1016/j.jbusres.2024.114874

Meriç, E., & Çakıcı, Z. (2024). From TikTok trends to pandemic essentials: A comparative analysis of the World Health Organization's health communication strategies on TikTok. In Gürkan, H., & Serttaş, A. (Eds.), *Transformed communication codes in the mediated world: A contemporary perspective* (pp. 1–23). IGI Global Scientific Publishing., DOI: 10.4018/979-8-3693-0896-7.ch001

Mhlanga, D. (2023). The value of open AI and chat GPT for the current learning environments and the potential future uses. Available at *SSRN* 4439267. DOI: 10.2139/ssrn.4439267

Miles, M. B., & Huberman, A. M. (1994). *Qualitative data analysis: An expanded sourcebook.* Sage Publications.

Mogavi, R. H., Wang, D., Tu, J., Hadan, H., Sgandurra, S. A., Hui, P., & Nacke, L. E. (2024). Sora OpenAI's Prelude: Social Media Perspectives on Sora OpenAI and the Future of AI Video Generation. *arXiv preprint* arXiv:2403.14665.

Molem, A., Makri, S., & Mckay, D. (2024, March). Keepin'it Reel: Investigating how short videos on TikTok and Instagram reels influence view change. *In Proceedings of the 2024 Conference on Human Information Interaction and Retrieval* (pp. 317-327).

Narimo, S., Mustofa, R. H., Anindhita, H., & Gano-an, J. C. (2023). Analysis of the utilization of TikTok as a financial and educational learning medium. *Journal of Higher Education Theory and Practice*, *23*(18), 35–45. DOI: 10.33423/jhetp.v23i18.6620

Ooi, K. H. (2022). How political parties used TikTok in the 2022 Malaysian general election. *RSIS Commentaries,* 133 (22), 1-4. https://hdl.handle.net/10356/165438
OpenAI (@OpenAI) / X. (2025, Şubat 27). X (Formerly Twitter). https://x.com/openai

Open, A. I. (@openai) Resmi | TikTok. (t.y.). Geliş tarihi 13 Mart 2025, gönderen https://www.tiktok.com/@openai

Open, A. I. Gönderiler | LinkedIn. (t.y.). Geliş tarihi 12 Mart 2025, gönderen https://www.linkedin.com/company/openai/posts/?feedView=all

Open, A. I. (t.y.). YouTube. Geliş tarihi 12 Mart 2025, gönderen https://www.youtube.com/channel/UCXZCJLdBC09xxGZ6gcdrc6A

Our approach to AI safety. (2024, Şubat 14). https://openai.com/index/our-approach-to-ai-safety/

Patton, M. Q. (2002). *Qualitative research and evaluation methods* (3rd ed.). Sage Publications.

Peng, H., Mengni, Z., Yang, L., Juanatas, R., Niguidula, J., & Huiliang, H. (2023). Research on brand marketing strategy on Tik Tok short video platform. *In SHS Web of Conferences* (Vol. 159, p. 02024). EDP Sciences.

Permatasari, P. A., Rech, M., Qohar, A., & Faizal, A. (2020). From web 1.0 to web 4.0: The digital heritage platforms for UNESCO's heritage properties in Indonesia. *Virtual Archaeology Review*, *11*(23), 75–93. DOI: 10.4995/var.2020.13121

Powell, T. C., & Dent-Micallef, A. (1997). Information technology as competitive advantage: The role of human, business, and technology resources. *Strategic Management Journal*, *18*(5), 375–405. DOI: 10.1002/(SICI)1097-0266(199705)18:5<375::AID-SMJ876>3.0.CO;2-7

Putriani, I., & Aras, M. (2022). Elevate Company's Social Media as a Diversity, Equity, and Inclusion Platform: Multi Cases Study to Digital Technology Companies. *Journal of Wood Science, 1*(10), 906–916. DOI: 10.36418/jws.v1i10.105

Rizomyliotis, I., Lin, C. L., Konstantoulaki, K., & Phan, T. (2024). TikTok short video marketing and Gen Z's purchase intention: Evidence from the cosmetics industry in Singapore. *Journal of Asia Business Studies, 18*(4), 930–945. DOI: 10.1108/JABS-04-2023-0138

Roumeliotis, K. I., & Tselikas, N. D. (2023). ChatGPT and Open-AI models: A preliminary review. *Future Internet, 15*(6), 192. DOI: 10.3390/fi15060192

Saifi, S., Tanveer, S., Arwab, M., Lal, D., & Mirza, N. (2025). Exploring the persistence of Open AI adoption among users in Indian higher education: A fusion of TCT and TTF model. *Education and Information Technologies, 30*(9), 12055–12078. Advance online publication. DOI: 10.1007/s10639-024-13282-x

Sheak, E., & Abdulrazak, S. (2023). The influence of social media marketing activities on TikTok in raising brand awareness. *Market-Tržište, 35*(1), 93–110. DOI: 10.22598/mt/2023.35.1.93

Taylor, S. H., & Brisini, K. S. C. (2024). Parenting the TikTok algorithm: An algorithm awareness as process approach to online risks and opportunities. *Computers in Human Behavior, 150*, 107975. DOI: 10.1016/j.chb.2023.107975

Vijay, D., & Gekker, A. (2021). Playing politics: How Sabarimala played out on TikTok. *The American Behavioral Scientist, 65*(5), 712–734. DOI: 10.1177/0002764221989769

Wang, S., & Ji, Q. (2015). Video affective content analysis: A survey of state-of-the-art methods. *IEEE Transactions on Affective Computing, 6*(4), 410–430. DOI: 10.1109/TAFFC.2015.2432791

Zeng, J., Abidin, C., & Schäfer, M. S. (2021). Research perspectives on TikTok & its legacy apps| research perspectives on TikTok and its legacy apps—Introduction. *International Journal of Communication, 15*, 12–12.

Zhang, W., Mei, J., Song, W., Evans, R., & Xiang, Y. (2021). Why do citizens engage with the TikTok accounts of public hospitals in China? *SAGE Open, 11*(4), 1–11. DOI: 10.1177/21582440211061568

Zou, W., Li, J., Yang, Y., & Tang, L. (2023). Exploring the Early Adoption of Open AI among Laypeople and Technical Professionals: An Analysis of Twitter Conversations on #ChatGPT and #GPT3. *International Journal of Human-Computer Interaction*, 1–12. PMID: 39989507

KEY TERMS AND DEFINITIONS

Artificial Intelligence: A multidisciplinary domain within computer science dedicated to the development of computational systems capable of simulating human cognitive processes. These systems exhibit advanced capabilities such as machine learning, natural language processing, visual perception, and autonomous decision-making, enabling AI to perform complex problem-solving tasks and adapt to evolving informational contexts.

OpenAI: A leading research institution specializing in the advancement of artificial intelligence technologies, with a strong emphasis on developing AI systems that are robust, interpretable, and ethically aligned with global societal values. Through interdisciplinary research and policy advocacy, OpenAI seeks to ensure the responsible deployment of AI, mitigating potential risks while maximizing its long-term benefits for humanity.

Social Media: A multifaceted digital ecosystem designed for the creation, dissemination, and interactive exchange of user-generated content. It serves as a dynamic platform for fostering virtual communities, facilitating multidirectional communication, and enabling the real-time sharing of information, opinions, and multimedia content across geographically dispersed networks.

Strategic Communication: A systematic and research-driven communication process aimed at achieving predefined organizational objectives through the deliberate design and dissemination of messages. This discipline integrates persuasive techniques, audience analysis, and media selection strategies to shape public discourse, influence stakeholder perceptions, and enhance organizational reputation within competitive environments.

TikTok: A short-form video-sharing platform that leverages an advanced algorithmic recommendation system to curate personalized content feeds, thereby optimizing content virality and user engagement. Its interactive features, including duets, filters, and audio overlays, contribute to the platform's role as a key driver of participatory culture and digital trends.

ENDNOTES

1 This paper was presented as an oral presentation at the NEWPATH Symposium: New Paradigms in Communication Technologies and Humanity, organized by Üsküdar University and UTAD, held online on March 6-7, 2025.

Chapter 11

Bridging the Digital Divide With AI to Enhance Access and Equity in Underserved Communities

Fransisco Chindongo Ntjamba
University of Namibia, Namibia

Daniel Opotamutale Ashipala
http://orcid.org/0000-0002-8913-056X
Univeristy of Namibia, Namibia

Medusalem Hangula Joel
http://orcid.org/0000-0003-3886-0367
Univeristy of Namiba, Namibia

ABSTRACT

Artificial Intelligence (AI) has a transformative force that has the potential to alleviate systemic disparities in rural regions through enhanced healthcare, education, and agriculture. Despite the potential of AI, obstacles such as infrastructure deficiencies, digital literacy gaps, and ethical concerns jeopardize its equitable adoption. By adopting a multi-stakeholder perspective, this study emphasizes the imperative of culturally sensitive AI solutions, moral governance, and public-private collaboration to facilitate ethical and inclusive AI uptake. By putting affordability, accessibility, and long-term viability, AI can promote the empowerment of marginalized communities,

DOI: 10.4018/979-8-3373-1509-6.ch011

build resilience, and forge a more equal future. This chapter explores how AI-driven innovations can transform these sectors through improved access to basic services, effective resource allocation, and empowering local communities.

INTRODUCTION

The digital divide refers to the socioeconomic disparity between individuals, families, communities, and places that have access to contemporary information and communication technologies (ICTs) and those that do not (Darus et al., 2021). This gap is evident in digital literacy, usage skills, and chances for significant interaction with technology, as well as in physical access to devices and high-speed internet. In its broader context, the digital divide reflects systemic inequalities, often rooted in poverty, rural underdevelopment, and marginalization that prevent underserved communities from fully benefiting from technological progress (Ndjama, 2025). This gap results in restricted access to electronic financial tools, e-governance platforms, telehealth services, and online education in remote and resource-poor settings. As a result, the digital divide worsens current social inequalities, impeding educational achievement, healthcare equity, and social mobility (Mishra, 2024). Crucially, the digital divide now encompasses not only who is connected, but also how and to what extent individuals can benefit from digital advancements. Those marginalized by the digital transition run the risk of becoming even more marginalized as artificial intelligence (AI) increasingly influences worldwide progress. Closing this gap requires more than just technological deployment by ensuring that technology is affordable, accessible, and culturally appropriate, as well as providing digital education support. The integration of AI into underserved regions has great potential however, such advancements could exacerbate rather than lessen inequality if deliberate measures are not taken to close the digital divide (Hussein & Safina, 2025).

AI has the potential to empower societies by helping to lessen disparities in economic growth, education, and health care (Dakanalis et al., 2023). AI has introduced a variety of machines and instruments in various fields but its use in underserved areas is insufficient (Abuagla, 2023). AI-led strategies could offer solutions to bridge the access gaps, enhance service delivery, and eliminate hiccups (Qwiumin, 2023). Despite the fact that (AI) has many uses, its implementation presents several difficulties that need strategic interventions and collaboration between the public and corporate sectors as well as the local population (Braun & Möller, 2024). The evolution of Artificial Intelligence (AI) has opened a new era of innovation and possibilities affecting everything humans do (Wang, 2025). With applications in healthcare, learning, farming and governance, and other areas, AI helps automate normal tasks to solve complex problems. Although much of this advancement has occurred in cities and developed countries, AI has a lot to offer to tackle challenges in remote and underserved areas (Dahlstrand-Rudin et al., 2023). On the whole, outlying areas confront extreme challenges to poverty with which urban centre's have little experience (Asiedu et al., 2024).

AI has shown that it can change many industries for the better. AI-based diagnostics and telemedicine platforms help support areas lacking medical professionals in the healthcare sector (Singh & Singh, 2024). In the field of education, adaptive learning technologies offer personalized learning experiences to children in resource-constrained environments. In agriculture, the optimal application of AI-based technologies can enhance food security by improving crop yield and allocation of resources (Hemmler & Ifenthaler, 2022). Despite advancements made in deploying AI application, various issues limit their usage in the real world. There are limited access to digital infrastructure, data privacy concerns and algorithmic biases. Moreover, adoption and trust of underprivileged communities could be hindered due to deficient culturally-relevant AI model (Dempere, 2024). AI can process a large amount of data, recognize patterns, and provide useful insights. This can help people solve some deep-rooted problems in poor communities. For example, AI-equipped diagnostic tools can allow early detection of disease in regions where health workers are not available (Kuska et al., 2024). AI learning programs can give an experience to people in far-flung places. In addition, predictive modelling can help with the wastage of agricultural products in floods or fixing human insulin for those suffering from diabetes-related complications. AI has the amazing ability to enhance its capacity to perform the same or similar tasks. AI solutions can be used by simply utilizing mobile phones or other low-cost devices (Goel & Dixit, 2023).

BACKGROUND

Access to healthcare remains an important issue in accessing health services in remote and underserved areas. Inadequate access to healthcare services in remote areas often happens because of a shortage of healthcare professionals and inadequate medical infrastructure (Pandey, 2024). AI can help improve the delivery of healthcare services in these areas via remote diagnosis, telemedicine, and predictive health analytics. Diagnostic equipment powered by AI, including a portable ultrasound machine with a machine-learning algorithm will allow community health workers to detect pneumonia or complications in pregnancy (Kambala, 2024). Further, AI-based telemedicine platforms can overcome geographic barriers, linking patients in remote localities with doctors in urban centers. With tools like these, not just lives but resource drain too, could be managed with a slight ease (Hswen & Schweitzer, 2025). Predictive analytics could help in disease prediction and management. Using wearable devices, mobile health apps, and electronic health records, AI can spot trends and predict disease outbreaks by processing the data (Yadav, 2023). AI algorithms have been employed to predict malaria outbreaks in sub-Saharan Africa, enabling government officials and NGOs to pre-position resources efficiently. These innovations can show how promising AI is going to be in health care for left-behind areas (Kosaraju, 2024).

Enhancing Learning Opportunities

The foundation of development is education, but many people in underprivileged areas still struggle to get access to high-quality education (Ye et al., 2023). Learning is significantly hampered by elements like a lack of educational resources, out-of-date curricula, and a teacher shortage. Enabling scalable and customized educational models, AI provides creative answers to these problems (Mishra, 2024). Learning platforms with AI capabilities, like adaptive tutoring programs, can offer students personalized instruction based on their unique requirements and learning style (Ragab, 2025). In order to provide more effective learning outcomes, these platforms employ machine learning algorithms to pinpoint students' problem areas and provide tailored interventions. AI can act as an additional resource in areas with a shortage of teachers, closing the gap and enhancing educational equity. Additionally, by making educational content accessible to non-native speakers, AI-driven language translation tools can promote inclusivity while preserving regional languages and cultures (Sol & Heng, 2024). Artificial intelligence (AI)-powered virtual reality (VR) and augmented reality (AR) technologies can also improve experiential learning by enabling students to investigate difficult physics, history, and geography concepts in an interesting and participatory way (Deng, 2024). AI has the potential to enable people in marginalized communities to escape poverty cycles and create better futures by democratizing access to high-quality education (Egunjobi & Adeyeye, 2024).

Revolutionizing Food Security

Agriculture serves as the driving force for many isolated and underdeveloped regions of the world. However, agricultural industry is exposed to the threats of climate changes, pests, and resource shortage. The implementation of AI in agriculture has the capability of transforming the entire industry and opening avenues through precision farming, better resource allocation, and increased crop production (Arya, 2025). Farmers can make use of AI-powered tools, such as drones equipped with computer vision technologies, to track the health of their crops, monitor pest activities, and analyze the condition of the soil (Selwyn et al., 2025). This means, farmers can now able to make decisions with useful information which enables them to minimize wastage while maximizing productivity. Moreover, farmers can time their activities using AI-powered weather forecasts to plan for unpredictable weather changes (Stamoulis & Giannopoulos, 2021)

In areas where market accessibility is low, AI can digitally link farmers to buyers. AI can help farmers set reasonable prices by analyzing market dynamics and demand shares, ensuring that intermediaries do not take advantage of farmers. With these implementations, food security is enhanced as well as the economic status of the farming community's core, which sustainably grows the economy (Ratcliffe, 2023).

Collaboration for Inclusive AI Deployment

The effective utilization of AI in underdeveloped regions necessitates the involvement of governments, private sector companies, NGOs, and local participants (Nour et al., 2025). Governments contribute by funding infrastructure projects, adopting favorable laws, and promoting public-private cooperation, which paves the way for the public sector (Oh & Ki, (2024). The private sector has the distinct advantage of being able to implement cutting-edge technologies and invest capital towards innovation, leading to large-scale solutions. While NGOs and local organizations assist the technical provider with culturally sensitive solutions, they can make sure that AI technologies align with local needs (Nour et al., 2025). Through multi-stakeholder partnerships, this can build an ecosystem that enables the use of AI in a manner that helps even the most marginalized regions which, in turn, increases the efficacy of the technology (Oh & Ki, (2024).

AI's Role in Community Empowerment

It is important to emphasize AI's wider role in empowering communities, even though its direct applications in healthcare, education, and agriculture are equally important (Nag et al., 2024). AI could democratize information and provide a voice to groups that are frequently left out of decision-making procedures. AI-driven platforms have the potential to magnify local issues and empower remote communities to advocate for their needs and take part in policy discussions (Mark & Bizz, 2025). AI promotes accountability and inclusion, two crucial tenets of sustainable development, by reducing the communication gap between underrepresented groups and decision-makers. Similarly, AI tools can improve financial inclusion by giving marginalized groups mobile platform access to banking, credit, and insurance services (Mhlanga, 2024). This lessens susceptibility to financial shocks, increases economic resilience, and encourages entrepreneurship. Additionally, by enabling communities to record and disseminate their distinctive customs and knowledge systems, AI-driven projects can aid in the preservation of cultural heritage (Mhlanga, 2024).

AI's function goes beyond solving problems; it becomes a force for structural change. Incorporating AI into a more comprehensive community empowerment this can guarantee and ensure that, its advantages are experienced both individually and collectively, promoting sustainable and all-encompassing growth (Mark & Bizz, 2025).

Fostering Resilience in Underserved Communities

The potential of AI to improve community resilience is a crucial component of its application in underprivileged areas. AI-powered solutions can improve disaster response and preparedness, which is crucial for remote regions that are frequently disproportionately hit by natural disasters. For example, predictive analytics tools can forecast natural disasters like hurricanes, droughts, and floods by analyzing historical data and weather patterns. These systems have the potential to save lives and reduce financial losses by enabling prompt evacuations and resource deployment through the provision of early warnings (Ruiz-Vanoye & Díaz-Parra, 2025). AI can enhance resilience in addition to disaster management by facilitating better access to resources and information in times of emergency. For instance, chatbots with natural language processing skills can communicate important information in local languages regarding emergency relief efforts, healthcare services, and safety precautions. Even the most remote communities can get essential assistance when they need it (Chakravarthy & K S, 2024). AI can help with climate change adaptation as well as resilience building. For example, in agriculture, machine learning algorithms can assist farmers in forecasting crop yields under different climate conditions, allowing them to make well-informed planting and harvesting decisions. Communities whose livelihoods are closely linked to natural resources will find these insights invaluable in preparing them to deal with the uncertainties of a changing environment (Goel et al., 2024).

Building Capacity through AI Education and Training

Building local capacity through training and education is crucial if underprivileged communities are to fully benefit from AI (Mathai, 2024). Programs for digital literacy that are adapted to the particular requirements of these groups can enable people to use AI technologies sensibly and with confidence. Such initiatives can remove adoption barriers and promote a sense of ownership over technological solutions by demythologizing and making AI accessible Additionally, AI-related skill-focused vocational training programs can give marginalized communities access to new economic opportunities (Trust & Maloy, 2024). To ensure the sustainability of these solutions, for example, training community members to operate and maintain AI-powered systems like agricultural drones or diagnostic tools can generate jobs (Voorneveld, 2024). Building local capacity through training and education is crucial if underprivileged communities are to fully benefit from AI. Programs for digital literacy that are adapted to the particular requirements of these groups can enable people to use AI technologies sensibly and with confidence. Such initiatives can remove adoption barriers and promote a sense of ownership over technological solutions by demythologizing and making AI accessible (Drumm & Sami, 2024). Additionally, AI-related skill-focused vocational training programs can give marginalized communities access to new economic opportunities. To ensure the sustainability of these solutions, for example, training community members to operate and maintain AI-powered systems like agricultural drones or diagnostic tools can generate jobs (Smt. Sumela Chatterjee, 2023). Furthermore, incorporating AI ideas into school curricula can motivate the upcoming generation of creative thinkers and prepare them to use the technology to advance their communities (Bieser, 2022). Building capacity can be greatly aided by cooperative efforts involving governments, academic institutions, and businesses. These stakeholders can create and carry out programs that specifically address the needs and goals of marginalized communities by combining their resources and expertise. These initiatives support the general resilience and self-sufficiency of these areas in addition to empowering individuals (Bieser, 2022).

Challenges and Ethical Considerations

While many parts of the world could benefit from Artificial Intelligence, their lack of infrastructure proves to be a critical challenge. For example, in certain regions where electricity and internet connectivity is unreliable, the use of AI technologies becomes difficult. A significant portion of the population who live in the hinterlands do not possess the basic skills required to use AI technologies which widens the gap even further (Alakitan & Makinde, 2024).

It is impossible to ignore the ethical ramifications of implementing AI in underprivileged areas. Data privacy and bias are two of the foremost ethical considerations that can come into play from the use of AI technologies. AI systems function well when trained on large datasets, but collecting data to train AI in underserved communities' raises concerns about consent (Ntjamba & Ashipala, 2024). It is crucial to protect data security and privacy, especially in areas with less developed regulatory frameworks (Ntjamba & Ashipala, 2024). To uphold people's rights and foster confidence in AI systems, transparent data collection procedures, informed consent procedures, and strong cybersecurity measures are crucial (Suresh et al., 2025). The possibility of algorithmic bias is another ethical issue. Biased results could be produced by AI systems trained on data that does not accurately represent the diversity of marginalized communities, thus sustaining current disparities (Sharma et al., 2025). An AI-powered diagnostic tool that was trained primarily on data from urban healthcare facilities, for example, might not be able to identify diseases that are common in rural areas. Diverse and representative datasets are necessary to address these biases, and AI systems must be continuously monitored and assessed to spot and address inequities (Kerketta & Balasundaram, 2024). Another crucial component of deploying AI ethically is accountability. Communities can hold stakeholders accountable for any negative effects and have a say in how AI technologies are used when there are clear procedures for oversight and grievance redress. Ensuring ethical issues to become a priority, this can ensures that, AI is a positive force that promotes sustainable and equitable development (Sidoo, 2023).

Moreover, if sufficient care is not taken during the design of AI algorithms, the pre-existing biases are more likely to be built upon rather than removed, further deepening the inequality gap. For instance, an AI trained on urban data may not understand rural needs creating poor results (Sharma & Chaudhary, 2025). Owing to the aforementioned challenges, a new approach focusing on stakeholder participation is essential in AI development and deployment processes (Gaurav et al., 2024). Employing the members of the community when implementing AI solutions ensures that their needs are adequately articulated. Moreover, initiatives such as building capacity through programs aimed at improving digital literacy enable individuals to make effective use of AI technologies (Malik & Sharma, 2023).

The Future of AI in Underserved Communities

Artificial intelligence has great potential to change underprivileged communities. From more effective algorithms to creative hardware solutions, advances in artificial intelligence technology are opening more opportunities for impact (Pagani & Champion, 2023). More open and responsible systems made possible by emerging disciplines including explainable artificial intelligence and ethical artificial intelligence also guarantee that technology advances human best interests (Kerrigan, 2022). Realizing this promise, though, calls for a group dedication to diversity, cooperation, and creativity. Working with governments, businesses, non-governmental organizations, and local communities can remove obstacles and guarantee that artificial intelligence technologies are sustainable, ethical, and accessible. Encouragement of a shared vision for AI-driven development can open new chances for underprivileged groups and can help to create a more fair and rich future (Lamuth, 2010). AI presents a transforming chance to close the disparity between underprivileged areas and basic services. From education and healthcare to agriculture and community empowerment, the possible uses are rather broad

(Shibuya, 2020). This will enable the power of artificial intelligence to produce significant and long-lasting transformation by tackling obstacles, giving ethics top priority, and encouraging teamwork so that none of anyone is left behind in the age of technology (Aljawawdeh et al., 2024).

A Future of Equitable Rural Healthcare

The adoption of AI technologies in rural healthcare systems offers a chance to eliminate existing obstacles while enhancing patient care results and promoting equitable health services (Malviya & Goyal, 2023). AI transforms rural healthcare delivery through wider medical service access and optimized resource distribution combined with improved diagnostic abilities and reinforced mental health support. The evolving technology can increasingly strengthen its impact on healthcare equality and rural community empowerment (Menona-Jacob, 2022) The possibility of AI transforming remote, underserved communities poses both an opportunity and a challenge due to the technological revolution (Nuka, 2024). Addressing systemic barriers in healthcare, education, agriculture, and other fields can enable marginalized groups to realize equity and sustainable development with the help of AI. But achieving this kind of social equity, it will demand great mobilization to overcome barriers and deal with ethical dilemmas concerning inclusivity and contextual relevance (Mariyono et al., 2024).This chapter seeks to provide evidence on the impact of AI technology in addressing gaps for underserved communities in healthcare and other services. It aims to motivate stakeholders for action and collaboration to build a future responsive to the needs of all people, regardless of their social status and origin on how AI can help low-income communities get the services they need.

Conceptual Framework of AI Implementation Barriers in Public Healthcare

The conceptual framework of the implementation barriers of AI in public healthcare illustrates the various tools and applications of AI, through which sustainable operational excellence can be obtained in the public healthcare systems (Joshi et al., 2022).

Figure 1. Conceptual framework of implementation barriers of AI in public healthcare enhancement

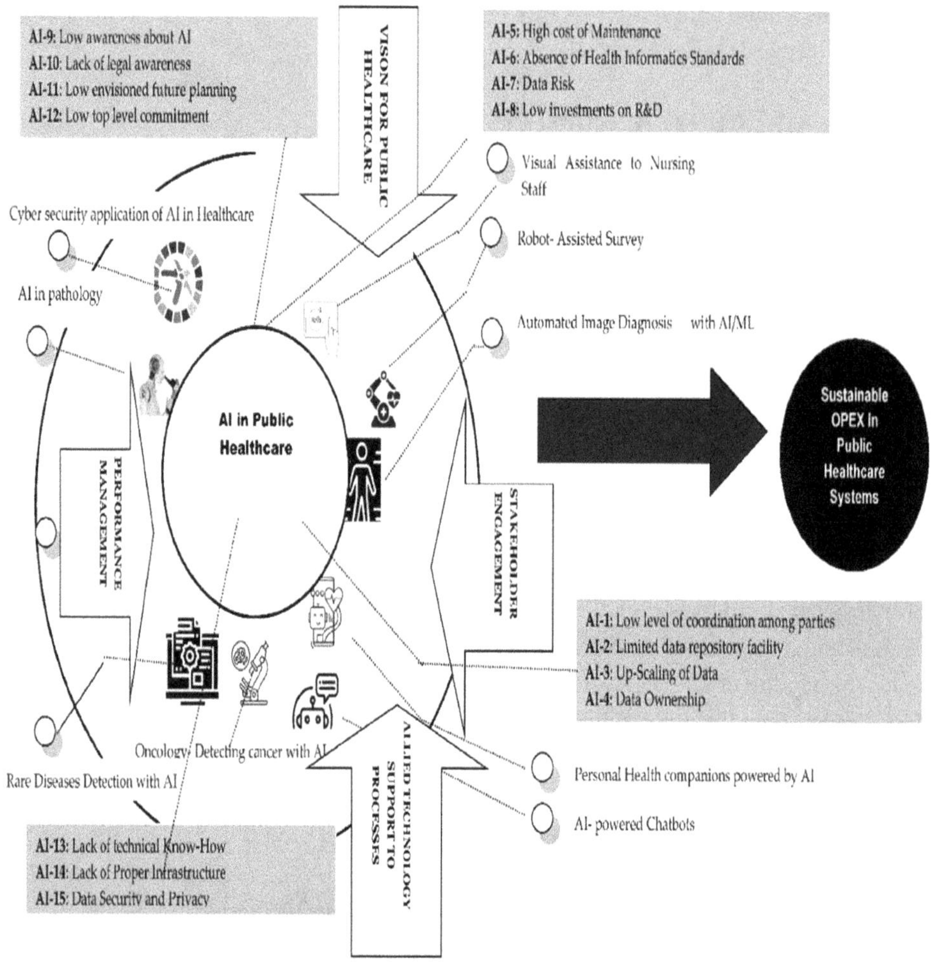

(Joshi et al., 2022).

Table 1. AI Implementation in Public Healthcare- Key Barriers, Applications, and Sustainable Solutions

Category	Key Aspects
Barriers to AI Implementation	- **Infrastructure Deficits:** Inconsistent electricity, unreliable internet, and insufficient digital infrastructure.
	- **Data Privacy and Security:** Concerns over patient confidentiality, data protection laws, and cybersecurity threats.
	- **Limited AI Literacy and Training:** Healthcare professionals and community health workers lack AI skills and knowledge.
	- **Regulatory and Ethical Challenges:** Unclear policies, algorithmic biases, and ethical concerns in AI-driven decision-making.
	- **Financial Constraints:** Limited public healthcare budgets, high costs of AI deployment, and lack of funding for AI research.
AI Tools and Applications	- **AI-Powered Diagnostics:** AI-driven tools for early disease detection in low-resource settings, reducing reliance on specialists.
	- **Predictive Analytics:** AI models for forecasting disease outbreaks, tracking patient trends, and optimizing resource allocation.
	- **Telemedicine & Virtual Assistants:** AI-powered chatbots, remote consultations, and mobile health apps to expand healthcare access.
	- **Robotic Process Automation (RPA):** AI automates data entry, patient scheduling, and hospital workflow management.
	- **Personalized Treatment Plans:** AI-based decision support tools tailored to individual patient needs in rural and underserved areas.
Pathway to Sustainable Excellence	- **Interdisciplinary Collaboration:** Partnerships among healthcare providers, AI developers, governments, and NGOs.
	- **Ethical & Inclusive AI Models:** Development of AI solutions that consider cultural, linguistic, and contextual diversity.
	- **Public-Private Partnerships (PPP):** Joint efforts between governments, private tech firms, and healthcare institutions to fund and scale AI solutions.
	- **Capacity Building & Workforce Training:** AI literacy programs, continuous training, and mentorship for healthcare professionals.
	- **Policy & Regulatory Frameworks:** Establishing robust, transparent policies to ensure ethical AI deployment and patient safety.
	- **Community Engagement:** Involving local communities in AI adoption to enhance trust, usability, and long-term sustainability.

(Joshi et al., 2022).

Table 1, ensures a strong focus on AI's role in underserved public healthcare systems and addresses real-world constraints with practical and scalable solutions.

REVIEW OF RELATED LITERATURE

Artificial intelligence (AI) has revolutionized many sectors, such as healthcare, education, farming, and governance. (Ojedokun, 2024). An exceptional chance to address persistent socioeconomic inequalities and systemic issues is provided by the use of AI in isolated and underprivileged communities. Because of their remote location, inadequate infrastructure, and shortage of qualified workers, these communities frequently face barriers to accessing healthcare, education, and economic opportunities (Ojedokun, 2024). AI-powered solutions have enormous potential to bridge these gaps, especially in the fields of precision agriculture, telemedicine, personalized learning, and diagnostics (Sinha, 2024). However, there are some obstacles to integrating AI in these settings, such as limited infrastructure, moral dilemmas, and digital literacy issues (Nienhaus, 2021). The impact, difficulties, and potential of artificial intelligence (AI) in underprivileged communities are reviewed in this review of the literature.

There has been a substantial increase in writing about artificial intelligence (AI) and its capacity to remedy systemic inequalities in underprivileged communities during the past few years. Especially in areas where traditional infrastructure is inadequate, scholars and practitioners alike highlight AI's potential to revolutionize access to vital services like healthcare, education, and agriculture (Dempere, 2024; Sinha, 2024). AI technologies, such as robotic process automation, natural language processing, machine learning, and predictive analytics, make it possible to enable financial inclusion mechanisms, intelligent agricultural systems, and virtual learning environments, thereby circumventing conventional obstacles (Goel & Dixit, 2023; Arya, 2025).

Nevertheless, it is not at all simple to implement these technologies in underprivileged situations. The promise of artificial intelligence should be tempered with an awareness of ingrained structural issues like digital illiteracy, algorithmic bias, and infrastructure shortfalls, as Nienhaus (2021) points out. Even the most sophisticated AI applications are still inaccessible or unworkable in regions with poor internet access or unreliable electricity supply (Alakitan & Makinde, 2024). In addition, data privacy issues and ethical concerns, especially in areas lacking robust legal frameworks, create significant issues regarding consent, surveillance, and the possible abuse of AI technologies (Ntjamba & Ashipala, 2024; Sharma et al., 2025).

According to Lourdu-Vesna (2025), AI will keep replicating current inequalities unless interventions tackle not just physical access but also digital skills, localized content, and linguistic inclusivity. As a result, researchers have supported community-driven artificial intelligence models, in which the technology is developed in partnership with local communities to guarantee cultural appropriateness and usability (Gaurav et al., 2024).

Conversely, the literature also provides evidence of effective AI-enabled interventions. In sub-Saharan Africa, AI-based telemedicine systems have been shown to enhance health outcomes, and adaptive learning platforms are assisting in closing the educational gap (Dahlstrand-Rudin et al., 2023; Krishna-Pasupuleti, 2024). In agriculture, precision farming technologies powered by AI have been successful in increasing productivity, especially when used with mobile platforms that are accessible to smallholder farmers (Agrawal & Arafat, 2024).

The role of multi-sector collaboration is another common theme in the literature. Effective AI adoption in underserved communities, according to Nour et al. (2025), necessitates coordinated action among local users, NGOs, technology developers, and governments. To close the digital divide and guarantee that AI benefits all segments of society, inclusive policy frameworks, capacity-building initiatives, and public-private partnerships (PPPs) are increasingly viewed as foundational pillars (Oh & Ki, 2024).

AI in Healthcare: Expanding Access and Improving Outcomes

Predictive analytics and diagnostics are two of the most well-known uses of AI in remote healthcare settings. Even in environments with limited resources, studies have shown how well AI-powered diagnostic tools can identify illnesses like malaria, pneumonia, and tuberculosis (Reethika & Priya, 2024). Early detection and prompt intervention are made possible by the ability of machine learning algorithms trained on large datasets to recognize patterns in medical images (Dooraki, 2024). AI-powered telemedicine platforms have also become more popular, enabling medical experts to conduct consultations and prescribe treatments remotely, relieving pressure on nearby healthcare facilities (Kambala, 2024). Another AI-driven healthcare innovation that has proven crucial in managing public health responses and predicting disease outbreaks is predictive analytics. To help with the strategic distribution of medical resources, AI models have been used, for example, to forecast the spread of infectious diseases like COVID-19 and Ebola (Ali, 2025). However, issues with algorithmic bias, data privacy, and the moral application of AI in decision-making continue to be major obstacles to its adoption (Byun, 2022).

AI in Education: Enhancing Learning in Resource-Limited Environments

Another area where AI has shown transformative potential is education. Learning platforms with AI capabilities use adaptive algorithms to offer individualized instruction based on each student's needs (Krishna-Pasupuleti, 2024). In order to make education more inclusive and effective—especially for students in remote areas with limited access to qualified teachers—these platforms analyze student performance data to identify areas of difficulty and modify content accordingly. Additionally, multilingual education has been made easier by AI-driven language translation tools, which have broken down linguistic barriers in diverse communities (Yoon, J. (2023). In science and vocational training, where experiential learning is crucial, virtual and augmented reality combined with artificial intelligence have also been utilized to create immersive learning environments (Sukhdeve, 2021). However, obstacles like digital literacy, internet connectivity, and reluctance to embrace new technologies prevent AI from being widely used in education.

Promoting Food Security and Sustainable Practices

AI-powered solutions have been implemented in agriculture, a vital industry for many marginalized communities, to increase productivity, optimize resource allocation, and reduce climate-related risks. By using AI-powered sensors, drones, and data analytics, precision agriculture has made it possible for farmers to keep an eye on crop health, anticipate pest infestations, and improve irrigation techniques (Agrawal & Arafat, 2024). Farmers have also benefited from machine learning models trained on historical soil and weather data, which have helped them make well-informed decisions regarding planting and harvesting schedules (Agrawal & Arafat, 2024). AI-powered supply chain management systems have not only increased yields but also made it easier for small-scale farmers to reach markets, which has decreased post-harvest losses and increased profitability (Palla, 2025). Notwithstanding these benefits, adoption of AI in agriculture is severely hampered by obstacles like high implementation costs, a lack of technical expertise, and slow internet penetration. AI-powered supply chain management systems have not only increased yields but also made it easier for small-scale farmers to reach markets, which has decreased post-harvest losses and increased profitability (Al-Jenaibi & Bulhoon, 2024). Notwithstanding these benefits, adoption of AI in agriculture is severely hampered by obstacles like high implementation costs, a lack of technical expertise, and slow internet penetration.

Challenges in AI Implementation: Ethical, Technical, and Societal Concerns

Although AI has great potential for underserved and remote communities, a number of obstacles stand in the way of its effective implementation. A significant issue is the "digital divide," which restricts access to AI-powered solutions because of poor internet connectivity, a lack of electricity, and a lack of digital literacy (Lourdu-Vesna, 2025). The fair application of AI technologies is also called into question by ethical issues pertaining to algorithmic bias, consent, and data privacy (Wang, 2024).

AI's dependence on massive datasets presents additional difficulties in areas with irregular or sparse data collection. AI models that contain bias due to non-representative training data may produce erroneous predictions and exacerbate already-existing inequities (Goankar et al., 2020). To address these problems, strong legal frameworks, inclusive AI development methods, and community involvement are necessary to make sure AI solutions suit regional requirements and ideals.

SOLUTIONS AND RECOMMENDATIONS

A concerted multi-stakeholder approach is required to optimize AI's potential in underprivileged communities. To invest in digital infrastructure, improve AI literacy, and create culturally relevant AI applications that meet local needs, governments, businesses, and communities must work together. Adoption of AI technologies can be accelerated while maintaining inclusivity by offering easily accessible training courses and online resources. To encourage the long-term adoption of AI-driven solutions in important industries like healthcare, education, and agriculture, public-private partnerships ought to be reinforced. These partnerships can spur AI advancements that tackle urgent issues in underprivileged areas by utilizing the financial and technical resources of both industries. Supporting the creation and application of open-source AI models can promote community-driven solutions and further democratize access to AI.

FUTURE RESEARCH DIRECTIONS

In order to ensure responsible AI deployment, future research should concentrate on developing affordable AI technologies tailored for resource-constrained environments. Transparent policies and ethical guidelines should be integrated and AI can be used as a transformative tool for social and economic development in underserved communities. Addressing ethical issues like algorithmic bias, data privacy, and equitable AI governance should be a top priority. Future research should explore interdisciplinary collaboration models, emphasizing partnerships between governments, technology developers, healthcare providers, and local communities to co-create AI-driven solutions that are ethical, effective, and sustainable.

CONCLUSION

In isolated and underprivileged areas, artificial intelligence (AI) holds the potential to transform healthcare, education, and agriculture, resolving long-standing inequalities and enhancing quality of life. However, overcoming technical, ethical, and infrastructure obstacles is necessary for its successful implementation. AI can be a potent instrument for social justice and sustainable development if it is developed inclusively, digital infrastructure is invested in, and ethical governance is maintained. Cooperative initiatives that put human-centric and culturally appropriate solutions first will be key to the future of AI in underprivileged communities, ultimately bridging the gap between technology and social good. Furthermore, ensuring AI solutions are designed with accessibility, affordability, and ethical considerations at their core is crucial for long-term success. Public-private partnerships, capacity-building initiatives, and strong regulatory frameworks must be established to create an ecosystem where AI can thrive without reinforcing existing disparities. Addressing challenges such as bias in AI models, data privacy concerns, and the digital divide will be essential to maximize AI's benefits while minimizing risks. Continuous research, stakeholder collaboration, and adaptive policy frameworks will be necessary to scale AI-driven solutions sustainably. By prioritizing inclusivity and ethical responsibility, AI can become a transformative force in underserved regions, empowering communities, improving service delivery, and fostering economic and social resilience. If leveraged effectively, AI can bridge healthcare and technology gaps, creating a more equitable and sustainable future for all.

REFERENCES

Abuagla, M. Y. (2023). Recent trends in the use of artificial intelligence techniques and their impact on investment, development and growth for developed and developing countries under corona pandemic. *The International Scientific Forum*, 136-162. DOI: 10.36772/rg11931.3

Agrawal, J., & Arafat, M. Y. (2024). Transforming farming: A review of AI-powered UAV technologies in precision agriculture. *Drones (Basel)*, *8*(11), 664. DOI: 10.3390/drones8110664

Al-Jenaibi, B., & Bulhoon, A. (2024). Investigating adoption determinants, obstacles, and interventions for AI implementation in Emirati media organizations. *South Eastern European Journal of Public Health*, 1901–1909. DOI: 10.70135/seejph.vi.2254

Alakitan, M., & Makinde, E. (2024). Where are the ethical guidelines? Examining the governance of digital technologies and AI in Nigeria. *Policy and Internet*, *17*(1), e416. Advance online publication. DOI: 10.1002/poi3.416

Ali, H. (2025). AI for pandemic preparedness and infectious disease surveillance: Predicting outbreaks, modeling transmission, and optimizing public health interventions. *International Journal of Research Publication and Reviews*, *6*(2), 4605–4619. DOI: 10.55248/gengpi.6.0225.0941

Aljawawdeh, H., Droubi, A., Mashaqbeh, D., & Alazzeh, H. (2024). Transforming healthcare in Jordan: Enhancing the patient journey with AI innovations using AI to enhance patient experiences and hospital efficiency. *2024 Global Digital Health Knowledge Exchange & Empowerment Conference (gDigiHealth.KEE)*, 1-8. DOI: 10.1109/gDigiHealth.KEE62309.2024.10761273

Arya, K. (2025). Machine learning for smart agriculture: Use of ML in precision farming, crop yield prediction, and resource optimization. DOI: 10.2139/ssrn.5025948

Asiedu, E., Henkel, O., & Horne-robinson, H. (2024). *AI, implementation and academic outcomes in poor resource settings: Experimental evidence from Ghana*. AEA Randomized Controlled Trials., DOI: 10.1257/rct.14476

Bieser, J. (2022). Creative through AI – How artificial intelligence can support the development of new ideas. DOI: 10.59986/CCHA2271

Brand, L. (2020). Why machines that talk still do not think, and why they might nevertheless be able to solve moral problems. *Artificial Intelligence*, 203–217. DOI: 10.30965/9783957437488_014

Braun, T., & Möller, R. (2024). AI in healthcare and the public sector: How to face the challenges of high-risk applications and what AI research can get out of it. *KI - Künstliche Intelligenz, 38*(3), 119-126. DOI: 10.1007/s13218-024-00853-w

Byun, S. (2022). The moral issues of AI bias in data ethics. *Value-Sharing between Humans and Robots*, 3-3. DOI: 10.13180/icres.2022.18-19.07.p01

Chakravarthy, S., & K S, D. S. (2024). Application of natural language processing for creating chatbots in healthcare. *International Journal of Research Publication and Reviews*, *5*(1), 1472–1481. DOI: 10.55248/gengpi.5.0124.0212

Dahlstrand-Rudin, A., Deserranno, E., & Guariso, A. (2023). *Remote care for remote areas: The impact of Telehealth in rural India.* AEA Randomized Controlled Trials., DOI: 10.1257/rct.10787

Dakanalis, A., Wiederhold, B. K., & Riva, G. (2023). Artificial intelligence: A game-changer for mental health care. DOI: 10.31234/osf.io/ubdxz

Darus, Mohamad Yusofet al. (2021). A case study on digital divide and access to information communication technologies (Icts) in pulau tuba, Langkawi, Malaysia. Turkish Journal of Computer and Mathematics Education (TURCOMAT), 12(5), 1710-1718. DOI: 10.17762/turcomat.v12i5.2166

Dempere, J. (2024). Deploying AI-driven natural language processing systems for public health crisis management. *Advancements in Health Sciences*, 131-153. DOI: 10.1007/978-3-031-57773-4_10

Deng, B. (2024). The application of AI video generation technology in virtual reality (VR) and augmented reality (AR). *2024 International Conference on Artificial Intelligence, Deep Learning and Neural Networks (AIDLNN)*, 168-173. DOI: 10.1109/AIDLNN65358.2024.00035

Dooraki, S. K. (2024). Early detection of celiac disease through its common symptoms using machine learning algorithms. *Journal of Clinical Images and Medical Case Reports*, *5*(3). Advance online publication. DOI: 10.52768/2766-7820/2915

Drumm, L., & Sami, A. (2024). Academic staff AI literacy development through LLM prompt training. *Effective Practices in AI Literacy Education: Case Studies and Reflections*, 41-49. DOI: 10.1108/978-1-83608-852-320241005

Egunjobi, D., & Adeyeye, O. J. (2024). Revolutionizing learning: The impact of augmented reality (AR) and artificial intelligence (AI on education. *International Journal of Research Publication and Reviews*, *5*(10), 1157–1170. DOI: 10.55248/gengpi.5.1024.2734

Farinu, U. (2025). Fairness, accountability, and transparency in AI: Ethical challenges in data-driven decision-making. DOI: 10.2139/ssrn.5128174

Gaurav, A., Gupta, B. B., & Castiglione, A. (2024). Ethical AI development and deployment. *Advances in Computational Intelligence and Robotics*, 115–144. DOI: 10.4018/979-8-3693-3860-5.ch004

Goankar, B., Cook, K., & Macyszyn, L. (2020). Ethical issues arising due to bias in training A.I. Algorithms in healthcare and data sharing as a potential solution. *AI Ethics Journal, 1*(2). Advance online publication. DOI: 10.47289/AIEJ20200916

Goel, P., & Dixit, A. (2023). Introduction to on-device AI and its applications in mobile phones and personal assistants. *AI in Wireless for Beyond 5G Networks*, 19-26. DOI: 10.1201/9781003303527-3

Goel, R., Mishra, S., & Srivastava, S. (2024). Applying machine learning algorithms for the scientific prediction of crop yields. SSRN *Electronic Journal*. DOI: 10.2139/ssrn.4925156

Hemmler, Y. M., & Ifenthaler, D. (2022). Indicators of the learning context for supporting personalized and adaptive learning environments. *2022 International Conference on Advanced Learning Technologies (ICALT)*, 61-65. DOI: 10.1109/ICALT55010.2022.00026

Hswen, Y., & Schweitzer, K. (2025). Researcher explores using AI to overcome language barriers with patients. *Journal of the American Medical Association, 333*(1), 9. DOI: 10.1001/jama.2024.23526 PMID: 39641948

Hussein, S., & Safina, S. S. Q. (2025). Analyzing how AI can both exacerbate and help overcome digital inequalities in education, especially in underserved regions. *The Critical Review of Social Sciences Studies, 3*(2), 143–157. DOI: 10.59075/4964v051

Jain, D. (2024). AI-driven climate solutions: Leveraging LEO satellites for sustainability. DOI: 10.58445/rars.1814

Joshi, S., Sharma, M., Das, R. P., Rosak-Szyrocka, J., Żywiołek, J., Muduli, K., & Prasad, M. (2022). Modeling conceptual framework for implementing barriers of AI in public healthcare for improving operational excellence: Experiences from developing countries. *Sustainability (Basel), 14*(18), 11698. DOI: 10.3390/su141811698

Kambala, M. (2024). AI-powered healthcare: Transforming patient outcomes with machine learning. *Journal of Medical Science and clinical Research, 12*(08), 34-47. DOI: 10.18535/jmscr/v12i08.07

Kerketta, A., & Balasundaram, S. (2024). “Leveraging AI tools to bridge the health-care gap in rural areas in India”. DOI: 10.1101/2024.07.30.24311228

Kerrigan, C. (2022). Explainable AI and responsible AI. *Artificial Intelligence*, 527–533. DOI: 10.4337/9781800371729.00043

Kosaraju, D. (2024). Predictive analytics in healthcare: Leveraging AI to anticipate disease outbreaks and enhance patient outcomes. *Galore International Journal of Health Sciences and Research*, *8*(3), 73–79. DOI: 10.52403/gijhsr.20230312

Krishna Pasupuleti, M. (2024). AI-powered learning: Transformative innovations in smart education technology. *Advanced AI-Driven EdTech: Transformative Innovations in Smart Learning*, 292-310. DOI: 10.62311/nesx/802874

Kuska, M., Wahabzada, M., & Paulus, S. (2024). Ai-chatbots for agriculture - Where can large language models provide substantial value? DOI: 10.2139/ssrn.4685971

Lamuth, J. (2010). *Ethical artificial intelligence*. SciVee., DOI: 10.4016/22623.01

Malik, R., & Sharma, A. (2023). Empowering customer experience with AI tools and technologies. *Building AI Driven Marketing Capabilities*, 33-45. DOI: 10.1007/978-1-4842-9810-7_3

Malviya, R., & Goyal, P. (2023). Obstacles and complexity in AI embedded remote health monitoring. *Remote Patient Monitoring: A Computational Perspective in Healthcare*, 47-64. DOI: 10.1201/9781032627991-4

Mariyono, D., Hidayatullah, A. N., & Alifatul Kamila, A. N. (2024). Navigating ethical dilemmas in AI-enhanced language education: Addressing bias and ensuring inclusivity. DOI: 10.2139/ssrn.4936530

Mark, B., & Bizz, P. (2025). Ethical implications of AI-driven decision making in healthcare information systems. DOI: 10.20944/preprints202503.0078.v1

Mathai, A. (2024). Enhancing education for underprivileged children through AI-powered native language learning inclusive education through AI-powered native language learning. DOI: 10.2139/ssrn.4899553

Menona Jacob, A. (2022). *Healthcare delivery systems in rural areas*. Rural Health., DOI: 10.5772/intechopen.98184

Mhlanga, D. (2024). Artificial intelligence (AI) financial inclusion and wealth creation in the fourth Industrial Revolution. *Financial Inclusion and Sustainable Development in Sub-Saharan Africa*, 80-95. DOI: 10.4324/9781003515715-6

Mishra, A. (2024). Scalable AI algorithms and models. *Scalable AI and Design Patterns*, 45-60. DOI: 10.1007/979-8-8688-0158-7_4

Mishra, D. (2024). Role of social media and AI in transforming healthcare communication. *Advances in Electronic Government, Digital Divide, and Regional Development*, 29–51. DOI: 10.4018/979-8-3693-5976-1.ch002

Nag, P. K., Bhagat, A., & Priya, R. V. (2024). Expanding AI's role in healthcare applications: A systematic review of emotional and cognitive analysis techniques. DOI: 10.36227/techrxiv.172297472.22422829/v1

Naudts, L. (2017). Fair or unfair algorithmic differentiation? Luck egalitarianism as a lens for evaluating algorithmic decision-making. SSRN Electronic Journal. DOI: 10.2139/ssrn.3043707

Ndjama, J. D. (2025). Bridging the digital divide in the access and usage of technology through digital literacy in rural vocational schools. Institutes of Higher Education (IHE) and Workforce Collaboration for Digital Literacy, 91-124. DOI: 10.4018/979-8-3373-0004-7.ch004

Nienhaus, V. (2021). Digital transformation of the world economy by AI: Some moral, ethical and Sharīʿah concerns., *16*(16), 245-277. DOI: 10.33001/M011020211591

Nour, S. M., Abdelwahab, M. A., Mahmoud, M. S., & Nabil, K. E. (2025). Innovation in medical application based on artificial intelligence (AI) by cutting-edge technologies. *Practice, Progress, and Proficiency in Sustainability*, 57–68. DOI: 10.4018/979-8-3693-7117-6.ch003

Ntjamba, F. C., & Ashipala, D. O. (2024). Impact on and ethical considerations of artificial intelligence on human healthcare. *Advances in Computational Intelligence and Robotics*, •••, 1–36. DOI: 10.4018/979-8-3693-9158-7.ch001

Nuka, T. F. (2024). AI-driven financial inclusion: The role of DEI in shaping equitable credit opportunities for underserved communities. *International Journal of Research Publication and Reviews*, *5*(11), 3824–3840. DOI: 10.55248/gengpi.5.1124.3306

Nuka, T. F. (2024). AI-driven financial inclusion: The role of DEI in shaping equitable credit opportunities for underserved communities. *International Journal of Research Publication and Reviews*, *5*(11), 3824–3840. DOI: 10.55248/gengpi.5.1124.3306

Oh, J., & Ki, E. (2024). Can we build a relationship through artificial intelligence (AI)? Understanding the impact of AI on organization-public relationships. *Public Relations Review*, *50*(4), 102469. DOI: 10.1016/j.pubrev.2024.102469

Ojedokun, S. A. (2024). Artificial intelligence (AI) and machine learning (ML); A revolutionary game changer in clinical laboratory diagnosis. *Clinical Pathology & Research Journal*, *8*(1), 1–4. DOI: 10.23880/cprj-16000194

Pagani, M., & Champion, R. (2023). Could artificial intelligence make us humans more creative? *Artificial Intelligence for Business Creativity*, 24-43. DOI: 10.4324/9781003287582-4

Palla, S. R. (2025). AI-powered supply chain innovation: From predictive analytics to warehouse automation. *INTERNATIONAL JOURNAL OF INFORMATION TECHNOLOGY AND MANAGEMENT INFORMATION SYSTEMS*, *16*(1), 888–901. DOI: 10.34218/IJITMIS_16_01_063

Pandey, M. (2024). Access to the healthcare in remote areas is support for poverty reduction. *International Journal of Social Science and Human Research*, *07*(02). Advance online publication. DOI: 10.47191/ijsshr/v7-i02-65

Qwiumin, s. (2023). Enhancing public service delivery with AI-driven solutions in developing countries. https://doi.org/DOI: 10.31219/osf.io/cwqz4

Ragab, K. (2025). Empowering STEM education with AI and prompt engineering. *Advances in Computational Intelligence and Robotics*, 175–188. DOI: 10.4018/979-8-3693-7332-3.ch011

Ratcliffe, J. (2023). Can AI help market your brand? *Journal of Aesthetic Nursing*, *12*(2), 98–100. DOI: 10.12968/joan.2023.12.2.98

Reethika, A., & Priya, P. K. (2024). Using AI-powered predictive analytics tools to identify students falling behind or dropping out. *Innovation in the University 4.0 System based on Smart Technologies*, 101-115. DOI: 10.1201/9781003425809-7

Ruiz-Vanoye, J., & Díaz-Parra, O. (2025). *Machine and deep learning solutions for achieving the sustainable development goals*. IGI Global. DOI: 10.4018/979-8-3693-8161-8

Selwyn, E. J., Tiwari, A., Pandurang, G. G., Das, S. S., & Vardhan, R. (2025). Healthsyncc: An AI-powered mobile application for health management. *Modern Technologies in Healthcare*, 241-266. DOI: 10.1201/9781003481959-11

Sharma, G. K., Tyagi, E., & Chaudhary, A. (2025). Ethical considerations in the use of AI and big data in corporate decision-making. *Digital Citizenship and the Future of AI Engagement, Ethics, and Privacy*, 467-532. DOI: 10.4018/979-8-3693-9015-3.ch017

Shibuya, K. (2020). Correction to: Digital transformation of identity in the age of artificial intelligence. *Digital Transformation of Identity in the Age of Artificial Intelligence*, C1-C1. DOI: 10.1007/978-981-15-2248-2_15

Sidoo, F. (2023). Ensuring ethical data collection and use in AI-driven development programs. DOI: 10.31219/osf.io/5mhcw

Singh, P., & Singh, P. (2024). AI strategies for delivering better value in the healthcare sector. *Advances in Medical Technologies and Clinical Practice*, 131–156. DOI: 10.4018/979-8-3693-3731-8.ch007

Sinha, S. (2024). Revolutionizing human health: The role of artificial intelligence in modern healthcare. DOI: 10.31224/4181

Smt. Sumela Chatterjee. (2023). Evaluating the effects of AI-powered training programs on skill development and career growth. *International Journal of Advanced Research in Science, Communication and Technology*, 941-945. DOI: 10.48175/IJARSCT-12743D

Sol, K., & Heng, K. (2024). AI-powered chatbots as personalized academic writing assistants for non-native English speakers. *Encyclopedia of Educational Innovation*, 1-5. DOI: 10.1007/978-981-13-2262-4_313-1

Stamoulis, D., & Giannopoulos, P. (2021). Pesonalising weather forecasts using AI techniques. DOI: 10.5194/ems2021-502

Suh, C. (2022). *3. Machine learning applications.* Convex Optimization for Machine Learning.

Sukhdeve, P. S. (2021). Implementing augmented reality into immersive virtual learning environments. *Advances in Educational Technologies and Instructional Design*, 102–118. DOI: 10.4018/978-1-7998-4222-4.ch006

Suresh, N. V., Selvakumar, A., Sridhar, G., & Catherine, S. (2025). Ethical considerations in AI implementation for patient data security and privacy. *Advances in Healthcare Information Systems and Administration*, 139–147. DOI: 10.4018/979-8-3693-7452-8.ch008

Trust, T., & Maloy, R. W. (2024). Generative AI writing technologies. *How to Use Digital Learning with Confidence and Creativity*, 74-84. DOI: 10.4337/9781035311293.00020

Vesna, L. (2025). Digital divide in AI-powered education: Challenges and solutions for equitable learning. *Journal of Information Systems Engineering & Management*, *10*(21s), 300–308. DOI: 10.52783/jisem.v10i21s.3327

Voorneveld, M. (2024). AI technology adoption & Sustainability improvement though cloud solutions. *Proceedings of the 26th International Conference on Enterprise Information Systems*, 675-687. DOI: 10.5220/0012612500003690

Wang, W. (2024). *Ethical issues in personalized medicine: Privacy, consent, and data sharing*. GMO Biosafety Research., DOI: 10.5376/gmo.2024.15.0002

Wang, X. (2025). Industry innovation in the era of artificial intelligence. DOI: 10.1201/9781003510604

Yadav, V. (2023). AI for predicting disease outbreaks: Developing AI systems capable of predicting disease outbreaks by analyzing global health data and trends. *Journal of Artificial Intelligence. Machine Learning and Data Science*, *1*(1), 946–953. DOI: 10.51219/JAIMLD/vivek-yadav/226

Ye, X., Zhai, M., & Feng, L. (2023). Improving access to high-quality teachers in underdeveloped areas: Evidence from free teacher education in China. DOI: 10.2139/ssrn.4503575

Yoon, J. (2023). Integrating AI tools and linguistic content in college education. *Convergence English Language & Literature Association*, *8*(3), 349–370. DOI: 10.55986/cell.2023.8.3.349

ADDITIONAL READING

Bartneck, C., Lütge, C., Wagner, A., & Welsh, S. (2020). *An introduction to ethics in robotics and AI*. Springer Nature.

Boylan, M., & Teays, W. (2022). *Ethics in the AI, technology, and Information Age*. Rowman & Littlefield. DOI: 10.5040/9798881836429

Dubber, M. D., Pasquale, F., & Das, S. (2020). *The Oxford handbook of ethics of AI*. DOI: 10.1093/oxfordhb/9780190067397.001.0001

Elmassarawy, A. (2025). *The transformative power of AI applications in healthcare*. Ayman Elmassarawy.

Hayes, L. D. (2024). *Talking machines: The rise of chatbots and virtual assistants*. eBookIt.com.

Hilty, D. M., Ahuja, S., Naslund, J. A., & Crawford, A. (2022). Approaches to virtual care in underserved communities and settings: Bridging the behavioral health-care gap. *Virtual Mental Health Care for Rural and Underserved Settings*, 101-129. DOI: 10.1007/978-3-031-11984-2_7

Maddela, S. (2025). Healthcare information technology: Bridging the healthcare access gap in underserved communities. *International Journal of Science and Research Archive, 14*(1), 323–329. DOI: 10.30574/ijsra.2025.14.1.0051

Majebi, N. L., & Hamza, O. (2022). Bridging the autism diagnosis gap through digital inclusion in underserved communities. *International Journal of Multidisciplinary Research and Growth Evaluation, 3*(1), 761–770. DOI: 10.54660/.IJMRGE.2022.3.1.761-770

Mehta, V. (2023). Artificial intelligence in medicine: Revolutionizing healthcare for improved patient outcomes. *Journal of Medical Research and Innovation, 7*(2), e000292. DOI: 10.32892/jmri.292

Morgan, S. (2024). LibriHouse.

Paliszkiewicz, J., & Gołuchowski, J. (2024). *Trust and artificial intelligence: Development and application of AI technology*. Taylor & Francis. DOI: 10.4324/9781032627236

Paliszkiewicz, J., & Gołuchowski, J. (2024). *undefined*. Taylor & Francis.

Pillay, R. (2024). *undefined*. Xlibris Corporation.

Spair, R. (n.d.). *The AI revolution: How artificial intelligence will reshape our lives, careers, and future*. Rick Spair.

Stahl, B. C. (2021). *Artificial intelligence for a better future: An ecosystem perspective on the ethics of AI and emerging digital technologies*. Springer Nature. DOI: 10.1007/978-3-030-69978-9

Tzimas, T. (2021). *Legal and ethical challenges of artificial intelligence from an international law perspective*. Springer Nature. DOI: 10.1007/978-3-030-78585-7

KEY TERMS AND DEFINITIONS

Algorithmic Bias: When AI decision-making processes contain systematic errors, it can produce unfair results, especially for underrepresented groups (Naudts, 2017).

Artificial Intelligence: (AI): The simulation of human intelligence in machines that are designed to think, learn, and solve problems (Brand, 2020).

Ethical AI: The process of creating and deploying AI systems that follow accountability, transparency, and fairness principles in order to avoid biases and unintended harm (Farinu, 2025).

Machine Learning: (ML): A branch of artificial intelligence that allows systems to learn from data and enhance their functionality without the need for explicit programming (Suh, 2022).

Sustainability: The ability of AI-driven solutions to continue to be useful, accessible, and effective over time in underprivileged communities (Jain, 2024).

Underserved Communities: Are groups of people who, frequently as a result of financial, infrastructural, or geographic constraints, do not have sufficient access to basic services like healthcare, education, and employment opportunities (Nuka, 2024).

Chapter 12
Synthaesthetics and the Future of Creative Expression

Justin Baird
http://orcid.org/0009-0006-1522-4738
Synthaesthetic.art, Singapore

ABSTRACT

Artificial Intelligence (AI) is transforming the creative landscape, challenging traditional notions of authorship, originality, and artistic collaboration. This chapter explores the multi-faceted impact of AI on art and creativity, beginning with a critical discussion on whether creativity is an exclusively human domain. It explores how AI is enabling new forms of artistic expression and how artists are adapting to collaborate with intelligent algorithms. The author also discusses the ethical considerations surrounding AI-generated art, including questions of authorship, fairness, and the economic impact on artists. Additionally, the chapter addresses intellectual property challenges – how laws around copyright and ownership are evolving (or struggling to evolve) in response to AI. Finally, it considers future directions, including global trends, the emergence of the Synthaesthetic art movement, and strategies for fostering a creative future where AI enhances rather than replaces human ingenuity.

INTRODUCTION

For most of history, creativity was seen as an exclusively human trait. Early computing pioneers like Ada Lovelace in 1843 argued that machines could not truly originate anything new. This stance, often called the "Lovelace Objection," held that

DOI: 10.4018/979-8-3373-1509-6.ch012

genuine creativity required human consciousness and experience. However, recent advances in artificial intelligence are challenging this human-centric definition of creativity. AI systems today can compose music, paint, and write, producing outputs that appear both novel and imaginative. If an AI produces an original, valuable piece of art, can we call the AI creative? Some researchers suggest that we may need to broaden our definition of creativity to include non-human agents (Wingström et al., 2022). Traditionally, creativity has been defined by elements like the creator, process, product, domain, and audience. AI introduces a non-human creator (the algorithm) and data-driven processes, forcing us to reconsider whether creativity is a property only of human minds.

AI'S CHALLENGE TO HUMAN-EXCLUSIVE CREATIVITY

Philosophical Challenges of AI Creativity

Sceptics argue that AI-generated works lack the intentionality and emotional depth of human art. AI can mimic style but does not have lived experience or feelings; it cannot truly understand its creations. For example, an AI might paint in Van Gogh's style, but it does not feel the sunflowers or the night sky. From this perspective, AI is an impressive imitator, not an originator. Without consciousness or self-awareness, an algorithm's output might be novel but not "creative" in the human sense. Human creativity, it is argued, is rooted in the intention and personal context behind art—qualities AI currently lacks.

On the other hand, human creativity itself often recombines existing ideas in ways not unlike what AI does. Psychologists and cognitive scientists distinguish between creativity that simply combines familiar ideas, creativity that explores a conceptual space deeply, and creativity that truly transforms that space. Much of human art and invention falls into the first two categories, while truly transformational creativity is rare. By this view, AI's ability to combine and explore within established styles is not categorically different from human creators building on prior art. If an algorithm produces a painting or story that audiences find meaningful, perhaps creativity can be seen as residing partly in the output and process, not only in the conscious intent of the maker.

A productive way to frame AI in art, then, is as a tool or collaborator rather than an independent artist. Many artists who use AI emphasise that it augments rather than replaces their creativity. AI can generate numerous ideas or variations that a human might not have imagined, effectively serving as a tireless assistant. For instance, an artist can input a concept into a generative model and receive dozens of varied outputs, some of which spark new inspiration. One digital artist noted that an unexpected AI output—or even an AI "mistake"—can become "a source of a new idea" for the artist (Silk, 2022), highlighting how a serendipitous error can inspire creativity. In this collaborative view, the human provides the vision, intuition, and critical judgment, while the AI provides suggestions or raw material. The artist remains the director of the process, curating and refining the AI's contributions and their own into a final work.

Figure 1. Interactive Augmented Violin, Anna Savery

(photograph supplied by author)

Early evidence suggests that treating AI as a creative partner can enhance creative productivity and expand the range of ideas explored during the artistic process. Recent research has found that generative AI can significantly increase creative productivity while supporting the exploration of novel creative directions when used collaboratively with human creators (Zhou & Lee, 2024). While relying heavily on the same algorithms can lead to stylistic convergence, skilled artists can harness these systems to push into new creative territory. In other words, AI can expand the exploratory phase of art-making, generating "happy accidents" and novel combinations that humans can then cultivate into finished pieces.

Principles of Synthaesthetic Art

This influence of technological shifts is creating a common aesthetic vision and artistic philosophy leading to a new direction in art, which we define as the Synthaesthetic Art Movement that explores machine-produced creativity and AI-driven artistic expression. Defining this term further, it is rooted in the fusion of synthesis (the artificial, computational, or generative process) and aesthetic (the principles of artistic beauty and expression).

The core principles of Synthaesthetic Art include:

- **Machine-Augmented Creativity** – Rather than replacing human artists, AI and generative algorithms serve as collaborators, expanding the boundaries of artistic expression.
- **Hybrid Intelligence** – Synthaesthetic art values the interplay between human intuition and machine computation, embracing co-creation over pure automation.
- **Temporal and Spatial** Fluidity – Digital tools allow Synthaesthetic works to be nonlinear, ever-evolving, and adaptable, often existing in multiple states or dimensions.
- **Sensory Fusion** – Inspired by synesthesia, the movement explores multisensory interactions, blending visual, auditory, and even tactile elements in novel ways.
- **Algorithmic Aesthetics** – The movement challenges traditional definitions of beauty by exploring patterns, randomness, and emergent complexity through AI-generated forms.
- **Decentralized Authorship** – Works within this movement often question traditional ideas of authorship and originality, embracing open-source and communal creative processes.

TECHNICAL FOUNDATIONS AND AI SYSTEMS

Synthaesthetic artists use a wide range of technological mediums to achieve their creative outputs. Their artistic expressions include AI-generated visuals, such as those produced by generative adversarial networks (GANs) – systems where two neural networks compete against each other, with one generating images and the other evaluating their quality until realistic outputs are achieved – and diffusion models, which create images by gradually removing noise from random data through a learned reverse process. They also create music and sound art driven by algorithmic improvisation, interactive installations that blend human and machine interaction, alongside generative poetry and literature crafted using large language models (LLMs) – AI systems trained on vast amounts of text to understand and generate human-like language. Additionally, these artists explore virtual and augmented reality to construct immersive, synesthetic environments that push the boundaries of perception and creativity.

Synthaesthetic art shares its roots with Surrealism, exploring the subconscious through AI, as well as Generative Art, which relies on rule-based creation, and Futurism, with its embrace of technology. However, unlike these movements, Synthaesthetic Art is not solely about technological advancement but also about its aesthetic and philosophical implications—how machines create beauty and meaning. This movement challenges the role of artistic creation in the age of artificial intelligence, raising profound questions: Can a machine be truly creative? How do we define beauty in an era shaped by generative models? What happens when art is no longer fixed but continuously evolving? By tackling these ideas, Synthaesthetic Art redefines the boundaries between human and machine-driven creativity.

As an example, the field of prompt engineering —the process of crafting precise inputs to guide AI models—has found a natural extension in the creative domain, where artists, musicians, and designers leverage AI as a collaborator rather than a mere tool. This practice can be framed as "Synthaesthetic Prompting." Unlike traditional prompt engineering, which is often focused on efficiency and accuracy in text generation, Synthaesthetic Prompting is inherently exploratory, embracing ambiguity, iteration, and unexpected outcomes. It is the deliberate act of sculpting prompts to evoke specific moods, styles, or emotional responses, much like an artist mixing pigments or a composer layering harmonies. By treating AI as an extension of creative intuition, Synthaesthetic Prompting transforms the act of prompting into an art form in itself, where the process of guiding machine creativity becomes a meaningful contribution to the final output.

AI is redefining human creativity. It challenges the notion that creativity stems from a mystical human spark alone, showing that some mechanics of creative production—combining elements, exploring variations, and generating aesthetically pleasing forms—can be performed by algorithms. Yet human intuition, emotion, and intentionality remain crucial in guiding the process and producing art with meaning. The intersection of AI and artistic creation is thus a space of both exciting possibility and ongoing reflection, where we are renegotiating what it means to be creative in the age of intelligent machines. This redefinition of creativity sets the stage for exploring how Synthaesthetics is actively generating novel forms of art, which is the focus of the next section.

AI AS A CATALYST FOR NEW CREATIVE FORMS

Artificial intelligence has emerged as a transformative force in the arts, acting as a catalyst for new creative forms that push the boundaries of traditional media. In contemporary practice, artists increasingly integrate AI not just as a tool but as a creative collaborator, yielding artworks that were previously inconceivable. This trend spans across visual art, music, performance, and beyond, giving rise to hybrid art forms that fuse technology with human imagination.

Global Perspectives on AI-Driven Art

One area of impact is new media art and installations. AI-driven installations can be dynamic and responsive, changing in real time based on sensory input or audience interaction. This marks a departure from traditional static art pieces. For example, some interactive art projects use generative AI algorithms to create evolving visuals and soundscapes that respond to viewers in the gallery space (Varvara & Mar, 2024). Such an installation becomes a "living" artwork—its form is not fixed, but continuously generated by the algorithm in collaboration with its environment. Artists like Refik Anadol have harnessed AI to produce immersive digital experiences: his works analyse vast datasets (such as city images or museum collections) and translate them into swirling, dynamic projections, enveloping audiences in what he calls "machine hallucinations" (Anadol, 2021).

Beyond Western contexts, artists globally are exploring AI's creative potential through their cultural lenses. In Nigeria, artist Malik Afegbua's "The Elder Series" used AI to create a fashion show featuring African senior citizens as models, challenging both ageism and the underrepresentation of African aesthetics in digital art (Kitongo & Tutton, 2023). In Japan, AI has been used to generate haiku poetry that follows traditional 5-7-5 syllable patterns while exploring contemporary themes, bridging ancient literary forms with modern technology. Korean artists like Sougwen Chung have created installations that weave traditional materials such as red thread—a symbol in Korean and Chinese culture—with AI-generated elements, reflecting on human-AI interconnection. In Brazil, the collective Beep Studio has used AI to generate music that incorporates traditional Brazilian rhythms with electronic elements, creating new forms of digital samba and bossa nova. These pioneering projects demonstrate how AI enables novel aesthetics and interactive experiences that were not possible before, while respecting and incorporating diverse cultural traditions.

Figure 2. Dvorak Dreams Art Installation by Refik Anadol at the Kennedy Center

(Laporte, 2024)

AI is also catalysing innovation in performing arts and film. In music, composers use AI systems to generate melodies or even entire songs in the style of various genres. The Flow Machines project, for instance, famously produced a pop song mimicking The Beatles' sound (Ferreira, 2016). Rather than replace human songwriters, these tools serve as a creative springboard—musicians can iterate on AI-generated riffs or harmonies to develop new pieces. In dance and theatre, choreographers have begun experimenting with AI to generate movement sequences. British choreographer Wayne McGregor's "Living Archive" project used machine learning to suggest original dance movements by analysing his past works (Leprince-Ringuet, 2018). The AI offered novel choreographic ideas that McGregor could then incorporate and refine, effectively acting as a high-tech muse. In cinema, AI has been used to aid scriptwriting and video editing; notably, the sci-fi short film Sunspring was written entirely by an AI trained on screenplays (Gallagher, 2016). While the result was surreal, it opened debate about machine creativity in storytelling. Major film productions are now adopting AI tools for visual effects and post-production, showing how the technology can contribute to new cinematic techniques rather than overtaking the director's vision.

AI's collaborative role has given rise to novel art genres that blend human and machine creativity. A striking example is the emergence of AI-generated fine art that makes its way into galleries and auctions. In 2018, an AI-created portrait titled Edmond de Belamy was auctioned at Christie's for over $400,000, signalling that AI art had entered the mainstream art market (Caselles-Dupré et al., 2018). The portrait was produced by a generative adversarial network trained on thousands of classic paintings. Its sale provoked discussion: who is the artist—the algorithm's creators, the algorithm itself, or the collective that framed and presented the piece? Similarly, an earlier project in the Netherlands used AI to generate a brand-new painting in Rembrandt's style ("The Next Rembrandt"), demonstrating how algorithms might extend the legacy of past masters. Such milestones suggest a future in which AI techniques like neural style transfer and generative adversarial networks become standard tools in artists' palettes across the globe. However, experts also caution that the "human touch" will remain a prized element – audiences may still crave the authenticity of human-made work even as AI art becomes more prevalent (Mineo, 2023). The future art world may reach an equilibrium where AI expands creative possibilities, while human creativity and intentionality gain new appreciation as the defining elements of art.

The embrace of AI in artistic creation is not without critical reflection. Artists and audiences are actively debating the conceptual implications of AI-driven art. Some critics question whether art generated by an algorithm can convey meaning or authentic emotion, while others see these new forms as expanding what art can be. Many of the artists pioneering AI-based art are mindful of these questions: they often present their work as an exploration of the relationship between human and machine, rather than pretending the machine operates in isolation. In doing so, they invite audiences to consider how art created with AI fits into the continuum of creative expression.

In summary, AI is acting as a catalyst for innovation across a wide range of artistic mediums. It enables interactive and generative artworks in visual media, provides inspiration and novel techniques in music and performance, and blurs the boundaries between disciplines. These new creative forms, born from human-AI collaboration, are expanding the artistic landscape. They demonstrate that when artists leverage AI thoughtfully, technology can open exciting frontiers in art-making—while also sparking valuable dialogue about the evolving role of the artist and the essence of creativity. In the following section, we turn to how human creators are adapting to the ambiguity and unpredictability inherent in AI-assisted art.

HUMAN ADAPTATION TO MACHINE AMBIGUITY

As artists integrate artificial intelligence into their creative workflows, they often encounter a new kind of uncertainty: the ambiguity and unpredictability of machine-generated output. Unlike traditional tools that reliably execute an artist's intent, advanced AI systems (such as generative neural networks) can produce results that are surprising, strange, or even seemingly nonsensical. This unpredictability—sometimes described as AI "hallucinations" when the AI output deviates from reality—poses a challenge: How do humans adapt to collaborating with a creative process that they cannot fully control or foresee?

Many artists have responded by embracing ambiguity as part of the creative journey. Instead of treating AI's unexpected outputs as errors, they treat them as happy surprises or starting points for iteration. In practice, working with generative AI often means surrendering some control and engaging in a dialogue with the machine's output. An artist might prompt an AI image generator with a concept and receive an image that only partly matches their intention or perhaps introduces an odd element. Rather than discarding it, the artist can interpret the oddity in a creative way or refine the prompt and try again. This iterative loop of prompt and output becomes a collaborative conversation between human and algorithm. A 2022 study of new media artists found that many view AI's randomness as a source of inspiration rather than a flaw (Wingström et al., 2022). By adopting a mindset of exploration, artists expand their creative repertoire—using AI to venture into visual or musical ideas they might not have arrived at alone.

The process of curation and editing has become a crucial human skill in the age of AI art. Creators often generate dozens or hundreds of outputs with an AI system and then curate the results, picking those that resonate and refining them further by hand. In this way, the role of the artist shifts somewhat: from solely generating content to filtering and shaping content produced in tandem with a machine. This requires developing new sensibilities—knowing how to "read" an AI's output, spot promising variations, and guide the system toward desired outcomes. Some have likened this to a gardener tending a wild garden: the AI sprouts many possibilities, and the human selectively cultivates certain growths into a final artwork. The ambiguity of the machine's contributions means the artist must make interpretative decisions that imbue the work with meaning.

Human adaptation is also evident in how creators are learning new technical skills to better collaborate with AI. Beyond the technical craft of prompt engineering—the practice of designing inputs to guide AI effectively—a broader artistic movement has begun to emerge: Synthaesthetics. This term fuses "synthesis", representing AI's generative capabilities, with "aesthetic", the artistic principles shaping its output. Synthaesthetics is not just about crafting effective AI prompts but about exploring new forms of artistic expression through human-AI collaboration, where the process of guiding machine creativity becomes an art form in itself. At the core of this movement is Synthaesthetic Prompting, the practice of sculpting AI interactions to evoke specific moods, styles, and emotions. Artists who never coded before are now experimenting with phrasing, parameters, and iteration to refine their results, developing an intuitive understanding of AI's strengths and limitations—such as a visual model excelling at surreal compositions but struggling with anatomical precision. In response, art schools and creative workshops are incorporating these techniques into their curricula, reflecting a broader shift in artistic training where AI literacy and creative intuition intertwine, shaping the future of artistic innovation, Ho (2024).

Even audiences and consumers of art are adapting to this new creative paradigm. People encountering AI-generated art must adjust their interpretation frameworks. When we know a piece was produced with the assistance of an algorithm, we might view it through a different lens—perhaps focusing more on the concept or the human choices behind the algorithm, rather than on technical skill alone. Some art observers find that AI-generated works prompt them to ask new questions: for example, what part of this do I appreciate as the human artist's vision versus the algorithm's contribution? Exhibitions of Synthaesthetic art often include explanations of the human–machine collaboration to help audiences adapt their expectations and appreciate the work's context.

This adaptation is not always smooth or universally accepted. Some artists initially feel threatened by AI's unpredictability and the need to yield creative control. There can be a sense of loss of authorship, as if the machine is "intruding" on the creative process. However, many who experiment with these tools report that they eventually find the unpredictability liberating. One artist described the AI as a "playful partner" that brings a sense of discovery back into her studio practice. By letting go of total control, she felt more like an explorer, encountering unexpected imagery and then responding to it creatively. Similarly, in writing, some authors have used AI to generate text and then edited it, effectively acting as co-authors. A notable experiment by researcher Almira Osmanovic Thunström involved asking the GPT-3 language model to write an academic paper about itself; she then submitted the result for publication (Osmanovic Thunström, 2022). Exercises like this force human creators to adapt by taking on the role of editors, moderators, and critics of AI-produced content rather than sole originators.

Notably, the idea of introducing chance into art is not new. Surrealist painters and poets, for instance, practiced automatic techniques to let the unconscious and randomness influence their work, and composer John Cage famously incorporated chance operations into music. In a sense, AI provides a high-tech form of serendipity that artists can incorporate and guide, much like these earlier creative experiments with randomness. By adapting to machine ambiguity, human creators are also spurring changes in artistic norms and definitions.

Overall, human adaptation to AI's unpredictability involves a combination of technical learning, mindset shifts, and cultural changes in how we evaluate art. Artists who embrace AI's surprises can transform potential pitfalls into creative opportunities. By learning to navigate and even welcome the ambiguity introduced by machine partners, humans are extending their creative capabilities. They remain essential as sense-makers and storytellers who give AI-generated material a coherent vision and purpose. In effect, the emergence of AI in art is prompting artists to evolve – to become adept at collaboration, curation, and new forms of craftsmanship – ensuring that human creativity continues to drive the process, even as it now intertwines with the generative power of machines.

ETHICAL CONSIDERATIONS AND IMPLICATIONS

The integration of AI into artistic creation has ignited complex ethical debates. Because AI blurs the lines between human and machine contributions, it raises questions about authorship, ownership, fairness, and the future role of human creators. This section examines some of the key ethical considerations facing the art world in the age of AI.

Authorship and Ownership: One pressing issue is who should be credited as the creator of AI-generated art, and who (if anyone) owns the rights to such works. Traditional copyright law has long required a human author for a work to be protected, which means purely AI-created images or music would typically lack copyright protection. Ethically, this brings up conflicting views. Some argue that if an artist uses AI as a tool (providing the ideas, prompts, and curation), that artist should still be considered the author of the result. Others counter that if the AI's contribution is dominant, calling a human the "author" is a legal fiction. This debate has practical consequences: for example, in 2022 an AI-generated painting won first place in a state art competition, sparking controversy among human artists who felt it was unfair to compete with a largely machine-made piece (Kenney, 2023). In a landmark 2023 U.S. case, a court denied copyright for an image created solely by an AI, reaffirming that human creativity is a prerequisite for legal authorship (Thaler v. Perlmutter, 2023). The judge underscored that no U.S. court has ever recognised copyright in a work originating from a non-human entity. While laws are beginning to address these questions, the ethical crux is balancing innovation with respect for human creators.

Impact on Artists' Livelihoods: Another concern is the potential displacement of human artists and the appropriation of their work. Generative AI can produce art, illustrations, designs, music, and written content at a fraction of the time and cost that a human might require, raising the spectre of automation displacing creative jobs. Many artists worry that AI systems have been trained on their past works without permission, and now those systems can generate new images in the artists' styles – effectively competing with them using their own creative "DNA." A vivid example is digital illustrator Greg Rutkowski, whose distinctive fantasy art style was mimicked tens of thousands of times by users prompting AI image generators with his name. Rutkowski found that online searches for his name turned up mostly AI-made images imitating his work, rather than his original art. From an ethical standpoint, this feels like a violation: the AI was built on his labour, yet he receives no credit or compensation as it floods the internet with copycats. Indeed, one open letter in 2023 – signed by over 3,500 artists and supporters – denounced AI art generators for "mass theft" of human artists' works ("Cancel the Christie's AI Art Auction," 2025). In late 2022, thousands of artists protested on ArtStation (a popular portfolio site) by posting "No AI Art" images to demand that their work not be used to train AI models (Edwards, 2022). Similarly, in 2023 a group of visual artists filed a class-action lawsuit against AI firms Stability AI, Midjourney, and DeviantArt, accusing them of infringing copyrights on millions of artworks used in training (Brittain, 2023). The broader fear is that if left unchecked, AI could undercut the economic livelihood of artists – companies might opt for cheap AI-generated visuals instead of hiring illustrators, for instance. There have already been incidents fuelling these anxieties: Netflix Japan's 2023 release of an anime short that used AI-generated backgrounds (justified by a supposed "labour shortage" of animators) sparked outrage among artists, who pointed out that the issue was not a true lack of human animators but rather the studio's unwillingness to pay for human talent (Chen, 2023).

On the other hand, some argue that AI will not so much replace artists as change the nature of their work. Optimists point out that new technologies in art (photography, digital tools, etc.) have historically created new opportunities and even new art forms, rather than eliminating creativity. AI might automate certain tedious tasks (like colouring, cleanup, or generating base illustrations), allowing artists to focus on higher-level creative decisions. Entirely new roles – such as AI art directors, Synthaesthetic prompt designers, or AI ethics consultants in creative teams – are emerging. The World Economic Forum has even coined the term "FOBO" (fear of being obsolete) to describe workers' anxiety about AI, noting that while certain tasks will be automated, AI is also likely to augment existing jobs and create new roles in the creative economy (World Economic Forum, 2024). For example, while demand may decline for routine graphic design work that AI can handle, new opportunities are arising for artists who can craft effective Synthaesthetic prompts, curate AI outputs, or blend AI with hand-made techniques. From this perspective, AI could be a creative multiplier—increasing productivity and enabling solo artists or small studios to accomplish what used to require larger teams. Ethically, the challenge is ensuring that this transition is fair: that artists are not left behind or forced into exploitative situations. Proposals to address these concerns include implementing compensation mechanisms for artists whose works help train AI (such as royalties or licensing fees), and establishing norms for transparency and consent (e.g. clearly labelling AI-generated art and obtaining permission to use artists' works in training datasets). The outcome will depend on choices by industry and policymakers; as AI tools become widespread, there is a collective responsibility to support human artists in adapting to these changes rather than simply replacing them.

Bias and Representation: AI-generated art can also reflect and amplify societal biases present in training data, raising concerns about fairness and representation. If a generative model is trained on a dataset that skews toward certain cultures or demographics, its outputs may be biased. Studies have found, for example, that when asked to draw a picture of "a person," some AI image models produced predominantly light-skinned male figures, underrepresenting women and people of colour (Milne, 2023). Similarly, these models have been observed reinforcing stereotypes – for instance, depicting nurses as female or CEOs as male – mirroring the imbalances in their training data. In an artistic context, such biases might mean AI art unintentionally marginalises certain groups or presents a narrow worldview. Ethically, creators using AI need to be aware of these issues. There is a push for more inclusive training datasets and for technical methods to mitigate bias, so that AI art can reflect diverse perspectives. International organisations like UNESCO have highlighted the importance of addressing bias in creative AI (Bailleul, 2023). Some AI art platforms now implement diversity filters or allow users to specify demographic details in prompts (for example, asking for a variety of ethnic backgrounds in generated characters) to counteract default biases. While technical fixes can help, they rely on human values: it takes conscious effort by developers and artists to ensure the content AI produces is respectful and representative of different communities. This is an ongoing ethical commitment to equity in the creative sector.

Human Creativity: Augmentation vs. Automation: A deeper philosophical debate underpins many of these issues: Is AI in art enhancing human creativity or threatening to automate it? Optimists see AI as a powerful tool – a kind of creative amplifier that can expand an artist's reach and repertoire. They argue that, much like the camera didn't destroy painting (but changed it), AI will push artists toward new modes of expression while taking over some lower-level tasks. For example, AI can quickly generate concept art or musical sketches that a human artist can then elaborate, potentially accelerating the creative process. Pessimists, however, worry about a future where human creativity is devalued. They fear an abundance of AI-generated content could saturate culture with derivative, soulless work – noise that drowns out human voices. Ethically, society must decide what role we want AI to play in the arts. Many advocate for a model where AI augments human creativity but humans remain at the centre of artistic decision-making. In this view, AI is a collaborator or Synthaesthetic instrument, and the art arises from how humans use it. Maintaining this balance means nurturing human creativity (through education, practice, and cultural support) even as we embrace new AI-driven possibilities.

Public Reaction and Cultural Shifts: The controversies and conversations around AI art indicate that we are in a period of cultural adaptation. There have been high-profile incidents – from an AI artwork winning a prize, to a robot artist like Ai-Da addressing the U.K. Parliament – that grab headlines and provoke public debate. Not all reactions are adversarial: some artists have found creative ways to critique AI through art itself (for example, generating deliberately absurd AI images to point out flaws, or creating artworks that mix AI and manual techniques in commentary). Such meta-critiques are part of the evolving dialogue. Meanwhile, creative professionals have been actively pushing back to assert the value of human creativity and control. In 2023, for instance, the Writers' Guild and Screen Actors' Guild in Hollywood went on strike partly over AI issues, leading to new contract provisions that limit the use of AI without consent. These disputes, while often heated, are driving increased public awareness about how AI works and what it means for the future of creativity. In that sense, they have an educational value: society at large is now grappling with questions that were once niche hypotheticals. The ethical choices we make now – about ownership, attribution, fairness, and the role of the human creator – will help determine the trajectory of art in the years to come. Will AI be remembered as just another tool that artists mastered and assimilated, or as a disruptive force that forever changed artistic production and our definition of art? The answer depends on how we navigate these dilemmas today – whether we can integrate AI in ways that uphold the value of human creativity, ensure equitable treatment of artists, and broaden rather than narrow the diversity of artistic expression.

INTELLECTUAL PROPERTY AND COPYRIGHT CHALLENGES

AI's ability to create art, text, and music has posed unprecedented challenges for intellectual property (IP) law. Around the world, legal systems are grappling with how to fit AI-generated content into copyright frameworks that were built for human creators. Can an AI be an author? If not, can the human who uses the AI claim authorship of AI-produced material? And is using existing artworks to train AI models a form of infringement or a permissible practice?

Authorship of AI-Created Works: In most jurisdictions, the answer so far is that only humans can be authors of copyrightable works. Copyright law traditionally requires a human creative spark. For instance, the European Union (through court decisions) emphasises that a work needs a human's creative choices to merit protection; if content is produced with minimal or no human input, it likely isn't a protected "work." Similarly, Japan's Copyright Act defines a "work" as an expression of human thought or emotion, inherently referring to a human creator. In the United Kingdom however, the law has taken a distinct approach by explicitly addressing "computer-generated" works. Under the UK Copyright, Designs and Patents Act 1988, if a literary, artistic, musical, or dramatic work "is generated by computer in circumstances such that there is no human author," then the person who undertook the arrangements necessary for creation of the work is deemed the author (UK Copyright, Designs and Patents Act 1988). In practice, this assigns authorship (and thus copyright) to the human who set up or programmed the AI to produce the work. The UK grants a shorter duration of protection (50 years from creation, instead of the usual life of the author plus 70 years) for these computer-generated works, reflecting a compromise on how strongly such works should be protected. Some Commonwealth jurisdictions influenced by UK law, such as India and Hong Kong, have analogous provisions or interpretations that attribute authorship to the person orchestrating the AI. However, even these solutions are somewhat internally inconsistent: if originality in copyright requires human intellectual creation (as it does in most countries), assigning a human author by legal fiction doesn't change the fact that the actual creative process was autonomous. In practice, even under UK law, there is debate about what level of human involvement is needed for a work to be considered "computer-assisted" (and thus protectable) rather than truly "computer-generated."

So far, no country has given legal authorship status to an AI itself. When AI acts with substantial human guidance, any resulting copyright is typically attributed to the human contributors; but if an AI ever achieved completely autonomous creativity with no human input, current laws suggest such works would fall outside traditional copyright protection. The U.S. has been unequivocal on this point: the U.S. Copyright Office's guidance and a 2023 federal court ruling (Thaler v. Perlmutter) affirmed that works with no human creator cannot be copyrighted (even acknowledging the grey areas when there is some human alteration or selection). Another example came with a 2022 comic book Zarya of the Dawn, which used AI-generated imagery. The U.S. Copyright Office initially granted it a registration, then later determined that copyright protection applied only to the human-authored portions of the work and not to the AI-generated images while maintaining it for the human-written text and overall story compilation. This illustrates how hybrid AI-human works are being treated: the human contributions remain protected, but the purely AI parts do not. Practically speaking, if an artwork is generated mostly or entirely by AI without a significant human contribution, it may fall into the public domain by default. Some creators see this as a problem – for instance, an artist who spends effort curating AI outputs might want copyright protection for the final product. Without clear protection, they have less legal control or incentive to commercialise it. Others view a public domain outcome as beneficial, preventing companies from monopolising

AI-generated content. Legal scholars and policymakers are actively debating these issues, and while minor reforms have been proposed (like shorter-duration rights or clarifications in law for AI-assisted works), there is not yet a consensus on granting AI any form of authorship or copyright.

Use of Existing Works in AI Training: Another major copyright issue is the use of copyrighted material to train AI models. Generative AIs learn by analysing large datasets of human-created images, text, music, and so on—often scraped from the internet, which means they include countless copyrighted works. This raises the question: is using someone's artwork or writing as fodder to train an AI a violation of the creator's rights? Artists and content owners have begun to challenge this practice. In early 2023, a group of visual artists sued the makers of AI art tools Stable Diffusion and Midjourney, alleging that the training of these models on billions of online images (including the artists' own work) was unlawful (Brittain, 2023). The legal arguments here are novel: AI developers argue that the training process is transformative and non-expressive, somewhat analogous to a human learning by observing art, and that the AI does not store or output exact copies of training images (aside from rare overfitting cases). Opponents argue that the scale of this unpermitted use, and the direct commercial competition from the outputs, make it far from fair use.

Different jurisdictions are approaching this issue in varying ways. The European Union's 2019 copyright directive introduced text-and-data mining exceptions that allow AI training on copyrighted content for certain purposes: one provision lets researchers mine data from copyrighted works for non-commercial research by default, and another allows others (including commercial entities) to mine data unless the rights-holder has opted out (e.g., via a notice on a website). Japan went even further; since 2018, Japanese law broadly permits the use of copyrighted works for information analysis (which covers AI training) regardless of purpose, without requiring permission, if the output does not simply replicate protected expression (Japanese Copyright Act, 2018). This wide exception legally cleared the way for AI developers in Japan to train models on virtually any data. In contrast, the UK initially considered a sweeping exception to allow text and data mining of any copyrighted material for AI development but backed off in 2023 after pushback from authors and publishers. In the United States, which relies on the flexible fair use doctrine, the question of whether AI training is fair use is now in the courts. The outcomes of the artists' class action will likely set important precedents. If courts rule that using data for AI training is fair use, AI companies will have more freedom to continue current practices; if the ruling goes the other way, we may see requirements for licenses or new opt-out mechanisms at scale.

Emerging Policies and Debates: Policymakers and international bodies are starting to respond to these challenges, though no country has fully overhauled its IP laws for AI yet. Instead, we see cautious, piecemeal approaches. Copyright offices and ministries in various countries have held consultations and issued guidance. For example, the U.S. Copyright Office in 2023 released guidelines instructing that when registering a work, any AI-generated portions must be disclosed and will be excluded from the claim unless the AI contributions are modified by a human author. This effectively operationalises the principle that you can only claim the parts of a work that you (a human) actually created. The World Intellectual Property Organization (WIPO) has convened discussions among member states on AI and IP since 2019, revealing a range of opinions but a general reluctance to grant AI legal personhood or authorship rights. Most policymakers seem inclined to adjust current laws with clarifications (e.g. requiring transparency and attribution of AI involvement, or defining the threshold of human contribution for a work to be protected) rather than introduce entirely new intellectual property categories for AI-generated works at this time.

One idea floated in academic and industry circles is a new category of rights for AI-generated content – a sui generis (unique) IP right that isn't full copyright but would give some limited protection to those who generate or invest in AI-produced works. The rationale is to incentivise the creation of valuable AI outputs by granting a short-term exclusive right, without stretching traditional copyright to non-human creators. However, this idea has not gained much traction yet, as lawmakers are cautious about unintended consequences (for example, flooding the market with lightly "monopolised" AI content or undermining human creators further). As of 2025, no country has implemented such a scheme.

Another angle is ensuring proper attribution and transparency. While copyright covers economic rights, moral rights (like attribution) are also in play. Some jurisdictions place strong emphasis on the author's right to be credited. Major content platforms are considering or implementing policies to label AI-generated media. This could help human creators distinguish their work in the marketplace and inform consumers about what they are viewing. It might also become standard to credit the human who operated the AI ("Artwork by Jane Doe using DALL-E") or even the model itself as a tool, akin to how one might credit software.

In summary, the legal landscape for AI-generated art and media is still taking shape. The clear trend is to keep humans at the centre of IP law – AI is treated as a tool, not an author. Those humans who conceive of, guide, or significantly shape the output can often claim authorship, provided their contribution meets the originality threshold in that jurisdiction. When an AI acts with more autonomy, the resulting work may be unprotected, which raises further economic and ethical questions as discussed. Additionally, the massive use of existing creative works to fuel AI brings to the forefront a clash between innovation and the rights of original creators. Lawmakers are trying to balance these interests by updating regulations (like data mining exceptions) and relying on courts to apply old doctrines (like fair use) to new situations. The coming years will likely see more legal precedents and possibly legislative action to clarify how creators can safely innovate with AI without undermining the creative community that feeds these algorithms. It's a delicate balance that jurists worldwide are now beginning to navigate.

THE FUTURE OF AI IN ARTISTIC CREATION

Looking ahead, AI is poised to redefine artistic boundaries in profound ways. Experts worldwide predict that AI will become an increasingly ubiquitous collaborator in the arts, rather than a wholesale replacement for human creativity. In virtually every creative field – from visual arts and literature to music, film, and performance – AI tools are expected to become part of the standard toolkit. This evolution carries both exciting opportunities and important responsibilities for the artistic community.

AI as a Creative Partner: In the coming years, AI will likely be deeply integrated into creative workflows, often handling routine or technical tasks while humans focus on high-level artistic direction. In film and television, for example, AI systems are beginning to streamline editing and visual effects. Industry analysts predict that by the mid-2020s, lower-level production tasks in media will increasingly be automated, freeing human creators to concentrate on storytelling and design (AlixPartners, 2025). Notably, the Oscar-winning film Everything Everywhere All at Once leveraged AI-based tools (such as generative image software from Runway) to create its imaginative multiverse visuals on a tight budget, demonstrating how human vision and AI assistance can combine to produce acclaimed art. Rather than replacing the film's artists, these tools served as supporting technology – much as earlier generations used computer-generated imagery (CGI) or digital editing software. This aligns with a stance emerging in creative communities that "AI should serve as a tool to support, not replace, human talent," a principle underscored by recent agreements in the entertainment industry. Similarly, in music, AI-driven software is expected to be embedded in music production globally, assisting musicians with tasks from generating melodies and beats to mastering tracks. Songwriters might use AI to explore novel combinations of chords or instantly orchestrate a demo in different styles, speeding up the creative cycle. Rather than rendering composers obsolete, AI could serve as a creative partner: the human sets the vision and emotional tone, while the AI offers a plethora of riffs and variations to choose from. In live performance and dance, choreographers have begun using AI to generate new movement ideas. For example, Wayne McGregor's experiment with AI "predicting" choreography from his past work hints at a future where dancers and theatre directors collaborate with algorithms to devise sequences that a human alone might not have imagined. Even in literature, AI's role is expanding – large language models have co-written short stories and poems. In Japan, an AI-coauthored short novel made it past the first round of a national literary competition, hinting at how storytelling might evolve with algorithms as collaborators. These developments illustrate a common prediction: AI will increasingly handle generative and iterative aspects of artistic labour, while humans provide overarching vision, emotional context, and critical curation. As one humanoid "artist" – the robot Ai-Da – remarked to the U.K. House of Lords, technology in art is both a threat and an opportunity; it will continue to grow as artists find new ways to use it to express themselves (Lawson-Tancred, 2022). The global consensus is that AI's creative potential will be harnessed to augment human artistry, opening exciting new frontiers, so long as we navigate the journey thoughtfully.

Not all prognostications are rosy or uniform, however. Some sceptics argue that the current AI art boom could plateau or fizzle out like past tech hype cycles if audiences tire of algorithmic aesthetics once the novelty wears off. Others fear a dystopian scenario where AI becomes so dominant in content creation that human artists struggle to compete. While these extreme outcomes are possible, they are generally viewed as unlikely extremes. A more moderate path – with AI deeply integrated into creative industries but coexisting with human creators – is considered the most plausible future (Zaber, 2024). Early evidence supports this: when AI-generated art and music are presented to the public, people still often assign greater value to human-made work, suggesting an inherent bias favouring the human touch (Mineo, 2023). This bias may diminish as society becomes more accustomed to AI art, or it may persist as a reminder that art is ultimately a human conversation. Either way, the coming years will test our definitions of creativity and force a reckoning with how we value art made with the assistance of intelligent machines. The trajectory will likely vary globally – some regions and industries may embrace AI faster than others – but overall AI is on track to become an indispensable creative partner. As YouTube's CEO Neal Mohan noted in 2023, "AI should not be a replacement for human creativity. It should be a tool used to enhance all our creativity."

Democratising Creativity and Broadening Access: One of the most frequently cited advantages of AI in the arts is its potential to democratise creativity and broaden access to artistic expression on a global scale. Traditionally, the ability to create high-quality art or media has been limited by technical skill, training, or access to certain resources and networks. AI tools are lowering many of these barriers. Text-to-image generators, music composition AIs, and language models allow people with minimal formal art education – or those who lack specific physical abilities – to produce compelling artworks by simply describing their ideas. This means individuals who might have been excluded from artistic creation in the past can now participate. A recent analysis on Latin American art noted that AI can help overcome linguistic, cultural, and geographic barriers that often prevent artists from accessing the means to create and distribute their work (LatAm Arte, 2023). AI-driven translation and style transfer tools allow creators to draw on influences from many cultures, blending motifs in ways that transcend their own linguistic or regional limitations. The result could be a richer, more cosmopolitan global arts scene (Elliott, 2024).

Crucially, AI is also empowering artists with disabilities and those facing physical or technical constraints. Adaptive AI interfaces (such as voice-controlled painting apps or gesture-based music tools) can help creators with limited mobility or vision execute their ideas in ways not possible with traditional methods. As one ethics report observed, AI can "make art creation more inclusive… letting more people express themselves artistically" (Clancy & Tweed, 2022). This democratisation through technology ensures that creative talent is not squandered due to disability or disadvantage. In the coming years, we can expect improved AI accessibility features – from intelligent prosthetics for musical performance to AI tutors that guide beginners – further levelling the playing field. The net effect could be an art community that is more inclusive and globally representative than ever before. Granting access to artists who were previously marginalised or silenced will mean a more diverse and vibrant tapestry of creative expression. Moreover, diversity in creators can feed back into the AI systems themselves: a wider range of users and inputs can help mitigate biases in algorithms, making the tools more culturally aware and effective for everyone.

AI is also changing the notion of artistic authorship in ways that might empower collectives over individuals. By its nature, AI art is often a product of collaboration – not only between a human artist and a machine, but among the many humans whose data or contributions inform the AI. This challenges the age-old idea of the solitary genius artist and points toward a model of distributed creativity. The process of training an AI on thousands of images, songs, or texts is essentially a form of crowdsourced input. AI art projects often bring together software developers, data curators, and artists in interdisciplinary teams to produce a final work. The outcome can be seen as a joint creation – a convergence of human and machine contributions. If managed ethically, such AI-assisted collective creation could amplify the voices of underrepresented groups by explicitly incorporating their aesthetic perspectives into algorithms. For example, Stephanie Dinkins' "Not the Only One N'TOO" project transformed multigenerational oral histories into a conversational deep learning 3D avatar that enables the user to explore the stories, myths and perspectives of Black American culture. In N'TOO, storytelling, art, technology, and social engagement combine to create a new artificially intelligent narrative form. (Art Papers, 2023)

Going forward, the global art community faces the challenge of maximising AI's empowering potential while minimising its downsides. On balance, the trajectory so far suggests that many more people will enter creative expression through AI. The key will be ensuring fair access to these technologies (through open-source models, public programs, or affordable tools) and protecting artists' rights so that democratisation doesn't come at the expense of those who have devoted their lives to art. If these issues are addressed, AI could indeed usher in a new era of creative democratisation: one where a teenager in a developing country, a differently abled creator, or an indigenous collective can all equally harness algorithmic tools to share their culture and imagination with the world. In such a future, the arts would truly become a global dialogue, enriched by diverse participation and made possible by intelligent technology.

Ensuring Human Oversight and Ethical Use: As AI becomes more entwined with creative practices, human oversight and strong ethical guidelines will be essential to steer this integration responsibly. The future of AI in art raises several ethical considerations that must be addressed through proactive policies and industry best practices. Key among these are issues of intellectual property, authenticity and attribution, bias and cultural sensitivity, and the broader impact on creative labour. Envisioning the next phase of AI-assisted art also means devising frameworks to ensure technology serves human creativity – and not the other way around – echoing the principle that technology should serve the needs of people.

Intellectual property and authorship stand out as urgent concerns. Current AI models are often trained on vast datasets of human-created art, text, or music, much of it scraped from the internet without explicit permission. This has triggered backlash from artists who discover their works were used to teach algorithms that now generate "new" art in their style. Legal systems worldwide are grappling with questions of who owns AI-generated content and whether using copyrighted material to train AI is fair use or infringement. For instance, in the music industry, songwriters and composers' guilds have argued that feeding music into AI without consent is tantamount to theft of their creative work. In 2023, the Writers' Guild of America went on strike in part to demand protections against such unlicensed AI usage, leading to contract provisions that explicitly forbid studios from using AI to undermine writers' compensation and credit. This set an important precedent: any AI-generated material provided to human creators (a screenplay draft, for example) cannot be used to deny those humans fair payment or acknowledgement. Similar debates are occurring in visual arts – as noted, several lawsuits have been filed by artists against AI image-generator firms for allegedly infringing copyrights by training on millions of online images. Moving forward, a likely scenario is the development of new copyright frameworks or licenses specific to AI. These might require AI developers to obtain opt-in permission or pay royalties when training on an artist's works, thus ensuring artists are remunerated when their creations fuel an algorithm.

Another pillar of oversight is addressing bias and cultural sensitivity. AI systems carry the values and biases of their training data; if that data lacks diversity, the AI's outputs can inadvertently reinforce stereotypes or exclude certain perspectives. This has already been observed in creative AI: for example, early versions of popular image generators struggled to accurately portray people from underrepresented ethnic groups or would default to Western-centric imagery. To counter this, there is a concerted effort to diversify training datasets and involve creators from different backgrounds in AI development. International collaborations and programs emphasise incorporating non-Western perspectives so that the future of AI-art isn't dominated by a single cultural narrative (Elliott, 2024). The goal is an AI that respects and reflects global cultural diversity, rather than acting as a homogenising force. Ethically, this ties into the concept of cultural IP and avoiding appropriation: AI designers should be careful that their models don't unknowingly plagiarise indigenous designs or sacred symbols, which would be disrespectful and harmful. Therefore, a key recommendation is ongoing human oversight of AI outputs – a "human in the loop" review process where culturally sensitive content is evaluated by knowledgeable humans. In the long run, embedding ethical constraints into the AI (such as filters to recognise and avoid biased or culturally inappropriate content) will also help. As one group of global experts summed up, we must strive to build AI that "accurately represents all global cultures" and maintains inclusion and diversity as core design tenets (Bailleul, 2023).

Another aspect of oversight is maintaining the integrity and emotional authenticity of art. If AI-generated works proliferate, audiences and creators alike will wrestle with questions of authenticity: Does a symphony composed by an AI evoke the same depth of feeling as one composed by a person who poured their life experiences into it? Many artists argue that genuine creativity requires consciousness, emotion, and lived experience – qualities AI doesn't possess. For example, Meredith Stiehm, president of the Writers Guild of America, noted during a writers' strike that "writing, like any art form, is based on a lived human experience… AI can't bring that" into the creative process (Haskins, 2024). From this view, human oversight is needed to instil artworks with contextual meaning and emotional resonance that an algorithm alone might miss. In practical terms, this could mean an editor shaping an AI-written story to ensure it conveys real human truths, or a painter adding hand-painted flourishes to an AI-generated canvas to imbue it with personal expression.

International and cross-industry collaboration on ethical guidelines is ramping up. UNESCO, for instance, convened a high-level global discussion in 2023 on the impact of AI in culture and media. Stakeholders from different countries concurred on principles such as informed consent, transparency, fair remuneration for artists, and inclusive development of AI (Bailleul, 2023). They called for better governance of AI in creative sectors, recognising that self-regulation alone may not suffice in protecting artists and audiences. A notable quote from that forum was: "We are not anti-technology. We are pro-people. And technology should serve the needs of people, not the other way around." This ethos – putting human welfare at the centre – should guide policies on AI in art. Concretely, we may see new laws that require AI training data to be obtained ethically, or that grant creators a right to opt out of AI datasets. We might also see industry standards emerging; for example, coalitions of artists and AI developers agreeing on certain best practices (such as watermarking AI art or sharing a portion of AI-generated art sale proceeds with the original source artists). Museums and galleries are beginning to develop curatorial standards for AI art, weighing how to credit algorithms and programmers alongside human artists. Tech companies, for their part, have a responsibility to implement ethical guard-rails – such as content filters to avoid obscene or prejudiced outputs, and features that allow artists to remove their works from training sets. Some experts advocate establishing audit trails for AI art: keeping logs of what source material influenced a given AI creation, which could be useful for both attribution and accountability.

Human oversight in the creative process itself will remain vital. AI can produce a plethora of options, but humans must decide which options are meaningful, appropriate, or resonant. In the future, artists might frequently take on the role of Synthaesthetic directors, guiding the machine's output in line with an artistic vision. This symbiosis will work best if artists are literate in how AI operates – pointing to a need for education and training so that creators worldwide can confidently and critically use these tools. It also suggests that the arts community should be involved in AI development from the ground up. By having artists, including those from diverse cultures, at the table when new creative AIs are designed, we can bake ethical and aesthetic values into the technology early on.

Ensuring a positive future for AI in artistic creation means establishing a robust ethical framework now. Safeguards around IP will protect artists' livelihoods; guide-lines on diversity and content will prevent cultural biases; transparency measures will maintain trust with audiences; and human involvement at every stage will keep art connected to lived experience. This is a collective global effort – no single country or sector can address all these issues in isolation. If stakeholders uphold the principle that AI is a tool for artists rather than a replacement for artists, the creative industries can navigate the ethical challenges and use AI in a way that re-spects human creativity. The balance we strike will determine whether the story of AI and art is one of empowerment and innovation, or one of unchecked disruption.

Pioneering AI-Driven Art Projects Shaping Tomorrow: A glimpse of the future can often be seen in the experimental projects of today. Around the world, artists and technologists are engaging in groundbreaking AI-driven art projects that hint at how artistic creation might evolve. These case studies – spanning different disciplines and cultures – illustrate the cutting edge of AI creativity and provide inspiration (as well as cautionary lessons) for the road ahead.

One striking example comes from Nigeria, where artist Malik Afegbua garnered global attention with "The Elder Series," an AI-generated fashion show featuring African senior citizens as models on the runway. (Kitongo & Tutton, 2023) Afegbua used AI tools to create vividly realistic images of elegant older adults in high-fashion attire. The series immediately went viral for its positive portrayal of aging in a context (runway fashion) that usually celebrates youth. The project's impact was twofold: it demonstrated AI's power to visualise concepts that have been underrepresented in mainstream art, and it sparked conversation about inclusion (both of the elderly and of African aesthetics) in the digital art world. Afegbua's work, supported by Africa's first AI art incubator (Gallery of Code in Lagos), shows how artists in the Global South are not just consuming AI tech but actively shaping it to tell their own stories. By blending traditional African textile patterns with contemporary digital techniques, The Elder Series points toward a future where AI art can challenge stereotypes and broaden societal narratives. It also illustrates how local context matters: the piece was celebrated for reflecting an African perspective in AI art, something historically scarce. As more community-rooted AI artworks proliferate, we can expect a more pluralistic AI art scene – one that doesn't emanate solely from Western labs, but from diverse societies each using AI in line with their local values and creative needs.

In East Asia, we see the rise of AI in literary and visual arts. In Japan, the experiment of submitting an AI-written short novel to a literary contest (where it passed initial screening) showed that algorithms can produce narrative structures that pass muster, if not win prizes. In South Korea, artists like Sougwen Chung (who has Korean-Chinese heritage, working in the West) are exploring installations that weave traditional materials (like red thread, a symbol in Korean and Chinese culture) with AI-generated elements, reflecting on human-AI interconnection. Seoul's art institutions (e.g. Art Center Nabi) have been nurturing tech-based art for years, and new galleries are now focusing specifically on AI art. The Korean AI art community's experience, including dealing with biases in tools like Midjourney (which initially struggled with Asian facial features), has underscored the importance of cultural nuance in AI – leading to improved AI models that better grasp local aesthetics. This regional case demonstrates how feedback from artists to AI developers can rapidly shape the evolution of creative tools. By voicing the gaps (such as the misrepresentation of non-Western cultures), artists help drive updates that make AI more inclusive.

Creative technologist teams in France and the U.S. have used AI to resurrect artistic styles or even artists from the past. The Next Rembrandt project in the

Netherlands (a collaboration involving Microsoft, Delft University, and museums) analysed Rembrandt's body of work to algorithmically generate a new "Rembrandt painting" in 2016, which was then 3D-printed in layers of paint to mimic the Dutch master's brush strokes. This hyper-real synthesis of past and present technology was a glimpse of a future where AI might routinely assist in restoration, conservation, or even extension of artistic legacies. In France, the art collective Obvious used a generative adversarial network to create a series of "Portraits of the Belamy Family," one of which (Edmond de Belamy, mentioned earlier) became the first AI art piece sold at a major auction. That sale in 2018 was a wake-up moment for the art market, raising questions about how to value AI-created work and who the true "artist" was (the programmers? the algorithm? the 18th-century portrait data it trained on?). The Belamy project was deliberately framed as a commentary on AI and art – the team even signed the canvas with part of the algorithm's formula – highlighting that AI can itself be the subject of art, not just a tool.

Figure 3. "Edmond de Belamy, from La Famille de Belamy" generative adversarial network print, on canvas, 2018, signed with GAN model loss function in ink by the publisher

(Obvious Art, 2018)

In the U.S., tech companies like Google have run artist residencies (e.g. the Magenta program) to experiment with AI in music and visual art. One outcome has been AI-curated music albums and the use of AI by mainstream musicians to co-produce songs or create AI "duets." The singer Taryn Southern, for instance, produced an album (I AM AI, 2018) using AI composition software for the instrumentals, positioning herself as one of the first pop artists to explicitly work with AI as a creative partner.

What all these projects indicate is a future artistic landscape brimming with hybrid creations: part human, part machine. They also show that different regions are contributing unique flavours to that future. In Africa and Latin America, AI art is intertwined with themes of social change and cultural identity. In Asia, AI art often engages with tradition – whether by attempting classical poetry or by integrating ancient symbols into digital art – thus ensuring continuity even in innovation. In Europe and North America, much AI art probes philosophical questions about creativity and authorship and often has the backing of galleries, auction houses, and tech giants, which helps bring AI art into public awareness quickly. The convergence of these threads globally is shaping a multifaceted future – one where AI is used not just to entertain or astonish, but also to preserve languages (there are projects using AI to revitalise endangered languages through music and VR), to educate, to critique society, and to collaborate across vast distances.

A likely scenario in the next decade is that what is experimental today will become commonplace. The avant-garde AI choreography seen in McGregor's work may lead to dance companies regularly employing AI to generate initial movement ideas. The AI painting techniques pioneered by tech-art teams could become standard in digital art software used by millions. AI-written content might populate our games, interactive stories, and even journalism in more refined ways. There will also be entirely new art forms born from AI: imagine personalised virtual reality experiences that evolve via AI based on each viewer's reactions, or artworks derived from other creative inputs. Pioneering artists are already toying with these possibilities. For example, Tesseract.art produces acrylic and watercolour paintings from the real-time analysis of a live music performance, Savery & Baird (2024). AI art is not confined to a single medium or style – it spans from fine art to performance to functional design.

Figure 4. Black Sheep performance, musicians claiming authorship

(photographs supplied by author)

These case studies also highlight challenges that persist into the future. Questions of credit and ownership surface: Tesseract.art postulates that the musicians who interacted with the robotic instrument own the copyright of the resulting artwork. In the Belamy portrait sale, the AI couldn't sign its own name (hence the algorithm equation as signature), a witty reminder of AI's legal invisibility. As AI art becomes more routine, the novelty will wear off, and works will likely be judged more on merit and meaning. At that stage, human artists who wield AI creatively will probably stand out more than AI working in isolation. The human element remains the heart of art, even in these forward-looking projects. A curator of a recent AI art exhibit noted that despite the excitement over algorithmic art, people "do seem to want to see evidence of the human hand" in creative works – a reminder that the dialogue between tradition and technology will continue.

FUTURE DIRECTIONS

The future of AI in artistic creation is being charted in real time by innovators around the globe. Their experiments serve as both inspiration and guideposts, showing us what is possible when human creativity and artificial intelligence intersect. From Nigerian runways to Asian literature, from robot painters to algorithmic dancers, each development expands our imagination of what art can be. They collectively suggest that tomorrow's art world will be more inclusive, boundary-pushing, and interdisciplinary. It will be a world where an artist might be part-coder, a machine might be part-artist, and audiences are participants in ever-more interactive, personalised artistic experiences. The challenge and opportunity lie in navigating this transformation consciously – celebrating the new possibilities while ensuring we carry forward the depth, diversity, and humanity that art has always embodied. With balanced global collaboration, ethical foresight, and creative daring, AI's integration into art could herald a vibrant renaissance, one that truly redefines artistic boundaries.

Revisiting the Lovelace Objection

As we stand at this crossroads of human and machine creativity, we return to Ada Lovelace's 1843 assertion that machines could not truly originate anything new. The Lovelace Objection, which held that genuine creativity required human consciousness and experience, has been fundamentally challenged by the emergence of Synthaesthetic Art. While AI systems may not possess consciousness in the human sense, they have demonstrated the ability to generate novel combinations, explore aesthetic spaces, and produce works that audiences find meaningful and emotionally resonant.

The evidence presented throughout this chapter suggests that creativity may not be the exclusive domain of human consciousness, but rather a more distributed phenomenon that can emerge from the interaction between human intention and machine capability. Synthaesthetic Art represents a new paradigm where creativity is not diminished by machine participation but rather expanded and democratised. The collaborative model that has emerged—where humans provide vision, context, and curation while AI contributes generative power and novel combinations—points toward a future where the boundaries between human and machine creativity become increasingly fluid.

Perhaps most significantly, the global adoption of AI in artistic practice across diverse cultures demonstrates that this technological shift is not merely a Western phenomenon but a universal expansion of creative possibility. From Nigerian fashion shows to Japanese haiku generation, from Korean textile installations to Brazilian musical fusion, artists worldwide are finding ways to integrate AI that honour their cultural traditions while pushing creative boundaries.

The Lovelace Objection may have been correct in its time, but the emergence of Synthaesthetic Art suggests that creativity itself is evolving. Rather than replacing human creativity, AI is revealing new dimensions of what it means to create, collaborate, and express meaning through art. In this new landscape, the question is not whether machines can be creative, but how human and machine creativity can work together to produce art that neither could achieve alone. The future of artistic expression lies not in the supremacy of human or machine, but in the synthesis of both—a truly Synthaesthetic approach to creativity that honours the past while embracing the transformative potential of our technological present.

REFERENCES

Bailleul, C. (2023, November 8). UNESCO unites diverse perspectives to inform policies for AI in the creative sectors. UNESCO. https://www.unesco.org/en/articles/unesco-unites-diverse-perspectives-inform-policies-ai-creative-sectors

Brittain, B. (2023, February 6). Getty Images lawsuit says Stability AI misused photos to train AI. Reuters. https://www.reuters.com/legal/getty-images-lawsuit-says-stability-ai-misused-photos-train-ai-2023-02-06/

Cancel the Christie's AI Art Auction. (2025, February 8). Open Letter. https://openletter.earth/cancel-the-christies-ai-art-auction-f5135435

Caselles-Dupré, H.. (2018). *Portrait of Edmond de Belamy* [Auction lot notes]. Christie's.

Chen, M. (2023, February 6). Netflix Japan is drawing ire for using A.I. to generate the background art of its new anime short. Artnet News. https://news.artnet.com/art-world/netflix-japan-ai-anime-dog-and-boy-2251247

Clancy, M., & Tweed, R. (2022). The voice of the artist in the age of the algorithm. IEEE Global Initiative on Ethics of Autonomous and Intelligent Systems, 10–12. https://sagroups.ieee.org/global-initiative/wp-content/uploads/sites/542/2022/07/ead-artists.pdf

Dinkins, S. (2023). NTOO: Not the only one. Stephanie Dinkins. https://www.stephaniedinkins.com/ntoo.html

Edwards, B. (2022, December 14). ArtStation artists protest AI-generated art by adding 'No AI Art' images. Ars Technica. https://arstechnica.com/information-technology/2022/12/artstation-artists-stage-mass-protest-against-ai-generated-artwork/

Elliott, L. (2024, August 6). Non-western perspectives on AI: Art ecosystems in Africa, Americas, and Asia. Salzburg Global Seminar News. https://www.salzburgglobal.org/news/regions/article/non-western-perspectives-on-ai-art-ecosystems-in-africa-americas-and-asia

Ferreira, E. (2016, October 14). How Sony Flow Machines' AI takes itself for the Beatles. Makery. https://www.makery.info/en/2016/10/14/comment-lia-de-sony-flow-machines-se-prend-pour-les-beatles/

Gallagher, S. (2016, June 13). Sci-fi film 'Sunspring' is the world's first movie created entirely by artificial intelligence. HuffPost UK. https://www.huffingtonpost.co.uk/entry/artificial-intelligence-algorithm-writes-first-film-sunspring_uk_575e90eee4b014b4f2540cca

Haskins, C. (2024, September 5). Meredith Stiehm. Time. https://time.com/7012819/meredith-stiehm/

Japanese Copyright Act, art. 2(1)(i) (definition of "work"), and art. 30-4 (2018 amendment on information analysis).

Kenney, A. (2023, September 6). Jason Allen's AI Art Won the Colorado Fair — but Now the Feds Say It Can't Get a Copyright. Colorado Public Radio. https://www.cpr.org/2023/09/06/jason-allens-ai-art-won-colorado-fair-feds-deny-copyright-protection/

Kitongo, G., & Tutton, M. (2023, June 23). Nigerian AI artist creates a fashion show for elderly people. CNN. https://edition.cnn.com/style/article/malik-afegbua-elderly-fashion-ai-art-spc-intl/index.html

LatAm Arte. (2023, November 26). The Impact of AI on Creating Art. LatAm Arte, 87, 95. https://www.latamarte.com/en/articles/quE4/

Lawson-Tancred, J. (2022, October 12). The robot artist Ai-Da just addressed U.K. Parliament about the future of A.I. and 'terrified' the House of Lords. Artnet News. https://news.artnet.com/art-world/ai-da-robot-artist-parliament-2190611

Leprince-Ringuet, D. (2018, December 17). Google's latest experiment teaches AI to dance like a human. WIRED, 67, 75. https://www.wired.com/story/google-ai-wayne-mcgregor-dance-choreography/

Milne, S. (2023, November 29). AI image generator Stable Diffusion perpetuates racial and gendered stereotypes, study finds. UW News.

Mineo, L. (2023, August 15). If it wasn't created by a human artist, is it still art? Harvard Gazette, 238, 247. https://news.harvard.edu/gazette/story/2023/08/is-art-generated-by-artificial-intelligence-real-art/

Osmanovic Thunström, A. (2022, September 30). We asked GPT-3 to write an academic paper about itself. Scientific American. https://www.scientificamerican.com/article/we-asked-gpt-3-to-write-an-academic-paper-about-itself-mdash-then-we-tried-to-get-it-published/

Silk, M. S. W. (2022, November 30). Man v. Machine: The ethics of AI art. Prindle Institute. https://www.prindleinstitute.org/2022/11/man-v-machine-the-ethics-of-ai-art/

Thaler v. Perlmutter, No. 1:22-cv-01564 (D.D.C. Aug. 18, 2023).

The World's First AI-Generated Portrait Sold for $432,500. (2018, October 25). The New York Times. (Quoted in Clancy & Tweed, The Voice of the Artist, p. 15).

Wingström, R., (2022). Redefining creativity in the era of AI? Perspectives of computer scientists and new media artists. *Digital Creativity*, *33*(3), 183–201.

World Economic Forum. (2024, January). How is AI impacting and shaping the creative industries? (WEF Annual Meeting, Davos).

Zaber, M. A. (2024, October 10). The AI playbook: What other sectors can learn from the creative industry's fight against AI. RAND Corporation. https://www.rand.org/pubs/commentary/2024/10/the-ai-playbook-what-other-sectors-can-learn-from-the.html

Compilation of References

Abuagla, M. Y. (2023). Recent trends in the use of artificial intelligence techniques and their impact on investment, development and growth for developed and developing countries under corona pandemic. *The International Scientific Forum*, 136-162. DOI: 10.36772/rg11931.3

ACLU. (2020). Williams v. City of Detroit | Face Recognition False Arrest. American Civil Liberties Union. https://www.aclu.org/cases/williams-v-city-of-detroit-face-recognition-false-arrest

Adobe. (2023). Content Authenticity Initiative. https://contentauthenticity.org

African Development Bank (AfDB). (2021). *Taxation and technology in East Africa.*

African Union. (2024). Continental artificial intelligence strategy. https://au.int/en/documents/20240809/continental-artificial-intelligence-strategy

Agaba, E., & Aboagye, E. (2022). Blockchain for land administration in Africa: Lessons from Ghana. *African Journal of Science, Technology, Innovation and Development, 14*(3), 558–570.

Agrawal, J., & Arafat, M. Y. (2024). Transforming farming: A review of AI-powered UAV technologies in precision agriculture. *Drones (Basel), 8*(11), 664. DOI: 10.3390/drones8110664

Ahmadi, A. (2023). ChatGPT: Exploring the threats and opportunities of artificial intelligence in the age of chatbots. *Asian Journal of Computer Science and Technology, 12*(1), 25–30. DOI: 10.51983/ajcst-2023.12.1.3567

Ahuerma, F. M. (2023). Recommendation of the OECD Council on Artificial Intelligence: Inequality and Inclusion. *Research Gate*. https://www.researchgate.net/publication/374508986_Recommendation_of_the_OECD_council_on_artificial_intelligence_inequality_and_inclusion

Ahuja, A. S., Reddy, V. P., & Marques, O. (2020). Artificial Intelligence and COVID-19: A Multidisciplinary Approach. *Integrative Medicine Research, 9*(3), 100434. DOI: 10.1016/j.imr.2020.100434 PMID: 32632356

Aiken, M., & Lu, W. (1993). The role of accounting in early civilizations. *The Accounting Historians Journal, 20*(2), 163–186. DOI: 10.2308/0148-4184.20.2.163

Akbari, D. A., Jastacia, B., Setiawan, E., & Ningsih, D. W. (2022). The marketing power of TikTok: A content analysis in higher education. *Binus Business Review, 13*(2), 159–170. DOI: 10.21512/bbr.v13i2.8014

Alakitan, M., & Makinde, E. (2024). Where are the ethical guidelines? Examining the governance of digital technologies and AI in Nigeria. *Policy and Internet, 17*(1), e416. Advance online publication. DOI: 10.1002/poi3.416

Al-Awa, M. (1995). The role of Islamic accounting in commerce and civilization. *Arab Accounting Review, 15*(3), 45–67.

Alexiev, A. S., Janssen, M., & den Hertog, P. (2018). The moderating role of tangibility in synchronous innovation in services. *Journal of Product Innovation Management, 35*(5), 682–701. DOI: 10.1111/jpim.12459

Ali, H. (2025). AI for pandemic preparedness and infectious disease surveillance: Predicting outbreaks, modeling transmission, and optimizing public health interventions. *International Journal of Research Publication and Reviews, 6*(2), 4605–4619. DOI: 10.55248/gengpi.6.0225.0941

Ali, W., Khan, A. Z., Ahmad, F., & Mahmood, F. (2024). Critical artificial intelligence readiness factors in the context of public sector organizations: An expert opinion survey. *Journal of Business and Management Research, 3*(2), 14–29.

Aljawawdeh, H., Droubi, A., Mashaqbeh, D., & Alazzeh, H. (2024). Transforming healthcare in Jordan: Enhancing the patient journey with AI innovations using AI to enhance patient experiences and hospital efficiency. *2024 Global Digital Health Knowledge Exchange & Empowerment Conference (gDigiHealth.KEE)*, 1-8. DOI: 10.1109/gDigiHealth.KEE62309.2024.10761273

Al-Jenaibi, B., & Bulhoon, A. (2024). Investigating adoption determinants, obstacles, and interventions for AI implementation in Emirati media organizations. *South Eastern European Journal of Public Health*, 1901–1909. DOI: 10.70135/seejph.vi.2254

Amann, J., Blasimme, A., Vayena, E., Frey, D., & Madai, V. I. (2020). Explainability for artificial intelligence in healthcare: A multidisciplinary perspective. *BMC Medical Informatics and Decision Making, 20*(1), 310. DOI: 10.1186/s12911-020-01332-6 PMID: 33256715

Ammirato, S., Felicetti, A. M., Della Gala, M., Aramo-Immonen, H., Jussila, J. J., & Kärkkäinen, H. (2019). The use of social media for knowledge acquisition and dissemination in B2B companies: An empirical study of Finnish technology industries. *Knowledge Management Research and Practice*, *17*(1), 52–69. DOI: 10.1080/14778238.2018.1541779

Anderson, R. H., Bikson, T. K., Law, S. A., & Mitchell, B. M. (1995). Universal access to e-mail: Feasibility and societal implications. RAND Corporation. https://eric.ed.gov/?id=ED401884

Anderson, M., & Anderson, S. L. (2023). *Machine ethics: The design and governance of ethical AI systems*. MIT Press.

Angwin, J., Larson, J., Mattu, S., & Kirchner, L. (2016). Machine bias: There's software used across the country to predict future criminals. And it's biased against blacks. ProPublica. https://www.propublica.org/article/machine-bias-risk-assessments-in-criminal-sentencing

Arnold, M., Bellamy, R. K. E., Hind, M., Houde, S., Mehta, S., Mojsilović, A., & Varshney, K. R. (2019). FactSheets: Increasing trust in AI services through supplier's declarations of conformity. IBM Research. https://arxiv.org/abs/1808.07261

Arya, K. (2025). Machine learning for smart agriculture: Use of ML in precision farming, crop yield prediction, and resource optimization. DOI: 10.2139/ssrn.5025948

Asaad, C. T. (2020). Investor confidence: Are you your own worst enemy? In Financial Planning Review (Vol. 3, Issue 2). Wiley. https://doi.org/DOI: 10.1002/cfp2.1092

Asiedu, E., Henkel, O., & Horne-robinson, H. (2024). *AI, implementation and academic outcomes in poor resource settings: Experimental evidence from Ghana*. AEA Randomized Controlled Trials., DOI: 10.1257/rct.14476

Aytuna, E. N., Çakıcı, Z., & Özkaya, A. E. (2024). Entertainment and persuasion in online politics: A qualitativestudy of young voters' approach in Turkey's 2023 elections. In Al-Obaidi, J. (Ed.), *Changing globalmedia landscapes: Convergence, fragmentation, and polarization* (pp. 215–233). IGI Global., DOI: 10.4018/979-8-3693-3767-7.ch012

Bahls, C. (2024). In focus: AI and clinical trials—The beginning of a journey. *Applied Clinical Trials*, *33*(12). https://www.appliedclinicaltrialsonline.com/view/in-focus-ai-and-clinical-trials-the-beginning-of-a-journey-

Bailleul, C. (2023, November 8). UNESCO unites diverse perspectives to inform policies for AI in the creative sectors. UNESCO. https://www.unesco.org/en/articles/unesco-unites-diverse-perspectives-inform-policies-ai-creative-sectors

Bajwa, J., Munir, U., Nori, A., & Williams, B. (2021). Artificial Intelligence in Healthcare: Transforming the Practice of Medicine. *Future Healthcare Journal*, *8*(2), e188–e194. DOI: 10.7861/fhj.2021-0095 PMID: 34286183

Bal, R., & Gill, I. S. (2020). Policy Approaches to Artificial Intelligence Based Technologies in China, European Union and the United States. *SSRN*. https://dx.doi.org/DOI: 10.2139/ssrn.3699640

Banoczi, J., Edwards, S., Irrechukwu, C., Klosterman, J., Perper, H., Prince, S., Symington, S., & Wynne, D. (2017, August 31). Access Rights Management for the Financial Services Sector. Nist.gov. https://csrc.nist.gov/publications/detail/sp/1800-9/draft

Barnett, D. (2021, June 7). Security Tips for Online LGBTQ+ Dating. Electronic Frontier Foundation. https://www.eff.org/deeplinks/2021/06/security-tips-online-lgbtq-dating

Barocas, S., Hardt, M., & Narayanan, A. (2019). Fairness and machine learning. https://fairmlbook.org/

Barrington, S., & Farid, H. (2023). A Comparative Analysis of Human and AI Performance in Forensic Estimation of Physical Attributes. *Scientific Reports*, *13*(1), 4784. DOI: 10.1038/s41598-023-31821-3 PMID: 36959267

Barta, S., Gurrea, R., & Flavián, C. (2025). Augmented reality experiences: Consumer-centered augmented reality framework and research agenda. *Psychology and Marketing*, *42*(2), 634–650. DOI: 10.1002/mar.22143

Bartlett, R., Morse, A., Stanton, R., & Wallace, N. (2022). Escaping the Subprime Trap in Algorithmic Lending. arXiv. https://arxiv.org/abs/2502.17816

BBC. (2023, March 3). Mother sentenced in deepfake cheerleader case. https://www.bbc.com/news/technology-56404038

Bender, E. M., Gebru, T., McMillan-Major, A., & Shmitchell, S. (2021). On the dangers of stochastic parrots: Can language models be too big? In *Proceedings of the 2021 ACM Conference on Fairness, Accountability, and Transparency* (pp. 610–623). Association for Computing Machinery. https://doi.org/DOI: 10.1145/3442188.3445922

Benjamin, R. (2019). Race After Technology: Abolitionist Tools for the New Jim Code. Polity. https://www.researchgate.net/publication/351607959_Benjamin_Ruha_2019_Race_After_Technology_Abolitionist_Tools_for_the_New_Jim_Code_Medford_Polity_Press_172_pages_eISBN_9781509526437

Bettelheim, A. (2024). Philadelphia hospitals drop race-based algorithms.

Bharadiya, J. P. (2023). A comparative study of business intelligence and artificial intelligence with big data analytics. *American Journal of Artificial Intelligence, 7*(1), 24–30. DOI: 10.11648/j.ajai.20230701.14

Bieser, J. (2022). Creative through AI – How artificial intelligence can support the development of new ideas. DOI: 10.59986/CCHA2271

Binns, R. (2018). Fairness in machine learning: Lessons from political philosophy. *Proceedings of the 2018 Conference on Fairness, Accountability and Transparency*, 149–159. DOI: 10.1145/3287560.3287598

Binns, R., Veale, M., & Edwards, L. (2023). Automated decision-making and meaningful human control: A review of recent EU law and scholarship. *Computer Law & Security Review, 49*, 105738.

Blum, D. (2020). Institute Resilience Through Detection, Response, and Recovery. Rational Cybersecurity for Business, 259–295. https://doi.org/DOI: 10.1007/978-1-4842-5952-8_9

BoardPro. (2024). 7 cognitive biases in board decision-making and how to overcome them. https://www.boardpro.com/blog/cognitive-biases-in-board-decision-making

Boden, M. A. (Ed.). (1996). Artificial Intelligence. *Elsevier,* 161-182. https://core.ac.uk/download/pdf/82671696.pdf

Bond, S. (2024, May 23). A political consultant faces charges and fines for Biden deepfake robocalls. NPR. https://www.npr.org/2024/05/23/nx-s1-4977582/fcc-ai-deepfake-robocall-biden-new-hampshire-political-operative

Boudreaux, M. J. A. (2024). *Transformational Leadership in AI-Driven Industry 4.0: Cultivating Adaptive, Ethical, and Resilient Leaders.* ProQuest Dissertations Publishing.

Brand, L. (2020). Why machines that talk still do not think, and why they might nevertheless be able to solve moral problems. *Artificial Intelligence*, 203–217. DOI: 10.30965/9783957437488_014

Braun, T., & Möller, R. (2024). AI in healthcare and the public sector: How to face the challenges of high-risk applications and what AI research can get out of it. *KI - Künstliche Intelligenz, 38*(3), 119-126. DOI: 10.1007/s13218-024-00853-w

Brazilian Senate. (2024). Brazil AI Act: Bill No. 2338/2023. https://artificialintelligenceact.com/brazil-ai-act/

Brim, O. G.Jr. (1976). Theories of the male mid-life crisis. *The Counseling Psychologist*, *6*(1), 2–9. DOI: 10.1177/001100007600600102

Brittain, B. (2023, February 6). Getty Images lawsuit says Stability AI misused photos to train AI. Reuters. https://www.reuters.com/legal/getty-images-lawsuit-says-stability-ai-misused-photos-train-ai-2023-02-06/

Brookings Institution. (2024). *Corruption and the global financial crisis.* https://www.brookings.edu/opinions/corruption-and-the-global-financial-crisis/

Brunette, E. S., Flemmer, R. C., & Flemmer, C. L. (2009). A Review of Artificial Intelligence. *ResearchGate.* 385-392. https://dx.doi.org/DOI: 10.1109/ICARA.2000.4804025

Buczynski, W., Steffek, F., Cuzzolin, F., Jamnik, M., & Sahakian, B. J. (2024). Hard Law and Soft Law Regulations of Artificial Intelligence in Investment Management. *Cambridge Yearbook of European Legal Studies*, 262-293. https://papers.ssrn.com/sol3/papers.cfm?abstract_id=4786682

Buolamwini, J., & Gebru, T. (2018). Gender shades: Intersectional accuracy disparities in commercial gender classification. Proceedings of the 1st Conference on Fairness, Accountability and Transparency, 77–91. https://proceedings.mlr.press/v81/buolamwini18a.html

Burns-Stanning, K. (2020). Identity in communities and networks TikTok social networking site empowering youth civic engagement. 27, 1-11.

Busacca, A., & Monaca, M. A. (2025). Who's afraid of "big bad" generative AI? Brief notes on the IDPA provision against OpenAI ChatGPT. In Marino, D., & Monaca, M. A. (Eds.), *Generative artificial intelligence and fifth industrial revolution. Lecture Notes in Networks and Systems* (Vol. 880). Springer., DOI: 10.1007/978-3-031-73880-7_9

Business Insider. (2023, August). In every reported false arrest based on facial recognition, that person has been Black. https://www.businessinsider.com/in-every-reported-false-arrests-based-on-facial-recognition-that-person-has-been-black-2023-8

Busuioc, M. (2021). Accountable Artificial Intelligence: Holding Algorithms to Account. *Public Administration Review, 81*(5), 825–836. DOI: 10.1111/puar.13293 PMID: 34690372

Byun, S. (2022). The moral issues of AI bias in data ethics. *Value-Sharing between Humans and Robots*, 3-3. DOI: 10.13180/icres.2022.18-19.07.p01

Çakıcı, Z. (2024). *Türk basınında düzensiz göçün görsel temsili: Taliban yönetimi sonrası Afgan göçü* (Yayımlanmamış doktora tezi). Galatasaray Üniversitesi.

Çakıcı, Z. (2025). Tiktok as a political communication medium: An examination of the United Nations. *İletişim Kuram ve Araştırma Dergisi* (69), 136-156. DOI: 10.47998/ikad.1580671

Çakıcı, Z. (2024). Cyberstalking practices among youth in Türkiye: Motivational drivers, digital avenues, and perceptual paradigms. *Türkiye İletişim Araştırmaları Dergisi*, (46), 1–22. DOI: 10.17829/turcom.1488278

Çakıcı, Z., & Meriç, E. (2024). Exploring United Nations High Commissioner for Refugees' (UNHCR) TikTok landscape: Insights into migration representations. In Kir Elitaş, S. (Ed.), *Media representation of migrants and refugees* (pp. 278–295). IGI Global., DOI: 10.4018/979-8-3693-3459-1.ch017

Camilleri, M. A. (2023). Artificial intelligence governance: Ethical considerations and implications for social responsibility. *Expert Systems: International Journal of Knowledge Engineering and Neural Networks*, *40*(5), e13406. Advance online publication. DOI: 10.1111/exsy.13406

Camillo, M. (2017). Cybersecurity: Risks and management of risks for global banks and financial institutions. https://doi.org/DOI: 10.69554/EPYV4777

Campbell, J. L., Quincy, C., Osserman, J., & Pedersen, O. K. (2013). Coding in-depth semi-structured interviews: Problems of unitization and intercoder reliability and agreement. *Sociological Methods & Research*, *42*(3), 294–320. DOI: 10.1177/0049124113500475

Cancel the Christie's AI Art Auction. (2025, February 8). Open Letter. https://openletter.earth/cancel-the-christies-ai-art-auction-f5135435

Capital, F. M. (2018). *Kenya's mobile commerce transactions value hits Sh1.1 trillion.* https://www.capitalfm.co.ke/business/2018/04/kenyas-mobile-commerce-transactions-value-hit-sh1-1trillion/

Carello, M. P., Spaccamela, A. M., Querzoni, L., & Angelini, M. (2023). A Systematization of Cybersecurity Regulations, Standards and Guidelines for the Healthcare Sector. ArXiv (Cornell University). https://doi.org//arxiv.2304.14955DOI: 10.48550

Caselles-Dupré, H.. (2018). *Portrait of Edmond de Belamy* [Auction lot notes]. Christie's.

Castle, S., Pervaiz, F., Weld, G., Roesner, F., & Anderson, R. (2016). Let's Talk Money. https://doi.org/DOI: 10.1145/3001913.3001919

Ceci, L. (2025). TikTok users by country 2025. Statista. https://www.statista.com/statistics/1299807/number-of-monthly-unique-tiktok-users/

Central Bank of Kenya (CBK). (2022). *Discussion paper on Central Bank Digital Currency.*

Cerrato, P. (2016). Reducing the Risk of a Data Breach. https://doi.org/DOI: 10.1016/B978-0-12-804392-9.00005-8

Cervi, L. (2023). TikTok use in municipal elections: From candidate-majors to influencer-politicians. *Más Poder Local,* (53), 8–29. DOI: 10.56151/maspoderlocal.175

Çetinkaya, A., & Dondurucu, Z. (2023). Haber üretim ve dağıtım pratikleri bağlamında TikTok kullanımı. *İletişim Kuram ve Araştırma Dergisi,* 61, 79-97.

Chakravarthy, S., & K S, D. S. (2024). Application of natural language processing for creating chatbots in healthcare. *International Journal of Research Publication and Reviews, 5*(1), 1472–1481. DOI: 10.55248/gengpi.5.0124.0212

Chen, M. (2023, February 6). Netflix Japan is drawing ire for using A.I. to generate the background art of its new anime short. Artnet News. https://news.artnet.com/art-world/netflix-japan-ai-anime-dog-and-boy-2251247

Chen, Y., Che, Z., & Wan, D. (2024). How does the financial technology innovation regulatory pilot influence financial regulation? *Finance Research Letters, 69,* 106255. DOI: 10.1016/j.frl.2024.106255

Chesney, R., & Citron, D. K. (2019). Deep fakes: A looming challenge for privacy, democracy, and national security. *California Law Review, 107*(6), 1753–1820. https://www.californialawreview.org/print/deep-fakes-a-looming-challenge-for-privacy-democracy-and-national-security

China Briefing. (2023, April 11). China releases new draft regulations on generative AI. https://www.china-briefing.com/news/china-releases-new-draft-regulations-on-generative-ai/

Choudhary, N., Gautam, C., & Arya, V. (2020). Digital marketing challenge and opportunity with reference to tiktok-a new rising social media platform. *Editorial Board, 9*(10), 189–197.

Clancy, M., & Tweed, R. (2022). The voice of the artist in the age of the algorithm. IEEE Global Initiative on Ethics of Autonomous and Intelligent Systems, 10–12. https://sagroups.ieee.org/global-initiative/wp-content/uploads/sites/542/2022/07/ead-artists.pdf

Clark, C. (2024, February). The Advisor's Guide to SEC Cybersecurity Compliance. Nitrogen Wealth. https://nitrogenwealth.com/blog/sec-cybersecurity-2024/

Coronado-Maldonado, I., & Benítez-Márquez, M. D. (2023). Emotional intelligence, leadership, and work teams: A hybrid literature review. *Heliyon*, *9*(12), e20356. DOI: 10.1016/j.heliyon.2023.e20356 PMID: 37790975

Cristea, L. M. (2020). Current security threats in the national and international context. *JOURNAL of ACCOUNTING and MANAGEMENT INFORMATION SYSTEMS*, *19*(2), 351–378. DOI: 10.24818/jamis.2020.02007

Cugurullo, F., & Xu, Y. (2025). When AIs Become Oracles: Generative Artificial Intelligence, Anticipatory Urban Governance, and the Future of Cities. *Policy and Society*, *44*(1), 98–115. DOI: 10.1093/polsoc/puae025

Culnan, M. J., McHugh, P. J., & Zubillaga, J. I. (2010). How large US companies can use Twitter and other social media to gain business value. *MIS Quarterly Executive*, *9*(4), 243–259.

Cyberspace Administration of China. (2022). Internet Information Service Algorithmic Recommendation Management Provisions. https://digichina.stanford.edu/work/translation-internet-information-service-algorithmic-recommendation-management-provisions-effective-march-1-2022/

Cyberspace Administration of China. (2023a). Provisions on the Administration of Deep Synthesis Internet Information Services. https://www.chinalawtranslate.com/en/deep-synthesis/

Cyberspace Administration of China. (2023b). Interim Measures for the Management of Generative Artificial Intelligence Services. https://www.chinalawtranslate.com/en/generative-ai-interim/

Czaja, S. J., Charness, N., Fisk, A. D., Hertzog, C., Nair, S. N., Rogers, W. A., & Sharit, J. (2006). Factors predicting the use of technology: Findings from the Center for Research and Education on Aging and Technology Enhancement (CREATE). *Psychology and Aging*, *21*(2), 333–352. DOI: 10.1037/0882-7974.21.2.333 PMID: 16768579

Dahlstrand-Rudin, A., Deserranno, E., & Guariso, A. (2023). *Remote care for remote areas: The impact of Telehealth in rural India.* AEA Randomized Controlled Trials., DOI: 10.1257/rct.10787

Dakanalis, A., Wiederhold, B. K., & Riva, G. (2023). Artificial intelligence: A game-changer for mental health care. DOI: 10.31234/osf.io/ubdxz

Damnjanović, A., Rašković, M., & Janković, D. (2024). Leadership approaches in managing AI-driven process automation for small and medium-sized enterprises. *SCIENCE International Journal*, *3*(4), 47–51. DOI: 10.35120/sciencej0304047d

Darus, Mohamad Yusofet al. (2021). A case study on digital divide and access to information communication technologies (Icts) in pulau tuba, Langkawi, Malaysia. Turkish Journal of Computer and Mathematics Education (TURCOMAT), 12(5), 1710-1718. DOI: 10.17762/turcomat.v12i5.2166

Dastin, J. (2018, October 10). Amazon scraps secret AI recruiting tool that showed bias against women. Reuters. https://www.reuters.com/article/us-amazon-com-jobs-automation-insight-idUSKCN1MK08G

De Cremer, D., & Narayanan, D. (2023). How AI tools can—and cannot—help organizations become more ethical. *Frontiers in Artificial Intelligence*, *6*, 1093712. DOI: 10.3389/frai.2023.1093712 PMID: 37426304

Deloitte Global Boardroom Program. (2022). Digital frontier: A technology deficit in the boardroom. Deloitte Touche Tohmatsu Limited. https://www.deloitte.com/us/en/insights/topics/leadership/digital-transformation-topics-for-corporate-technology-leadership.html

Deloitte Global Boardroom Program. (2024). Governance of AI: A critical imperative for today's boards. Deloitte Touche Tohmatsu Limited. https://www2.deloitte.com/content/dam/insights/articles/glob187394_global-boardroom-program_ai-boardroom/DI_Governance-of-AI_A-critical-imperative-for-todays-boards.pdf

Demirel, S., Bulur, N., & Çakıcı, Z. (2024). Utilizing artificial intelligence for text classification in communication sciences: Reliability of ChatGPT models in Turkish texts. In Darwish, D. (Ed.), *Design and development of emerging chatbot technology* (pp. 218–235). IGI Global., DOI: 10.4018/979-8-3693-1830-0.ch013

Demirel, S., Çakıcı, Z., & Bulur, N. (2024). Social perception of artificial intelligence on Twitter: A comparative study on global south and global north countries. *The Russian Sociological Review*, *23*(4), 48–79. DOI: 10.17323/1728-192x-2024-4-48-79

Demirel, S., Çakıcı, Z., & Bulur, N. (2025). Exploring reflections of the 2023 Kahramanmaraş earthquake on X: A computational study on Türkiye and Syria. *Akademik İncelemeler Dergisi*, *20*(1), 143–171. DOI: 10.17550/akademikincelemeler.1593038

Dempere, J. (2024). Deploying AI-driven natural language processing systems for public health crisis management. *Advancements in Health Sciences*, 131-153. DOI: 10.1007/978-3-031-57773-4_10

Deng, B. (2024). The application of AI video generation technology in virtual reality (VR) and augmented reality (AR). *2024 International Conference on Artificial Intelligence, Deep Learning and Neural Networks (AIDLNN)*, 168-173. DOI: 10.1109/AIDLNN65358.2024.00035

Deng, W. H., Yildirim, N., Chang, M., & Eslami, M. (2023). Investigating practices and opportunities for cross-functional collaboration around AI fairness in industry practice. *Proceedings of the 2023 CHI Conference on Human Factors in Computing Systems*, 1-16. DOI: 10.1145/3593013.3594037

Dezuanni, M., & Osman, K. (2024). IFLA Trend Report (2024). Facing the future of information with confidence. International Federation of Library Associations and Institutions. https://repository.ifla.org/handle/20.500.14598/3496

Didenko, A. N. (2020). Cybersecurity regulation in the financial sector: prospects of legal harmonization in the European Union and beyond. Uniform Law Review, 25(1), 125–167. Does My Smart Device Provider Care About My Privacy? Investigating Trust Factors and User Attitudes in IoT Systems. https://doi.org/DOI: 10.1093/ulr/unaa006

Dinkins, S. (2023). NTOO: Not the only one. Stephanie Dinkins. https://www.stephaniedinkins.com/ntoo.html

Dixit, S., & Maurya, M. (2021). Equilibrating emotional intelligence and AI-driven leadership for transnational organizations. In 2021 International Conference on Advances in Computing, Communication, and Control (ICAC3) (pp. 1-6).

Dixon, R. B. L. (2022). Artificial Intelligence Governance: A Comparative Analysis of China, the European Union, and the United States. *University of Minnesota*. https://hdl.handle.net/11299/229505

Dlamini, S. T., & Botha, C. (2024). The use of augmented and virtual reality to enhance the customer experience. *African Journal of Science, Technology, Innovation and Development*, *16*(3), 399–409. DOI: 10.1080/20421338.2024.2326791

Dooraki, S. K. (2024). Early detection of celiac disease through its common symptoms using machine learning algorithms. *Journal of Clinical Images and Medical Case Reports*, *5*(3). Advance online publication. DOI: 10.52768/2766-7820/2915

Doshi-Velez, F., & Kim, B. (2017). Towards a rigorous science of interpretable machine learning. arXiv preprint arXiv:1702.08608. https://arxiv.org/abs/1702.08608

Drumm, L., & Sami, A. (2024). Academic staff AI literacy development through LLM prompt training. *Effective Practices in AI Literacy Education: Case Studies and Reflections*, 41-49. DOI: 10.1108/978-1-83608-852-320241005

Dwivedi, Y. K., Hughes, L., Ismagilova, E., Aarts, G., Coombs, C., Crick, T., Duan, Y., Dwivedi, R., Edwards, J., Eirug, A., Galanos, V., Ilavarasan, P. V., Janssen, M., Jones, P., Kar, A. K., Kizgin, H., Kronemann, B., Lal, B., Lucini, B., & Williams, M. D. (2021). Artificial Intelligence (AI): Multidisciplinary perspectives on emerging challenges, opportunities, and agenda for research, practice and policy. *International Journal of Information Management*, *57*, 101994. DOI: 10.1016/j.ijinfomgt.2019.08.002

Edwards, B. (2022, December 14). ArtStation artists protest AI-generated art by adding 'No AI Art' images. Ars Technica. https://arstechnica.com/information-technology/2022/12/artstation-artists-stage-mass-protest-against-ai-generated-artwork/

e-Estonia. (2023). *X-Road: Interoperability services*. https://e-estonia.com

Egunjobi, D., & Adeyeye, O. J. (2024). Revolutionizing learning: The impact of augmented reality (AR) and artificial intelligence (AI on education. *International Journal of Research Publication and Reviews*, *5*(10), 1157–1170. DOI: 10.55248/gengpi.5.1024.2734

Elia, G., Margherita, A., & Petti, C. (2016). An operational model to develop technology entrepreneurship "EGO-system.". *International Journal on Management of Innovation & Technology*, *13*(5), 1640008. Advance online publication. DOI: 10.1142/S0219877016400083

El-Kouedi, M., Richards, J., Shubart, S., Jam, A., & Jordan, M. (2024). Report to the President for the year ended June 30, 2024, Program in Media Arts and Sciences. Massachusetts Institute of Technology. https://dspace.mit.edu/bitstream/handle/1721.1/156178/MAS-annualreport-2024.pdf

Elliott, L. (2024, August 6). Non-western perspectives on AI: Art ecosystems in Africa, Americas, and Asia. Salzburg Global Seminar News. https://www.salzburgglobal.org/news/regions/article/non-western-perspectives-on-ai-art-ecosystems-in-africa-americas-and-asia

EMARKETER. (2023, January 2). Artificial Intelligence in Financial Services: Applications and benefits of AI in finance. EMARKETER. https://www.emarketer.com/insights/ai-in-finance/

Enyejo, J. O., Obani, O. Q., Afolabi, O., Igba, E., & Ibokette, A. I. (2024). Effect of Augmented Reality (AR) and Virtual Reality (VR) experiences on customer engagement and purchase behavior in retail stores. *Magna Scientia Advanced Research and Reviews*, *11*(2), 132–150. DOI: 10.30574/msarr.2024.11.2.0116

Equal Employment Opportunity Commission (EEOC). (2023). Technical assistance on AI and employment discrimination. https://www.eeoc.gov/meetings/meeting-january-31-2023-navigating-employment-discrimination-ai-and-automated-systems-new

Ersöz, B. (2020). Yeni nesil web paradigması-web 4.0. *Bilgisayar Bilimleri ve Teknolojileri Dergisi, 1*(2), 58–65.

Ertel, W. (2024). Introduction to Artificial Intelligence. *Springer Nature.* https://unidel.edu.ng/focelibrary/books/Introduction%20to%20Artificial%20Intelligence%20(%20PDFDrive%20).pdf

European Commission. (2024, August 1). AI Act enters into force. https://commission.europa.eu/news/ai-act-enters-force-2024-08-01_en

Exonum. (2024). *Blockchain land registry in Georgia.* https://exonum.com/story-georgia

EY. (2024, August 7). EY study: 82% of European financial services investors state that age diversity in the boardroom influences investment decisions. https://www.ey.com/en_ro/newsroom/2023/07/ey-study-82-of-european-financial-services-investors-state-tha

Facebook. (2020). Facebook's civil rights audit – Final report. https://about.fb.com/wp-content/uploads/2020/07/Civil-Rights-Audit-Final-Report.pdf

Fahad, M., Basri, T., Hamza, M. A., Faisal, S., Akbar, A., Haider, U., & Hajjami, S. E. (2024). The Benefits and Risks of Artificial General Intelligence (AGI). *ResearchGate*, 27-52. https://doi.org/DOI: 10.1007/978-981-97-3222-7_2

Fahmideh, M., Gunawardana, A., Chen, S., Shen, J., & Yecies, B. (2022). Blockchain Developments and Innovations – an analytical evaluation of software engineering approaches. In Lecture notes in computer science (pp. 58–76). DOI: 10.1007/978-3-030-96068-1_5

Fallows, D. (2002). Email at work. Pew Internet & American Life Project. Retrieved May 3, 2025, from https://www.pewresearch.org/internet/2002/12/08/email-at-work/

Fang. (2024, July 5). Finance and Technology: Why Fintech is the future of finance – A case study of Singapore's Financial Sector - UNICAF - scholarship programme. Unicaf - Scholarship Programme. https://www.unicaf.org/thesis-publications/finance-and-technology-why-fintech-is-the-future-of-finance-a-case-study-of-singapores-financial-sector/

Farinu, U. (2025). Fairness, accountability, and transparency in AI: Ethical challenges in data-driven decision-making. DOI: 10.2139/ssrn.5128174

Feldman, M., Friedler, S. A., Moeller, J., Scheidegger, C., & Venkatasubramanian, S. (2015). Certifying and removing disparate impact. *Proceedings of the 21th ACM SIGKDD International Conference on Knowledge Discovery and Data Mining*, 259–268. DOI: 10.1145/2783258.2783311

Ferreira, E. (2016, October 14). How Sony Flow Machines' AI takes itself for the Beatles. Makery. https://www.makery.info/en/2016/10/14/comment-lia-de-sony-flow-machines-se-prend-pour-les-beatles/

Feyen, E., Natarajan, H., & Saal, M. (2023). Fintech and the Future of Finance: Market and Policy Implications. https://doi.org/DOI: 10.1596/978-1-4648-1914-8

Fischer, E. A. (2014). Federal Laws Relating to Cybersecurity: Overview of Major Issues, Current Laws, and Proposed Legislation. UNT Digital Library. crs: R42114

Fisher, C. K. (2024). US FDA comments on Unlearn's PROCOVA methodology. Unlearn.AI. https://www.unlearn.ai/blog/us-fda-comments-on-unlearns-procova-methodology

Food Trust, I. B. M. (2020). *Improving supply chain transparency*. https://www.ibm.com/blockchain/solutions/food-trust

Fox, J., & Bartholomae, S. (2020). Household finances, financial planning, and COVID-19. In Financial Planning Review (Vol. 3, Issue 4). Wiley. DOI: 10.1002/cfp2.1103

Frank, D.-A., Chrysochou, P., & Mitkidis, P. (2023). The paradox of technology: Negativity bias in consumer adoption of innovative technologies. *Psychology and Marketing*, *40*(3), 554–566. DOI: 10.1002/mar.21740

Friedler, S. A., Scheidegger, C., & Venkatasubramanian, S. (2021). The (Im)possibility of fairness. *Communications of the ACM*, *64*(4), 136–143. DOI: 10.1145/3433949

Funahashi, S. (2017). Prefrontal contribution to decision-making under free-choice conditions. *Frontiers in Neuroscience*, *11*, 431. Advance online publication. DOI: 10.3389/fnins.2017.00431 PMID: 28798662

Gallagher, S. (2016, June 13). Sci-fi film 'Sunspring' is the world's first movie created entirely by artificial intelligence. HuffPost UK. https://www.huffingtonpost.co.uk/entry/artificial-intelligence-algorithm-writes-first-film-sunspring_uk_575e90eee4b014b4f2540cca

Garg, A., Degiovanni, R., Jimenez, M., Cordy, M., Papadakis, M., & Le Traon, Y. (2022). Learning from what we know: How to perform vulnerability prediction using noisy historical data. *Empirical Software Engineering*, *27*(7), 169. Advance online publication. DOI: 10.1007/s10664-022-10197-4

Gars, J., & Ward, P. S. (2019). Can differences in individual learning explain patterns of technology adoption? Evidence on heterogeneous learning patterns and hybrid rice adoption in Bihar, India. *World Development*, *115*, 178–189. DOI: 10.1016/j.worlddev.2018.11.014 PMID: 30828125

Gaurav, A., Gupta, B. B., & Castiglione, A. (2024). Ethical AI development and deployment. *Advances in Computational Intelligence and Robotics*, 115–144. DOI: 10.4018/979-8-3693-3860-5.ch004

Gellert, R. (2020). *The Risk-Based Approach to Data Protection*. Oxford University Press., DOI: 10.1093/oso/9780198837718.001.0001

Gelskov, S. V., Madsen, K. H., Ramsøy, T. Z., & Siebner, H. R. (2016). Aberrant neural signatures of decision-making: Pathological gamblers display cortico-striatal hypersensitivity to extreme gambles. *NeuroImage*, *128*, 342–352. DOI: 10.1016/j.neuroimage.2016.01.002 PMID: 26780575

Gheyle, N., & Jacobs, T. (2017). Content analysis: A short overview. *Internal research note,* 10, 1-17. DOI: 10.13140/RG.2.2.33689.31841

Ghosh, U. K. (2025). Transformative AI applications in business decision-making: Advancing data-driven strategies and organizational intelligence. In Ghosh, U. K. (Ed.), *AI-powered leadership*., DOI: 10.4018/979-8-3373-1687-1.ch001

Giachino, C., Cepel, M., Truant, E., & Bargoni, A. (2024). Artificial intelligence-driven decision making and firm performance: A quantitative approach. *Management Decision*. Advance online publication. DOI: 10.1108/MD-10-2023-1966

Giorgio Barba Navaretti, Giacomo Calzolari, José Manuel Mansilla-Fernandez, & Pozzolo, A. F. (2023). Open Banking's Promise of a Financial Revolution: Are We Falling Short? Social Science Research Network. https://doi.org/DOI: 10.2139/ssrn.4469320

Goankar, B., Cook, K., & Macyszyn, L. (2020). Ethical issues arising due to bias in training A.I. Algorithms in healthcare and data sharing as a potential solution. *AI Ethics Journal*, *1*(2). Advance online publication. DOI: 10.47289/AIEJ20200916

Goel, P., & Dixit, A. (2023). Introduction to on-device AI and its applications in mobile phones and personal assistants. *AI in Wireless for Beyond 5G Networks*, 19-26. DOI: 10.1201/9781003303527-3

Goel, R., Mishra, S., & Srivastava, S. (2024). Applying machine learning algorithms for the scientific prediction of crop yields. SSRN *Electronic Journal*. DOI: 10.2139/ssrn.4925156

Gouveia, S. S. (2022). *Qualia and information in philosophy and neuroscience*. Springer. DOI: 10.1007/978-3-030-95369-0_6

Government of Canada. (2025, March). AI strategy for the federal public service 2025–2027: Overview. https://www.canada.ca/en/government/system/digital-government/digital-government-innovations/responsible-use-ai/gc-ai-strategy-overview.html

Grother, P., Ngan, M., & Hanaoka, K. (2019). *Face recognition vendor test (FRVT) Part 3: Demographic effects (NIST Interagency/Internal Report 8280)*. National Institute of Standards and Technology., DOI: 10.6028/NIST.IR.8280

Grygar, F. (2014). *The complementary thought of Niels Bohr in the context of physics, philosophy and biology*. Pavel Mervart.

Guerrero, K. J. (2023). *An analysis of TikTok's digital marketing capabilities and its effectiveness for brands and businesses across different industries* (Master's thesis, Lindenwood University). Lindenwood University.

Gurkaynak, G., Yilmaz, I., & Taskiran, N. P. (2013). Governmental Efforts and Strategies to Reinforce Security in Cyberspace. *International Law Research*, *2*(1). Advance online publication. DOI: 10.5539/ilr.v2n1p185

Gu, Z., Chen, X., Wang, T., & Zhao, Y. (2023). Deepfake detection algorithm based on dual-branch data augmentation and modified attention mechanism. *Applied Sciences (Basel, Switzerland)*, *13*(14), 8313. DOI: 10.3390/app13148313

Haile, T. T. (2024). Web's progression: Moving from passive content consumption to active content creation and content validation. *International Journal of Business and Management*, *18*(6), 135–135. DOI: 10.5539/ijbm.v18n6p135

Hammond, J. S., Keeney, R. L., & Raiffa, H. (1998, September). The hidden traps in decision making. Harvard Business Review. https://hbr.org/1998/09/the-hidden-traps-in-decision-making-2

Haskins, C. (2024, September 5). Meredith Stiehm. Time. https://time.com/7012819/meredith-stiehm/

Hassan, M. A., Nguyen, C. T., Nguyen, C. H., Hoang, D. T., Nguyen, D. N., & Dutkiewicz, E. (2024). A novel blockchain based information management framework for Web 3.0. *arXiv preprint* arXiv:2401.14420. DOI: 10.1109/GLOBECOM52923.2024.10901512

Hemmler, Y. M., & Ifenthaler, D. (2022). Indicators of the learning context for supporting personalized and adaptive learning environments. *2022 International Conference on Advanced Learning Technologies (ICALT)*, 61-65. DOI: 10.1109/ICALT55010.2022.00026

Hermawan, S., Khoirunisa, Z. A., & Tejomurti, K. (2023). Triangular Insight on Open Banking in Indonesia, Singapore, and Australia. In International Journal of Legal Information (Vol. 51, Issue 3, p. 197). Cambridge University Press. https://doi.org/DOI: 10.1017/jli.2024.11

Hine, E. (2024). Governing Silicon Valley and Shenzhen: Assessing a New Era of Artificial Intelligence Governance in the United States and China. Digital Society. https://link.springer.com/article/10.1007/s44206-024-00138-7

Hong, J. (2012). The state of phishing attacks. Communications of the ACM, 55(1), 74. https://doi.org/10.1145/2063176.2063197

Hossin, M. A., Du, J., Mu, L., & Asante, I. O. (2023). Big Data-Driven Public Policy Decisions: Transformation Toward Smart Governance. *SAGE Open*, *13*(4), 21582440231215123. Advance online publication. DOI: 10.1177/21582440231215123

Hswen, Y., & Schweitzer, K. (2025). Researcher explores using AI to overcome language barriers with patients. *Journal of the American Medical Association*, *333*(1), 9. DOI: 10.1001/jama.2024.23526 PMID: 39641948

Huang, H., Lu, Y., & Wang, L. (2025). Board age diversity and digital transformation: Evidence from China. *International Review of Economics & Finance*, *100*, 104106. https://www.sciencedirect.com/science/article/pii/S1059056025002692. DOI: 10.1016/j.iref.2025.104106

Hunt Scanlon Media. (2024). Top 5 leadership trends set to define 2025. https://huntscanlon.com/top-5-leadership-trends-set-to-define-2025/

Hussein-Elhakim Al Issa, M. T. T., & Nguyen, H. (2024). A systematic mapping of social entrepreneurship education: A call for increased collaboration, ethics, and research frameworks. *International Journal of Management Education*, *22*(3), 101025. Advance online publication. DOI: 10.1016/j.ijme.2024.101025

Hussein, S., & Safina, S. S. Q. (2025). Analyzing how AI can both exacerbate and help overcome digital inequalities in education, especially in underserved regions. *The Critical Review of Social Sciences Studies*, *3*(2), 143–157. DOI: 10.59075/4964v051

Hutson, M. (2024). How AI is being used to accelerate clinical trials. *Nature*, *627*(8003), S2–S5. DOI: 10.1038/d41586-024-00753-x PMID: 38480968

Illinois General Assembly. (2020). Artificial Intelligence Video Interview Act, 820 ILCS 42. https://www.ilga.gov/legislation/ilcs/ilcs3.asp?ActID=4015&ChapterID=68

Innovation, Science and Economic Development Canada. (2022). Artificial Intelligence and Data Act (AIDA). https://ised-isde.canada.ca/site/innovation-better-canada/en/artificial-intelligence-and-data-act

International Journal of Innovative Research and Development (IJIIRD). (2024). *India's blockchain initiatives in land titles and agriculture.* https://ijiird.com/wp-content/uploads/030222.pdf

Issac, B., Chiong, R., & Seibu Mary Jacob. (2014). Analysis of Phishing Attacks and Countermeasures. https://doi.org/DOI: 10.48550/arxiv.1410.4672

Jain, D. (2024). AI-driven climate solutions: Leveraging LEO satellites for sustainability. DOI: 10.58445/rars.1814

Japanese Copyright Act, art. 2(1)(i) (definition of "work"), and art. 30-4 (2018 amendment on information analysis).

Jerahian, A. (2024). Leadership evolution in the age of artificial intelligence. *Journal of Digital Banking*, *9*(2), 145–156. DOI: 10.69554/HZLR7004

JetBrains. (2023). Python programming – The state of developer ecosystem in 2023. JetBrains s.r.o. https://www.jetbrains.com/lp/devecosystem-2023/python/

Jin, S., Chen, L., Ding, H., Wang, M., & Yu, L. (2024). Can artificial intelligence predict clinical trial outcomes? Klabunde, T. (2024). Digital "twinning": Clinical trials powered by AI. Sanofi. https://www.sanofi.com/en/magazine/our-science/digital-twinning-clinical-trials-ai

Jin, Q., Wang, Z., Floudas, C. S., Chen, F., Gong, C., Bracken-Clarke, D., Xue, E., Yang, Y., Sun, J., & Lu, Z. (2024). Matching patients to clinical trials with large language models. *Nature Communications*, *15*(1), 9074. Advance online publication. DOI: 10.1038/s41467-024-53081-z PMID: 39557832

Ji, Z., Lee, N., Frieske, R., Yu, T., Su, D., Xu, Y., & Yang, Y. (2023). Survey of hallucination in natural language generation. *ACM Computing Surveys, 55*(12), 1–38. Advance online publication. DOI: 10.1145/3571730

Joshi, D. (2024). AI governance in India – Law, Policy and Political Economy. *Communication Research and Practice, 10*(3), 328–339. DOI: 10.1080/22041451.2024.2346428

Joshi, S., Sharma, M., Das, R. P., Rosak-Szyrocka, J., Żywiołek, J., Muduli, K., & Prasad, M. (2022). Modeling conceptual framework for implementing barriers of AI in public healthcare for improving operational excellence: Experiences from developing countries. *Sustainability (Basel), 14*(18), 11698. DOI: 10.3390/su141811698

Kakareka, A. (2009). What Is Vulnerability Assessment? Computer and Information Security Handbook, 383–393. https://doi.org/DOI: 10.1016/B978-0-12-374354-1.00023-6

Kambala, M. (2024). AI-powered healthcare: Transforming patient outcomes with machine learning. *Journal of Medical Science and clinical Research, 12*(08), 34-47. DOI: 10.18535/jmscr/v12i08.07

Kandasamy, U. C. (2024). Ethical Leadership in the Age of AI Challenges, Opportunities and Framework for Ethical Leadership. arXiv preprint arXiv:2410.18095. https://arxiv.org/abs/2410.18095

Kaplan, J. M. (2016). Cybersecurity for Commercial Advantage. Elsevier EBooks, 97–111. https://doi.org/DOI: 10.1016/B978-0-12-803773-7.00006-1

Karmaza, O. O., Koroied, S. O., Makhinchuk, V. M., Strilko, V. Y., & Iosypenko, S. T. (2021). Artificial Intelligence in Justice. *Linguistics and Culture Review, 5*(S4), 1413–1425. DOI: 10.21744/lingcure.v5nS4.1764

Kaur, J. (2024). Responsible Artificial Intelligence (AI) Governance: Ethical Frameworks for Business Decision-Making. In *Ethical Quandaries in Business Practices: Exploring Morality and Decision Making*. IGI Global., https://www.igi-global.com/chapter/responsible-artificial-intelligence-ai-governance/356327 DOI: 10.4018/979-8-3693-3948-0.ch014

Kemp, S. (2025, February 5). Digital 2025 global overview report. We Are Social. https://wearesocial.com/uk/blog/2025/02/digital-2025/

Kemp, S. (2025, Şubat 5). Digital 2025: The essential guide to the global state of digital. *We Are Social UK.* https://wearesocial.com/uk/blog/2025/02/digital-2025-the-essential-guide-to-the-global-state-of-digital/

Kenney, A. (2023, September 6). Jason Allen's AI Art Won the Colorado Fair — but Now the Feds Say It Can't Get a Copyright. Colorado Public Radio. https://www.cpr.org/2023/09/06/jason-allens-ai-art-won-colorado-fair-feds-deny-copyright-protection/

Kerketta, A., & Balasundaram, S. (2024). "Leveraging AI tools to bridge the healthcare gap in rural areas in India". DOI: 10.1101/2024.07.30.24311228

Kerrigan, C. (2022). Explainable AI and responsible AI. *Artificial Intelligence*, 527–533. DOI: 10.4337/9781800371729.00043

Khan, H. U., Malik, M. Z., Alomari, M. K., Khan, S., Al-Maadid, A., Hassan, M. K., & Khan, K. (2022a). Transforming the Capabilities of Artificial Intelligence in GCC Financial Sector: A Systematic Literature Review. In Wireless Communications and Mobile Computing (Vol. 2022, p. 1). Wiley. https://doi.org/DOI: 10.1155/2022/8725767

Khan, N. F., Ikram, N., Saleem, S., & Zafar, S. (2022b). Cyber-security and risky behaviors in a developing country context: a Pakistani perspective. In Security Journal (Vol. 36, Issue 2, p. 373). Palgrave Macmillan. https://doi.org/DOI: 10.1057/s41284-022-00343-4

Khare, P. & Raghuwanshi, V. (2025). Analysis of Privacy Laws Impacting Healthcare Data Across Different Regions. *IGI Global Scientific Publishing*. 393-410. https://doi.org/DOI: 10.4018/979-8-3693-7630-0.ch016

Khare, P. & Raghuwanshi, V. (2025). Legal Frameworks Surrounding the Use of AI in Online Content Moderation. *IGI Global Scientific Publishing*, 235-260. https://doi.org/DOI: 10.4018/979-8-3693-9904-0.ch011

Khare, P. (2025). Investigating the Influence of AI in Social Media Governance. *IGI Global Scientific Publishing*. 365-388. https://doi.org/DOI: 10.4018/979-8-3693-9904-0.ch015

Khudhair, D. Z., Al-Zubaidi, F. K. A., & Raji, A. A. (2019). The effect of board characteristics and audit committee characteristics on audit quality. *Management Science Letters*, •••, 271–282. DOI: 10.5267/j.msl.2018.11.012

Kingston University. (2023). Kingston University renews calls for urgent action to prepare students for workplace increasingly dominated by AI [Press release]. https://www.kingston.ac.uk/news/article/2904/06-dec-2023-kingston-university-renews-calls-for-urgent-action-to-prepare-students-for-workplace-increasingly-dominated-by/

Kingston, J. K. (2016). Artificial Intelligence and Legal Liability. *Springer International Publishing*. 269-279. https://doi.org/DOI: 10.1007/978-3-319-47175-4_20

Kitces, M. (2014, July 21). Michael Kitces. Nerd's Eye View | Kitces.com. https://www.kitces.com/blog/fpa-rpi-studies-how-advisors-communicate-segment-clients-and-conduct-review-meetings/

Kitongo, G., & Tutton, M. (2023, June 23). Nigerian AI artist creates a fashion show for elderly people. CNN. https://edition.cnn.com/style/article/malik-afegbua-elderly-fashion-ai-art-spc-intl/index.html

Klug, D., Qin, Y., Evans, M., & Kaufman, G. (2021, June). Trick and please. A mixed-method study on user assumptions about the TikTok algorithm. *In Proceedings of the 13th ACM web science conference* 2021 (pp. 84-92). DOI: 10.1145/3447535.3462512

Kobie, N. (2024, February 6). Fake AI-generated news sites are spreading online. Wired. https://www.wired.com/story/ai-generated-fake-news-is-coming-to-an-election-near-you/

Kooij, D. T. A. M., de Lange, A. H., Jansen, P. G. W., Kanfer, R., & Dikkers, J. S. E. (2011). Age and work-related motives: Results of a meta-analysis. *Journal of Organizational Behavior*, *32*(2), 197–225. DOI: 10.1002/job.665

Korea Tech Today. (2024). *South Korea to invest $14.5M in blockchain innovation.* https://www.koreatechtoday.com/south-korea-to-invest-14-5m-in-blockchain-innovation-across-public-and-private-sectors/

Kosaraju, D. (2024). Predictive analytics in healthcare: Leveraging AI to anticipate disease outbreaks and enhance patient outcomes. *Galore International Journal of Health Sciences and Research*, *8*(3), 73–79. DOI: 10.52403/gijhsr.20230312

Kotter, J. P. (1996). *Leading change*. Harvard Business Review Press.

Krishna Pasupuleti, M. (2024). AI-powered learning: Transformative innovations in smart education technology. *Advanced AI-Driven EdTech: Transformative Innovations in Smart Learning*, 292-310. DOI: 10.62311/nesx/802874

Kshetri, N. (2023). 1.5 decades of blockchain research in business and management: A review. *Journal of Business Research*, *161*, 113839.

Kurshan, E., Mehta, D., Bruss, B., & Balch, T. (2024). AI versus AI in Financial Crimes and Detection: GenAI Crime Waves to Co-Evolutionary AI. ArXiv (Cornell University). https://doi.org//arxiv.2410.09066DOI: 10.1145/3677052.3698655

Kuska, M., Wahabzada, M., & Paulus, S. (2024). Ai-chatbots for agriculture - Where can large language models provide substantial value? DOI: 10.2139/ssrn.4685971

Lamuth, J. (2010). *Ethical artificial intelligence.* SciVee., DOI: 10.4016/22623.01

Lancaster, C., Schulenberg, K., Flathmann, C., & Freeman, G. Z. (2023). "It is everybody's role to speak up... but not everyone will": Understanding AI professionals' perceptions of accountability for AI bias mitigation. *ACM Journal on Responsible Computing*, *1*(1), 1–30. Advance online publication. DOI: 10.1145/3632121

Landers, R. N., & Behrend, T. S. (2023). Auditing the AI auditors: A framework for evaluating fairness and bias in high stakes AI predictive models. American Psychologist. https://psycnet.apa.org/fulltext/2022-30899-001.html

Landwehr, C. E. (2008). *Cyber security and artificial intelligence.*, DOI: 10.1145/1456377.1456391

Langeveldt, D. C. (2021). AI-driven leadership: A conceptual framework for educational decision-making in the AI era. *Journal of Educational Administration.*

LatAm Arte. (2023, November 26). The Impact of AI on Creating Art. LatAm Arte, 87, 95. https://www.latamarte.com/en/articles/quE4/

Laux, J., Wachter, S., & Mittelstadt, B. (2024). Three pathways for standardisation and ethical disclosure by default under the European union artificial intelligence act. *Computer Law & Security Review*, *53*, 105957. DOI: 10.1016/j.clsr.2024.105957 PMID: 41322429

LAWPRO. (2025, March 17). When a Law firm gets Hacked: A case study in Cybersecurity Risks and Recovery. practicePRO. https://www.practicepro.ca/2025/03/when-a-law-firm-gets-hacked-a-case-study-in-cybersecurity-risks-and-recovery/

Lawson-Tancred, J. (2022, October 12). The robot artist Ai-Da just addressed U.K. Parliament about the future of A.I. and 'terrified' the House of Lords. Artnet News. https://news.artnet.com/art-world/ai-da-robot-artist-parliament-2190611

Lazarsfeld, P. F., & Field, P. B. (1946). *The people look at radio.* University of North Carolina Press.

Leprince-Ringuet, D. (2018, December 17). Google's latest experiment teaches AI to dance like a human. WIRED, 67, 75. https://www.wired.com/story/google-ai-wayne-mcgregor-dance-choreography/

Li, G. (2024). A new regulatory road in clinical trials: Digital twins. Applied Clinical Trials. https://www.appliedclinicaltrialsonline.com/view/new-regulatory-road-clinical-trials-digital-twins

Li, S. (2024). How AI is being used to accelerate clinical trials. AAIATECH. https://www.aaiatech.org/how-ai-is-being-used-to-accelerate-clinical-trials/

Litwin, A. S., & Racabi, G. (2024). Varieties of AI Regulations: The United States Perspective. *Industrial & Labor Relations Review, 77*(5), 799–812. DOI: 10.1177/00197939241278956a

Liu, M., Zhang, H., & Sui, Y. (2024). Workplace Artificial Intelligence Regulation in China: Between Efficiency and Social Stability. *Industrial & Labor Relations Review, 77*(5), 813–824. DOI: 10.1177/00197939241278956b

Li, W., Shen, Z., & Wen, X. (2025). Access to banking and corporate dividend policy: Evidence from the distribution of bank branches in China. *Research in International Business and Finance, 75*, 102731. DOI: 10.1016/j.ribaf.2024.102731

Li, Y., Guan, M., Hammond, P., & Berrey, L. E. (2021). Communicating COVID-19 information on TikTok: A content analysis of TikTok videos from official accounts featured in the COVID-19 information hub. *Health Education Research, 36*(3), 261–271. DOI: 10.1093/her/cyab010 PMID: 33667311

Li, Y., & Xie, Y. (2020). Is a picture worth a thousand words? An empirical study of image content and social media engagement. *JMR, Journal of Marketing Research, 57*(1), 1–19. DOI: 10.1177/0022243719881113

López, C. (2023). Artificial intelligence and advanced materials. *Advanced Materials, 35*(23), 2208683. DOI: 10.1002/adma.202208683 PMID: 36560859

Loss aversion. (2025, May). In Wikipedia. Retrieved May 2025, from https://en.wikipedia.org/wiki/Loss_aversion

Lovejoy, K., & Saxton, G. D. (2012). Information, community, and action: How nonprofit organizations use social media. *Journal of Computer-Mediated Communication, 17*(3), 337–353. DOI: 10.1111/j.1083-6101.2012.01576.x

Lundberg, S. M., & Lee, S.-I. (2017). A unified approach to interpreting model predictions. Advances in Neural Information Processing Systems, 30. https://proceedings.neurips.cc/paper_files/paper/2017/file/8a20a8621978632d76c43dfd28b67767-Paper.pdf

Madanchian, M., Taherdoost, H., Vincenti, M., & Mohamed, N. (2024). Transforming Leadership Practices through Artificial Intelligence. *Procedia Computer Science, 235*, 2101–2111. DOI: 10.1016/j.procs.2024.04.199

Maity, S. (2022). Blockchain Basic. https://github.com/Subham-Maity/blockchain-basic

Ma, J. H., Erdogmus, E., Kangisser, S., & Yang, E. (2025). A comparative analysis of the effectiveness of immersive virtual reality on end-user design review. *Building and Environment*, *267*, 112237. DOI: 10.1016/j.buildenv.2024.112237

Malik, R., & Sharma, A. (2023). Empowering customer experience with AI tools and technologies. *Building AI Driven Marketing Capabilities*, 33-45. DOI: 10.1007/978-1-4842-9810-7_3

Malviya, R., & Goyal, P. (2023). Obstacles and complexity in AI embedded remote health monitoring. *Remote Patient Monitoring: A Computational Perspective in Healthcare*, 47-64. DOI: 10.1201/9781032627991-4

Manda, V. K., Christy, V., & Jitta, M. R. (2025). Ethical AI and Decision-Making in Management Leadership. In *Ethical Dimensions of AI Development*. IGI Global., https://www.igi-global.com/chapter/ethical-ai-and-decision-making-in-management-leadership/359644

Mantelero, A., Vaciago, G., Samantha Esposito, M., & Monte, N. (2021). The common EU approach to personal data and cybersecurity regulation. *International Journal of Law and Information Technology*, *28*(4), 297–328. Advance online publication. DOI: 10.1093/ijlit/eaaa021

Marinova, N. (2024). Effects of the Artificial Intelligence Ethical Regulations on Society. *IDEAS*, (1), 99-106. https://ideas.repec.org/a/nwe/iitfed/y2024i1p99-106.html

Mariyono, D., Hidayatullah, A. N., & Alifatul Kamila, A. N. (2024). Navigating ethical dilemmas in AI-enhanced language education: Addressing bias and ensuring inclusivity. DOI: 10.2139/ssrn.4936530

Mark, B., & Bizz, P. (2025). Ethical implications of AI-driven decision making in healthcare information systems. DOI: 10.20944/preprints202503.0078.v1

Markoš, A. (2015). Signs and meanings in evolution. Nová beseda.

Mathai, A. (2024). Enhancing education for underprivileged children through AI-powered native language learning inclusive education through AI-powered native language learning. DOI: 10.2139/ssrn.4899553

Matheny, M. E., Yang, J., Smith, J. C., Walsh, C. G., Al-Garadi, M. A., Davis, S. E., Marsolo, K. A., Fabbri, D., Reeves, R. R., Johnson, K. B., Dal Pan, G. J., Ball, R., & Desai, R. J. (2024). Enhancing postmarketing surveillance of medical products with large language models. *JAMA Network Open*, *7*(8), e2428276. Advance online publication. DOI: 10.1001/jamanetworkopen.2024.28276 PMID: 39150707

Matli, W. (2024). Integration of warrior artificial intelligence and leadership reflexivity to enhance decision-making. *Applied Artificial Intelligence*, *38*(1), 2411462. Advance online publication. DOI: 10.1080/08839514.2024.2411462

Mattessich, R. (2000). *The beginnings of accounting and accounting thought*. Taylor & Francis.

McCarthy, E. (2021). Cybersecurity for Retirement Plan Advisors. https://www.wealthmanagement.com/retirement-planning/cybersecurity-retirement-plan-advisors

McDermid, J. A., Jia, Y., Porter, Z., & Habli, I. (2021). Artificial Intelligence Explainability: The Technical and Ethical Dimensions. *Philosophical Transactions. Series A, Mathematical, Physical, and Engineering Sciences*, *379*(2207), 20200363. DOI: 10.1098/rsta.2020.0363 PMID: 34398656

McDonnell, E. J., Eagle, T., Sinlapanuntakul, P., Moon, S. H., Ringland, K. E., Froehlich, J. E., & Findlater, L. (2024, May 11-16). "Caption it in an accessible way that is also enjoyable": Characterizing user-driven captioning practices on TikTok. *In Proceedings of the CHI Conference on Human Factors in Computing Systems* (pp. 1-16).

McKinsey & Company. (2024). The state of AI: Global survey. https://www.mckinsey.com/capabilities/quantumblack/our-insights/the-state-of-ai

Meng, L. M., Kou, S., Duan, S., & Bie, Y. (2024). The impact of content characteristics of Short-Form video ads on consumer purchase Behavior: Evidence from TikTok. *Journal of Business Research*, *183*, 114874. DOI: 10.1016/j.jbusres.2024.114874

Menona Jacob, A. (2022). *Healthcare delivery systems in rural areas*. Rural Health., DOI: 10.5772/intechopen.98184

Meriç, E., & Çakıcı, Z. (2024). From TikTok trends to pandemic essentials: A comparative analysis of the World Health Organization's health communication strategies on TikTok. In Gürkan, H., & Serttaş, A. (Eds.), *Transformed communication codes in the mediated world: A contemporary perspective* (pp. 1–23). IGI Global Scientific Publishing., DOI: 10.4018/979-8-3693-0896-7.ch001

Mhlanga, D. (2024). Artificial intelligence (AI) financial inclusion and wealth creation in the fourth Industrial Revolution. *Financial Inclusion and Sustainable Development in Sub-Saharan Africa*, 80-95. DOI: 10.4324/9781003515715-6

Mhlanga, D. (2023). The value of open AI and chat GPT for the current learning environments and the potential future uses. Available at *SSRN* 4439267. DOI: 10.2139/ssrn.4439267

Mialon, G., ... (2023). Augmented language models: A survey. arXiv preprint. https://arxiv.org/abs/2302.07842

Miles, M. B., & Huberman, A. M. (1994). *Qualitative data analysis: An expanded sourcebook*. Sage Publications.

Milne, S. (2023, November 29). AI image generator Stable Diffusion perpetuates racial and gendered stereotypes, study finds. UW News.

Mineo, L. (2023, August 15). If it wasn't created by a human artist, is it still art? Harvard Gazette, 238, 247. https://news.harvard.edu/gazette/story/2023/08/is-art-generated-by-artificial-intelligence-real-art/

Mishra, A. (2024). Scalable AI algorithms and models. *Scalable AI and Design Patterns*, 45-60. DOI: 10.1007/979-8-8688-0158-7_4

Mishra, D. (2024). Role of social media and AI in transforming healthcare communication. *Advances in Electronic Government, Digital Divide, and Regional Development*, 29–51. DOI: 10.4018/979-8-3693-5976-1.ch002

MIT Sloan Office of Communications. (2024). Humans and AI: Do they work better together or alone? MIT Sloan School of Management. https://mitsloan.mit.edu/press/humans-and-ai-do-they-work-better-together-or-alone

MIT Technology Review. (2012, May 9). Are smart phones spreading faster than any technology in human history? MIT Technology Review. https://www.technologyreview.com/2012/05/09/186160/are-smart-phones-spreading-faster-than-any-technology-in-human-history/

Mitchell, M., Wu, S., Zaldivar, A., Barnes, P., Vasserman, L., Hutchinson, B., & Gebru, T. (2019). Model cards for model reporting. *Proceedings of the Conference on Fairness, Accountability, and Transparency*, 220–229. DOI: 10.1145/3287560.3287596

Mogavi, R. H., Wang, D., Tu, J., Hadan, H., Sgandurra, S. A., Hui, P., & Nacke, L. E. (2024). Sora OpenAI's Prelude: Social Media Perspectives on Sora OpenAI and the Future of AI Video Generation. *arXiv preprint* arXiv:2403.14665.

Mökander, J., & Floridi, L. (2023). Operationalising AI governance through ethics-based auditing: an industry case study. AI and Ethics. https://link.springer.com/article/10.1007/s43681-022-00171-7

Molem, A., Makri, S., & Mckay, D. (2024, March). Keepin'it Reel: Investigating how short videos on TikTok and Instagram reels influence view change. *In Proceedings of the 2024 Conference on Human Information Interaction and Retrieval* (pp. 317-327).

Monetary Authority of Singapore (MAS). (2024). *Project Ubin: Blockchain for interbank payments.* https://www.mas.gov.sg/schemes-and-initiatives/project-ubin

Moxley, J., Sharit, J., & Czaja, S. J. (2022). The factors influencing older adults' decisions surrounding technology adoption. *JMIR Aging*, *5*(4), e39890. Advance online publication. DOI: 10.2196/39890 PMID: 36416885

Mullard, A. (2018). How much do phase III trials cost? *Nature Reviews. Drug Discovery*, *17*(11), 777. DOI: 10.1038/nrd.2018.198 PMID: 30374183

Nag, P. K., Bhagat, A., & Priya, R. V. (2024). Expanding AI's role in healthcare applications: A systematic review of emotional and cognitive analysis techniques. DOI: 10.36227/techrxiv.172297472.22422829/v1

Narimo, S., Mustofa, R. H., Anindhita, H., & Gano-an, J. C. (2023). Analysis of the utilization of TikTok as a financial and educational learning medium. *Journal of Higher Education Theory and Practice*, *23*(18), 35–45. DOI: 10.33423/jhetp.v23i18.6620

National Institute of Standards and Technology (NIST). (2023). Artificial Intelligence Risk Management Framework (AI RMF 1.0). https://nvlpubs.nist.gov/nistpubs/ai/nist.ai.100-1.pdf

National Institute of Standards and Technology. (2022). Towards a Standard for Identifying and Managing Bias in Artificial Intelligence (NIST Special Publication 1270). https://nvlpubs.nist.gov/nistpubs/SpecialPublications/NIST.SP.1270.pdf

Naudts, L. (2017). Fair or unfair algorithmic differentiation? Luck egalitarianism as a lens for evaluating algorithmic decision-making. SSRN Electronic Journal. DOI: 10.2139/ssrn.3043707

Ndjama, J. D. (2025). Bridging the digital divide in the access and usage of technology through digital literacy in rural vocational schools. Institutes of Higher Education (IHE) and Workforce Collaboration for Digital Literacy, 91-124. DOI: 10.4018/979-8-3373-0004-7.ch004

Neal, R. (2020). Tech Survey 2020: Advisors losing faith in planning software. https://www.financial-planning.com/news/financial-planning-tech-survey-advisors-lose-faith-in-planning-software

Neumeister, L. (2023, June 22). Lawyers submitted bogus case law created by ChatGPT. A judge fined them $5,000. AP News. https://apnews.com/article/artificial-intelligence-chatgpt-fake-case-lawyers-d6ae9fa79d0542db9e1455397aef381c

New York City Council. (2021). Int. No. 1894-A: Automated employment decision tools. https://legistar.council.nyc.gov/LegislationDetail.aspx?ID=4344524&GUID=B051915D-A9AC-451E-81F8-6596032FA3F9

New York City Council. (2021). Local Law Int. No. 1894-A. https://legistar.council.nyc.gov/LegislationDetail.aspx?GUID=B051915D-A9AC-451E-81F8-6596032FA3F9&ID=4344524

New York State Department of Financial Services. (2021, March). Report on Apple Card Investigation. https://www.dfs.ny.gov/reports_and_publications/202103_report_apple_card_investigation

Newswire, P. R. (2025). Clinical trials market size is estimated to be US$ 149.58 billion by 2034. BioSpace. https://www.biospace.com/press-releases/clinical-trials-market-size-is-estimated-to-be-us-149-58-billion-by-2034

Ng, S. C. H., Ho, G. T. S., & Wu, C. H. (2023). Blockchain-IIoT-big data aided process control and quality analytics. *International Journal of Production Economics*, *261*, 108871. DOI: 10.1016/j.ijpe.2023.108871

Nienhaus, V. (2021). Digital transformation of the world economy by AI: Some moral, ethical and Sharīʿah concerns., *16*(16), 245-277. DOI: 10.33001/M011020211591

Nierengarten, N. (2006). A Plan for Wealth Managers to Reduce the Risk of Cyber Threats. *The Journal of Wealth Management*, *9*(3), 24–30. DOI: 10.3905/jwm.2006.661429

Nour, S. M., Abdelwahab, M. A., Mahmoud, M. S., & Nabil, K. E. (2025). Innovation in medical application based on artificial intelligence (AI) by cutting-edge technologies. *Practice, Progress, and Proficiency in Sustainability*, 57–68. DOI: 10.4018/979-8-3693-7117-6.ch003

Ntjamba, F. C., & Ashipala, D. O. (2024). Impact on and ethical considerations of artificial intelligence on human healthcare. *Advances in Computational Intelligence and Robotics*, •••, 1–36. DOI: 10.4018/979-8-3693-9158-7.ch001

Nuka, T. F. (2024). AI-driven financial inclusion: The role of DEI in shaping equitable credit opportunities for underserved communities. *International Journal of Research Publication and Reviews*, *5*(11), 3824–3840. DOI: 10.55248/gengpi.5.1124.3306

Obermeyer, Z., Powers, B., Vogeli, C., & Mullainathan, S. (2019). Dissecting racial bias in an algorithm used to manage the health of populations. *Science*, *366*(6464), 447–453. https://www.science.org/doi/10.1126/science.aax2342. DOI: 10.1126/science.aax2342 PMID: 31649194

Oh, J., & Ki, E. (2024). Can we build a relationship through artificial intelligence (AI)? Understanding the impact of AI on organization-public relationships. *Public Relations Review, 50*(4), 102469. DOI: 10.1016/j.pubrev.2024.102469

Ojedokun, S. A. (2024). Artificial intelligence (AI) and machine learning (ML); A revolutionary game changer in clinical laboratory diagnosis. *Clinical Pathology & Research Journal, 8*(1), 1–4. DOI: 10.23880/cprj-16000194

Olcott, D. (2024). Leadership in the era of AI. *E-mentor, 5*(107), 5–9.

Oliva, P., Jack, B. K., Bell, S., Severen, C., & Walker, E. (2020). Technology adoption under uncertainty: Take-up and subsequent investment in Zambia. *The Review of Economics and Statistics, 102*(3), 617–632. DOI: 10.1162/rest_a_00823

Ooi, K. H. (2022). How political parties used TikTok in the 2022 Malaysian general election. *RSIS Commentaries,* 133 (22), 1-4. https://hdl.handle.net/10356/165438
OpenAI (@OpenAI) / X. (2025, Şubat 27). X (Formerly Twitter). https://x.com/openai

Open, A. I. (@openai) Resmi | TikTok. (t.y.). Geliş tarihi 13 Mart 2025, gönderen https://www.tiktok.com/@openai

Open, A. I. (2023). Usage policies. https://openai.com/policies/usage-policies

Open, A. I. (t.y.). YouTube. Geliş tarihi 12 Mart 2025, gönderen https://www.youtube.com/channel/UCXZCJLdBC09xxGZ6gcdrc6A

Open, A. I. Gönderiler | LinkedIn. (t.y.). Geliş tarihi 12 Mart 2025, gönderen https://www.linkedin.com/company/openai/posts/?feedView=all

Osmanovic Thunström, A. (2022, September 30). We asked GPT-3 to write an academic paper about itself. Scientific American. https://www.scientificamerican.com/article/we-asked-gpt-3-to-write-an-academic-paper-about-itself-mdash-then-we-tried-to-get-it-published/

Our approach to AI safety. (2024, Şubat 14). https://openai.com/index/our-approach-to-ai-safety/

Oyetade, K. E., Zuva, T., & Harmse, A. (2020). Technology adoption in education: A systematic literature review. *Advances in Science, Technology and Engineering Systems Journal, 5*(6), 108–112. https://www.astesj.com/publications/ASTESJ_050611.pdf. DOI: 10.25046/aj050611

Pagani, M., & Champion, R. (2023). Could artificial intelligence make us humans more creative? *Artificial Intelligence for Business Creativity*, 24-43. DOI: 10.4324/9781003287582-4

Palla, S. R. (2025). AI-powered supply chain innovation: From predictive analytics to warehouse automation. *INTERNATIONAL JOURNAL OF INFORMATION TECHNOLOGY AND MANAGEMENT INFORMATION SYSTEMS*, *16*(1), 888–901. DOI: 10.34218/IJITMIS_16_01_063

Pandey, M. (2024). Access to the healthcare in remote areas is support for poverty reduction. *International Journal of Social Science and Human Research*, *07*(02). Advance online publication. DOI: 10.47191/ijsshr/v7-i02-65

Paschen, U., Pitt, C., & Kietzmann, J. (2020). Artificial intelligence: Building Blocks and an Innovation Typology. *Business Horizons*, *63*(2), 147–155. DOI: 10.1016/j.bushor.2019.10.004

Pasquale, F. (2023). A rule of persons, not machines: The limits of legal automation. *Yale Journal of Law & Technology*, *25*(2), 110–152.

Patil, D. (2024). Ethical challenges in industrial artificial intelligence applications: Bias, privacy, and accountability. https://www.researchgate.net/publication/385890075_Ethical_challenges_in_industrial_artificial_intelligence_applications_Bias_Privacy_and_Accountability

Patton, M. Q. (2002). *Qualitative research and evaluation methods* (3rd ed.). Sage Publications.

Payne, P., & Asebedo, S. (2019). Two-factor risk preference for investment market and credit card risk. In *Financial Planning Review* (Vol. 2). Wiley., DOI: 10.1002/cfp2.1062

Peng, H., Mengni, Z., Yang, L., Juanatas, R., Niguidula, J., & Huiliang, H. (2023). Research on brand marketing strategy on Tik Tok short video platform. *In SHS Web of Conferences* (Vol. 159, p. 02024). EDP Sciences.

Perc, M., Ozer, M., & Hojnik, J. (2019). Social and Juristic Challenges of Artificial Intelligence. *Palgrave Communications*, *5*(1), 61. DOI: 10.1057/s41599-019-0278-x

Perez-Lopez, R., Ghaffari Laleh, N., Mahmood, F., & Kather, J. N. (2024). A guide to artificial intelligence for cancer researchers. *Nature Reviews. Cancer*, *24*(6), 427–441. DOI: 10.1038/s41568-024-00694-7 PMID: 38755439

Permatasari, P. A., Rech, M., Qohar, A., & Faizal, A. (2020). From web 1.0 to web 4.0: The digital heritage platforms for UNESCO's heritage properties in Indonesia. *Virtual Archaeology Review*, *11*(23), 75–93. DOI: 10.4995/var.2020.13121

Pesapane, F., Volonte, C., Codari, M., & Sardanelli, F. (2018). Artificial Intelligence as a Medical Device in Radiology: Ethical and Regulatory Issues in Europe and the United States. *Insights Into Imaging*, *9*(5), 745–753. DOI: 10.1007/s13244-018-0645-y PMID: 30112675

Pew Research Center. (2014). The web at 25 in the U.S. https://www.pewresearch.org/internet/2014/02/27/the-web-at-25-in-the-u-s/

Pew Research Center. (2025). Internet/broadband fact sheet. https://www.pewresearch.org/internet/fact-sheet/internet-broadband/

Pew Research Center. (2025). Mobile fact sheet. https://www.pewresearch.org/internet/fact-sheet/mobile/

Phan, H. P., & Ngu, B. H. (2021). A case for cognitive entrenchment: To achieve optimal best, taking into account the importance of perceived optimal efficiency and cognitive load imposition. *Frontiers in Psychology*, *12*, 662898. Advance online publication. DOI: 10.3389/fpsyg.2021.662898 PMID: 34385950

Pilipiec, P., Liwicki, M., & Bota, A. (2022a). Using machine learning for pharmacovigilance. *Systematic Reviews in Pharmacy*, *14*(2), 266. Advance online publication. DOI: 10.3390/pharmaceutics14020266 PMID: 35213998

Pizza, G. (2016). Growth hacking: A new marketing mindset [Master's thesis, LUISS Guido Carli University]. https://tesi.luiss.it/16597/1/655681.pdf

Popescu, C. R., & Yu, P. (Eds.). (2024). *Intersecting Environmental Social Governance and AI for Business Sustainability*. IGI Global., DOI: 10.4018/979-8-3693-1151-6

Powell, T. C., & Dent-Micallef, A. (1997). Information technology as competitive advantage: The role of human, business, and technology resources. *Strategic Management Journal*, *18*(5), 375–405. DOI: 10.1002/(SICI)1097-0266(199705)18:5<375::AID-SMJ876>3.0.CO;2-7

PricewaterhouseCoopers (PwC). (2024). EU's AI Act: What regulators should know. https://www.pwc.com/us/en/services/consulting/cybersecurity-risk-regulatory/library/tech-regulatory-policy-developments/eu-ai-act.html

Putriani, I., & Aras, M. (2022). Elevate Company's Social Media as a Diversity, Equity, and Inclusion Platform: Multi Cases Study to Digital Technology Companies. *Journal of Wood Science*, *1*(10), 906–916. DOI: 10.36418/jws.v1i10.105

Qwiumin, s. (2023). Enhancing public service delivery with AI-driven solutions in developing countries. https://doi.org/DOI: 10.31219/osf.io/cwqz4

Radicati Group. (2022). *Email market, 2022–2026* [Executive summary]. The Radicati Group, Inc.

Ragab, K. (2025). Empowering STEM education with AI and prompt engineering. *Advances in Computational Intelligence and Robotics*, 175–188. DOI: 10.4018/979-8-3693-7332-3.ch011

Raghuwanshi, V., Khare, P., & Sharma, P. (2025). Global AI Governance and Collaboration: The Role of International Bodies in AI Regulation. *IGI Global Scientific Publishing*, 225-250. https://doi.org/DOI: 10.4018/979-8-3373-1210-1.ch009

Rajkomar, A., Oren, E., Chen, K., Dai, A. M., Hajaj, N., Hardt, M., Liu, P. J., Liu, X., Marcus, J., Sun, M., Sundberg, P., Yee, H., Zhang, K., Zhang, Y., Flores, G., Duggan, G. E., Irvine, J., Le, Q., Litsch, K., & Dean, J. (2018). Scalable and accurate deep learning with electronic health records. npj. *Digital Medicine*, *1*(1), 18. Advance online publication. DOI: 10.1038/s41746-018-0029-1 PMID: 31304302

Rana, J., & Daultani, Y. (2023). Mapping the role and impact of artificial intelligence and machine learning applications in supply chain digital transformation: A bibliometric analysis. *Operations Management Research : Advancing Practice Through Research*, *16*(4), 1641–1666. DOI: 10.1007/s12063-022-00335-y

Ratcliffe, J. (2023). Can AI help market your brand? *Journal of Aesthetic Nursing*, *12*(2), 98–100. DOI: 10.12968/joan.2023.12.2.98

Reethika, A., & Priya, P. K. (2024). Using AI-powered predictive analytics tools to identify students falling behind or dropping out. *Innovation in the University 4.0 System based on Smart Technologies*, 101-115. DOI: 10.1201/9781003425809-7

ReliefWeb. (2024). *Finland's blockchain experiment for public records*. https://reliefweb.int/report/finland/how-finland-using-blockchain-revolutionise-financial-services-refugees

Remolina, N. (2019, October 24). Open Banking: Regulatory Challenges for a New Form of Financial Intermediation in a Data-Driven World. Papers.*ssrn*.com. https://papers.ssrn.com/sol3/papers.cfm?abstract_id=3475019

Reuters. (2023, February 1). ChatGPT sets record for fastest-growing user base – analyst note. Reuters. https://www.reuters.com/technology/chatgpt-sets-record-fastest-growing-user-base-analyst-note-2023-02-01/

Ribeiro, M. T., Singh, S., & Guestrin, C. (2016). "Why should I trust you?": Explaining the predictions of any classifier. Proceedings of the 22nd ACM SIGKDD International Conference on Knowledge Discovery and Data Mining, 1135–1144. DOI: 10.1145/2939672.2939778

Rizomyliotis, I., Lin, C. L., Konstantoulaki, K., & Phan, T. (2024). TikTok short video marketing and Gen Z's purchase intention: Evidence from the cosmetics industry in Singapore. *Journal of Asia Business Studies, 18*(4), 930–945. DOI: 10.1108/JABS-04-2023-0138

Roberts, K., Anderson, M., & Lee, S. (2023). Leadership development for AI-integrated organizations. *Management Development Review, 31*(3), 178–195.

Rodgers, W., Murray, J. M., Stefanidis, A., & Li, T. (2023). An artificial intelligence algorithmic approach to ethical decision-making in human resource management processes. *Human Resource Management Review, 33*(3), 100932. DOI: 10.1016/j.hrmr.2022.100925

Rogers, E. M. (2003). Diffusion of innovations (5th ed.). Free Press. Mueller, S. M., & Brand, M. (2018). Approximate number processing skills contribute to decision making under objective risk: Interactions with executive functions and objective numeracy. Frontiers in Psychology, 9, Article 1202. https://doi.org/DOI: 10.3389/fpsyg.2018.01202

Romano, A. (2023, April 5). ChatGPT lied about a law professor, and he's not the only one. The Washington Post. https://www.washingtonpost.com/technology/2023/04/05/chatgpt-lies/

Roumeliotis, K. I., & Tselikas, N. D. (2023). ChatGPT and Open-AI models: A preliminary review. *Future Internet, 15*(6), 192. DOI: 10.3390/fi15060192

Royal Commission into the Robodebt Scheme. (2023). Report of the Royal Commission into the Robodebt Scheme. https://robodebt.royalcommission.gov.au/publications/report

Roy, P., Ghose, B., Singh, P. K., Tyagi, P. K., & Vasudevan, A. (2025). Artificial Intelligence and Finance: A bibliometric review on the Trends, Influences, and Research Directions. *F1000 Research, 14*, 122. DOI: 10.12688/f1000research.160959.1 PMID: 39911727

Ruiz-Vanoye, J., & Díaz-Parra, O. (2025). *Machine and deep learning solutions for achieving the sustainable development goals*. IGI Global. DOI: 10.4018/979-8-3693-8161-8

Rukadikar, A., & Khandelwal, K. (2024). Leadership development through self-upskilling: Role of generative artificial intelligence. *Development and Learning in Organizations, 38*(4), 27–30. Advance online publication. DOI: 10.1108/DLO-01-2024-0005

Sabry, F. (2022). *Artificial general intelligence: Machines may eventually be as clever as humans, and possibly even smarter, but the game is still far from over.* PublishDrive KFT.

Sadoian, L. (2024, January 2). Ultimate List of Cybersecurity Regulations by Industry | UpGuard. Www.upguard.com. https://www.upguard.com/blog/cybersecurity-regulations-by-industry

Saheb, T., & Saheb, T. (2024). Mapping Ethical Artificial Intelligence Policy Landscape: A Mixed Method Analysis. *Science and Engineering Ethics*, *30*(2), 9. https://10.0.3.239/s11948-024-00472-6. DOI: 10.1007/s11948-024-00472-6 PMID: 38451328

Saifi, S., Tanveer, S., Arwab, M., Lal, D., & Mirza, N. (2025). Exploring the persistence of Open AI adoption among users in Indian higher education: A fusion of TCT and TTF model. *Education and Information Technologies*, *30*(9), 12055–12078. Advance online publication. DOI: 10.1007/s10639-024-13282-x

Salinger, T. (2022). What advisory firms should be doing on cybersecurity. https://www.financial-planning.com/news/how-rias-and-financial-advisors-can-bulk-up-on-cybersecurity

Sanford, C. (2022). European Medicines Agency qualifies Unlearn's AI-powered method for running smaller, faster clinical trials.

Scherer, M. U. (2015). Regulating Artificial Intelligence Systems: Risks, Challenges, Competencies, and Strategies. *SSRN*, *29*(2), 353–400. DOI: 10.2139/ssrn.2609777

Schwartz, C. (1991). CHECKING OUT FINANCIAL PLANNERS. *The Bottom Line (New York, N.Y.)*, *4*(1), 33–33. DOI: 10.1108/eb025274

ScienceDaily. (2018, July 10). Artificial intelligence helps researchers predict drug combinations' side effects. https://www.sciencedaily.com/releases/2018/07/180710072013.htm

SEC. (2022). SEC.gov | SEC Proposes Cybersecurity Risk Management Rules and Amendments for Registered Investment Advisers and Funds. SEC. https://www.sec.gov/newsroom/press-releases/2022-20 Sen, K. (2024). Top 9 Cybersecurity Regulations for Financial Services. https://www.upguard.com/blog/cybersecurity-regulations-financial-industry

Selbst, A. D., & Powles, J. (2017). Meaningful information and the right to explanation. *International Data Privacy Law*, *7*(4), 233–242. DOI: 10.1093/idpl/ipx022

Seltzer, B., & Lewek, T. (2019). Information Security for Academics: Small steps to Safeguard Your data – profession. Retrieved April 19, 2025, from https://profession.mla.org/information-security-for-academics-small-steps-to-safeguard-your-data/

Selwyn, E. J., Tiwari, A., Pandurang, G. G., Das, S. S., & Vardhan, R. (2025). Healthsyncc: An AI-powered mobile application for health management. *Modern Technologies in Healthcare*, 241-266. DOI: 10.1201/9781003481959-11

Shah, N. (2024). Artificial Intelligence and Leadership: How Artificial Intelligence is Changing the Leadership Role. Remittances Review. https://remittancesreview.com/menu-script/index.php/remittances/article/view/1487

Sharma, G. K., Tyagi, E., & Chaudhary, A. (2025). Ethical considerations in the use of AI and big data in corporate decision-making. *Digital Citizenship and the Future of AI Engagement, Ethics, and Privacy*, 467-532. DOI: 10.4018/979-8-3693-9015-3.ch017

Shaw, R. S., Chen, C. C., Harris, A. L., & Huang, H.-J. (2009). The impact of information richness on information security awareness training effectiveness. *Computers & Education*, *52*(1), 92–100. DOI: 10.1016/j.compedu.2008.06.011

Sheak, E., & Abdulrazak, S. (2023). The influence of social media marketing activities on TikTok in raising brand awareness. *Market-Tržište*, *35*(1), 93–110. DOI: 10.22598/mt/2023.35.1.93

Shibuya, K. (2020). Correction to: Digital transformation of identity in the age of artificial intelligence. *Digital Transformation of Identity in the Age of Artificial Intelligence*, C1-C1. DOI: 10.1007/978-981-15-2248-2_15

Shin, S., Cunningham, J., Ryoo, J., & Tucci, J. E. (2009). Authentication and protection for e-finance consumers: the dichotomy of cost versus ease of use. In International Journal of Electronic Finance (Vol. 3, Issue 1, p. 31). Inderscience Publishers. https://doi.org/DOI: 10.1504/IJEF.2009.024268

Shinde, N., & Kulkarni, P. (2021). Cyber incident response and planning: A flexible approach. *Computer Fraud & Security*, *2021*(1), 14–19. DOI: 10.1016/S1361-3723(21)00009-9

Shrestha, Y. R., Ben-Menahem, S. M., & von Krogh, G. (2019). Organizational decision-making structures in the age of artificial intelligence. *California Management Review*, *61*(3), 57–82. DOI: 10.1177/0008125619862257

Sidoo, F. (2023). Ensuring ethical data collection and use in AI-driven development programs. DOI: 10.31219/osf.io/5mhcw

Silk, M. S. W. (2022, November 30). Man v. Machine: The ethics of AI art. Prindle Institute. https://www.prindleinstitute.org/2022/11/man-v-machine-the-ethics-of-ai-art/

Sillince, J. A., Macdonald, S., Lefang, B., & Frost, B. (1998). Email adoption, diffusion, use and impact within small firms: A survey of UK companies. *International Journal of Information Management*, *18*(4), 231–242. DOI: 10.1016/S0268-4012(98)00012-7

Singh, P., & Singh, P. (2024). AI strategies for delivering better value in the healthcare sector. *Advances in Medical Technologies and Clinical Practice*, 131–156. DOI: 10.4018/979-8-3693-3731-8.ch007

Sinha, S. (2024). Revolutionizing human health: The role of artificial intelligence in modern healthcare. DOI: 10.31224/4181

Skadden. (2023). EU Artificial Intelligence Act – Overview and impact. https://www.skadden.com/insights/publications/2023/06/quarterly-insights/ai-and-the-workplace

Slack, D., Hilgard, S., Jia, E., Singh, S., & Lakkaraju, H. (2020). Fooling LIME and SHAP: Adversarial attacks on post hoc explanation methods. In Proceedings of the AAAI/ACM Conference on AI, Ethics, and Society (pp. 180–186). Association for Computing Machinery. DOI: 10.1145/3375627.3375830

Smt. Sumela Chatterjee. (2023). Evaluating the effects of AI-powered training programs on skill development and career growth. *International Journal of Advanced Research in Science, Communication and Technology*, 941-945. DOI: 10.48175/IJARSCT-12743D

Sol, K., & Heng, K. (2024). AI-powered chatbots as personalized academic writing assistants for non-native English speakers. *Encyclopedia of Educational Innovation*, 1-5. DOI: 10.1007/978-981-13-2262-4_313-1

Spencer Stuart. (2023). 2023 U.S. Spencer Stuart board index (38th ed.). https://www.spencerstuart.com/-/media/2023/september/usbi/2023_us_spencer_stuart_board_index.pdf

Sriharan, A., Sekercioglu, N., Mitchell, C., Senkaiahliyan, S., Hertelendy, A., Porter, T., & Banaszak-Holl, J. (2024). Leadership for AI transformation in health care organization: A scoping review. *Journal of Medical Internet Research*, *26*(1), e54556. DOI: 10.2196/54556 PMID: 39009038

Stafford, T. F., & Treiblmaier, H. (2020). Characteristics of a Blockchain Ecosystem for Secure and Sharable Electronic Medical Records. *IEEE Transactions on Engineering Management*, *67*(4), 1340–1362. DOI: 10.1109/TEM.2020.2973095

Stamoulis, D., & Giannopoulos, P. (2021). Pesonalising weather forecasts using AI techniques. DOI: 10.5194/ems2021-502

Suh, C. (2022). *3. Machine learning applications.* Convex Optimization for Machine Learning.

Sukhdeve, P. S. (2021). Implementing augmented reality into immersive virtual learning environments. *Advances in Educational Technologies and Instructional Design*, 102–118. DOI: 10.4018/978-1-7998-4222-4.ch006

Sullivan, Y., & Fosso Wamba, S. (2024). Artificial intelligence and adaptive response to market changes: A strategy to enhance firm performance and innovation. *Journal of Business Research, 174*, 114500. DOI: 10.1016/j.jbusres.2024.114500

Suresh, N. V., Selvakumar, A., Sridhar, G., & Catherine, S. (2025). Ethical considerations in AI implementation for patient data security and privacy. *Advances in Healthcare Information Systems and Administration*, 139–147. DOI: 10.4018/979-8-3693-7452-8.ch008

Svítek, M. (2024). Integrated information assessment using geometric algebra. Neural Network World, 3. Ashby, W. R. (1956). Introduction to cybernetics. J. Wiley.

Svítek, M. (2021). *Information physics - Physics-information analogies for complex systems modelling*. Elsevier.

Svítek, M., & Žák, L. (2022). *Behind the mirror = Za zrcadlem.* Svítek & Žák.

Svítek, M., & Žák, L. (2024). *In mysterious ways = Nevyzpytatelnými cestami = Неизноведимимы путьями. Svítek & Žák. Mařík, V., Maříková, T., & Svítek, M. (2024). Essays on consciousness - Towards artificial intelligence.* Pavel Mervart.

Svítek, M., & Žák, L. (2024). *Viability and living*. Svítek & Žák.

Taleb, N. N. (2007). *The black swan: The impact of the highly improbable*. Random House.

TalktoAngel. (2024). Fluid and crystallized intelligence in older adults. https://www.talktoangel.com/blog/fluid-and-crystallized-intelligence-in-older-adults

Tarafder, A., & Vadlamani, A. (2024). Will the EU AI Regulations Give Rise to Another 'Brussels Effect'? Lessons from the GDPR. *Journal of Development Policy and Practice, 10*(1), 45–60. DOI: 10.1177/24551333241247670

Taylor, S. H., & Brisini, K. S. C. (2024). Parenting the TikTok algorithm: An algorithm awareness as process approach to online risks and opportunities. *Computers in Human Behavior, 150*, 107975. DOI: 10.1016/j.chb.2023.107975

Thaler v. Perlmutter, No. 1:22-cv-01564 (D.D.C. Aug. 18, 2023).

The Corporate Governance Institute. (2024). A board member's guide to digital transformation. https://www.thecorporategovernanceinstitute.com/insights/guides/a-board-members-guide-to-digital-transformation/

The Guardian. (2024). *Blockchain's role in reducing corruption.* https://guardian.ng/features/executive-briefs/records-and-information-management-play-a-critical-role-in-fighting-corruption/

The Moscow Times. (2017). *Russia's blockchain platform for public services.* https://www.themoscowtimes.com/2017/10/19/russian-government-announces-first-blockchain-project-for-moscows-property-registries-a59321

The White House. (2022, October 4). Blueprint for an AI Bill of Rights. Biden White House Archives. https://bidenwhitehouse.archives.gov/ostp/ai-bill-of-rights/

The White House. (2023a, February 16). Executive order on further advancing racial equity and support for underserved communities through the federal government. Biden White House Archives. https://bidenwhitehouse.archives.gov/briefing-room/presidential-actions/2023/02/16/executive-order-on-further-advancing-racial-equity-and-support-for-underserved-communities-through-the-federal-government/

The White House. (2023b, October 30). Executive order on the safe, secure, and trustworthy development and use of artificial intelligence. Biden White House Archives. https://bidenwhitehouse.archives.gov/briefing-room/statements-releases/2023/10/30/fact-sheet-president-biden-issues-executive-order-on-safe-secure-and-trustworthy-artificial-intelligence/

The World's First AI-Generated Portrait Sold for $432,500. (2018, October 25). The New York Times. (Quoted in Clancy & Tweed, The Voice of the Artist, p. 15).

thelittlewonder. (2018). GitHub - thelittlewonder/Blockchain-and-Cryptocurrencies: Notes for the course IBCC630E. GitHub. https://github.com/thelittlewonder/Blockchain-and-Cryptocurrencies

Tips to Tighten Your Online Security. (2022). https://www.mindbodyonline.com/business/education/blog/how-to-tighten-online-security

Treasury Board of Canada Secretariat. (2020). Directive on Automated Decision-Making. https://www.canada.ca/en/government/system/digital-government/digital-government-innovations/responsible-use-ai/algorithmic-impact-assessment.html

Trottman-Adewumi, Y., Kelley, D., Smuglin, L., & Markovich, G. (2017). Assessing cybersecurity risks and practices in the broker-dealer ...: Ingenta Connect. Ingentaconnect.com. https://www.ingentaconnect.com/content/hsp/jsoc/2017/00000009/00000004/art00003

Truby, J., Brown, R. D., Ibrahim, I. A., & Parellada, O. C. (2022). A Sandbox Approach to Regulating High-risk Artificial Intelligence Applications. *European Journal of Risk Regulation*, *13*(2), 270–294. DOI: 10.1017/err.2021.52

Trunk, A., Birkel, H., & Hartmann, E. (2020). On the current state of combining human and artificial intelligence for strategic organizational decision-making. *Business Research*, *13*(2), 345–374. DOI: 10.1007/s40685-020-00133-x

Trust, T., & Maloy, R. W. (2024). Generative AI writing technologies. *How to Use Digital Learning with Confidence and Creativity*, 74-84. DOI: 10.4337/9781035311293.00020

Tuliao, D. (2014). Online Shopping? Five Security Tips. https://www.attorneyatwork.com/online-shopping-security-tips/

Turk, Z. (2024). Regulating Artificial Intelligence: A Technology-independent Approach. *European View*, *23*(1), 87–93. DOI: 10.1177/17816858241242890

Twitter. (2021). Twitter's first algorithmic bias bounty challenge. https://blog.twitter.com/engineering/en_us/topics/insights/2021/algorithmic-bias-bounty-challenge

U. S. (2013). Computer Security Incident Handling Guide. https://www.amazon.com/Computer-Security-Incident-Handling-Guide/dp/1494726378

U4 Anti-Corruption Resource Centre. (2019). *Anti-corruption reforms successful in Georgia—Blockchain stealing limelight*. Chr. Michelsen Institute. https://www.u4.no/blog/anti-corruption-reforms-successful-in-georgia-blockchain-stealing-limelight

Uddin, A. S. M. (2023). The Era of AI: Upholding Ethical Leadership. Open Journal of Leadership. https://www.scirp.org/journal/paperinformation?paperid=129401

Uddin, M. H., Ali, M. H., & Hassan, M. K. (2020). Cybersecurity hazards and financial system vulnerability: a synthesis of literature. In Risk Management (Vol. 22, Issue 4, p. 239). Palgrave Macmillan. https://doi.org/DOI: 10.1057/s41283-020-00063-2

Umoga, J., & Oluwademilade, E. (2024). A critical review of emerging cybersecurity threats in financial technologies. *International Journal of Science and Research Archive*, *11*(1), 1810–1817. DOI: 10.30574/ijsra.2024.11.1.0284

UN ESCAP. (2024). Asia-Pacific tech monitor (Vol. 43, No. 1, Jan-Mar 2024). United Nations Economic and Social Commission for Asia and the Pacific. https://repository.unescap.org/handle/20.500.12870/7421

UNESCO. (2023). *Artificial intelligence in education: Challenges and opportunities.*

Unlearn.AI. (2024). How AI-generated digital twins are speeding up clinical development. https://www.unlearn.ai/

Upadhyay, N.K., & Romashkin, S. (2019). Using Artificial Intelligence to Address Criminal Justice Needs. *Legal Analytics*, 21-34. https://doi.org/DOI: 10.1201/9781003215998-3

van der Veen, B. (2024). Bridging the generative leap: Exploring factors determining generative AI-readiness in organizations [master's thesis]. Eindhoven University of Technology.https://pure.tue.nl/ws/portalfiles/portal/340287031/Master_Thesis_Bas_van_der_Veen.pdf

Vedral, V. (2010). Introduction to quantum information science. Oxford Academic. Hoffmann, D. W. (2024). Gödel's incompleteness theorems - A guided tour through Kurt Gödel's historic proof. Springer.

Vesna, L. (2025). Digital divide in AI-powered education: Challenges and solutions for equitable learning. *Journal of Information Systems Engineering & Management, 10*(21s), 300–308. DOI: 10.52783/jisem.v10i21s.3327

Vidè, F., & Giacomelli, G. (2024). How new competencies drive innovation. Performance Appraisals Balancing Innovation and Tradition. https://www.torrossa.com/gs/resourceProxy?an=5867867&publisher=FM2570

Vijay, D., & Gekker, A. (2021). Playing politics: How Sabarimala played out on TikTok. *The American Behavioral Scientist, 65*(5), 712–734. DOI: 10.1177/0002764221989769

Vishnu Priya, N. (2022). Merck KGaA collaborate to expedite immunology trials. Clinical Trials Arena. https://www.clinicaltrialsarena.com/news/unlearn-ai-merck-immunology-trials/

Voorneveld, M. (2024). AI technology adoption & Sustainability improvement though cloud solutions. *Proceedings of the 26th International Conference on Enterprise Information Systems*, 675-687. DOI: 10.5220/0012612500003690

Wachter, S., Mittelstadt, B., & Russell, C. (2017). Counterfactual explanations without opening the black box. *SSRN, 31*(2), 841–887. DOI: 10.2139/ssrn.3063289

Walsh, B. (2020, August 19). How an AI grading system ignited a national controversy in the U.K. Axios. https://www.axios.com/2020/08/19/england-exams-algorithm-grading

Wang, X. (2025). Industry innovation in the era of artificial intelligence. DOI: 10.1201/9781003510604

Wang, S., & Ji, Q. (2015). Video affective content analysis: A survey of state-of-the-art methods. *IEEE Transactions on Affective Computing*, *6*(4), 410–430. DOI: 10.1109/TAFFC.2015.2432791

Wang, W. (2024). *Ethical issues in personalized medicine: Privacy, consent, and data sharing*. GMO Biosafety Research., DOI: 10.5376/gmo.2024.15.0002

Wang, X., Chen, M., & Chen, N. (2024). How Artificial Intelligence Affects the Labour Force Employment Structure from the Perspective of Industrial Structure Optimisation. *Heliyon*, *10*(5), e26686. DOI: 10.1016/j.heliyon.2024.e26686 PMID: 38434398

Weidinger, L., Uesato, J., Rauh, M., Griffin, C., Huang, P.-S., Mellor, J., Glaese, A., Cheng, M., Balle, B., Kasirzadeh, A., Biles, C., Brown, S., Kenton, Z., Hawkins, W., Stepleton, T., Birhane, A., Hendricks, L. A., Rimell, L., Isaac, W., & Gabriel, I. (2022). Ethical and social risks of harm from language models. *ACM FAccT*, *2022*, 214–229. Advance online publication. DOI: 10.1145/3531146.3533088

Wikipedia contributors. (2020, December 4). Robodebt scheme. Wikipedia. https://en.wikipedia.org/wiki/Robodebt_scheme

Wikipedia contributors. (2024, February 14). Ofqual exam results algorithm. Wikipedia. https://en.wikipedia.org/wiki/Ofqual_exam_results_algorithm

Wilcox, S. (2024, December 2). Leveraging tech to help manage clients in uncertain times - Professional Paraplanner. Professional Paraplanner. https://professionalparaplanner.co.uk/development/leveraging-tech-to-help-manage-clients-in-uncertain-times/

Wingström, R., (2022). Redefining creativity in the era of AI? Perspectives of computer scientists and new media artists. *Digital Creativity*, *33*(3), 183–201.

Wong, C. H., Siah, K. W., & Lo, A. W. (2019). Estimation of clinical trial success rates and related parameters. *Biostatistics (Oxford, England)*, *20*(2), 273–286. DOI: 10.1093/biostatistics/kxx069 PMID: 29394327

Wood, T. (2024). How can AI be used in clinical trials? Fast Data Science. https://fastdatascience.com/ai-in-pharma/ai-in-clinical-trials/

World Economic Forum & ZS. (2024). Intelligent clinical trials: Using generative AI to fast-track therapeutic innovations. Medidata. https://www.medidata.com/en/life-science-resources/medidata-blog/generative-ai-in-clinical-development/

World Economic Forum. (2024). *Chile initiates blockchain for public procurement.* https://www3.weforum.org/docs/WEF_Blockchain_Government_Transparency_Report_Supplementary%20Research.pdf

World Economic Forum. (2024, December). 5 ways generative AI could transform clinical trials. https://www.weforum.org/stories/2024/12/genai-could-transform-clinical-research-trials/

World Economic Forum. (2024, January). How is AI impacting and shaping the creative industries? (WEF Annual Meeting, Davos).

World Health Organization (WHO). (2021). *Counterfeit drugs in Africa.*

Xi, W. (2024). Regulatory Changes and Compliance Challenges. Emerald Publishing Limited EBooks, 119–134. https://doi.org/DOI: 10.1108/978-1-83608-106-720241008

Yadav, N., Pandey, S., Gupta, A., Dudani, P., Gupta, S., & Rangarajan, K. (2023). Data privacy in healthcare: In the era of artificial intelligence. *Indian Dermatology Online Journal*, *14*(6), 788–792. DOI: 10.4103/idoj.idoj_543_23 PMID: 38099022

Yadav, V. (2023). AI for predicting disease outbreaks: Developing AI systems capable of predicting disease outbreaks by analyzing global health data and trends. *Journal of Artificial Intelligence. Machine Learning and Data Science*, *1*(1), 946–953. DOI: 10.51219/JAIMLD/vivek-yadav/226

Yeske, D., & Buie, E. (2014). Policy-Based Financial Planning: Decision Rules for a Changing World (p. 189). https://doi.org/DOI: 10.1002/9781118813454.ch11

Ye, X., Zhai, M., & Feng, L. (2023). Improving access to high-quality teachers in underdeveloped areas: Evidence from free teacher education in China. DOI: 10.2139/ssrn.4503575

Yin, R. K. (2018). Case Study Research and Applications: Design and Methods (6th ed.). *SAGE Publications*,Inc.Retrieved from https://uk.sagepub.com/en-gb/eur/case-study-research-and-applications/book250150

Yoon, J. (2023). Integrating AI tools and linguistic content in college education. *Convergence English Language & Literature Association*, *8*(3), 349–370. DOI: 10.55986/cell.2023.8.3.349

Yu, P., Gong, R., & Sampat, M. (2022). Blockchain Technology in China's Digital Economy: Balancing Regulation and Innovation. In Global, I. G. I. (Ed.), *Blockchain Technology in China's Digital Economy: Balancing Regulation and Innovation* (pp. 132–157). IGI Global., DOI: 10.4018/978-1-7998-7927-5.ch007

Yu, P., Mulli, J., Syed, Z., & Umme, L. (Eds.). (2024). *Facilitating Global Collaboration and Knowledge Sharing in Higher Education with Generative AI*. IGI Global., DOI: 10.4018/979-8-3693-0487-7

Yu, P., Wong, S., & Prabhakar, A. (Eds.). (2025). *AI Strategies for Social Entrepreneurship and Sustainable Economic Development*. IGI Global., DOI: 10.4018/979-8-3693-6392-8

Zaber, M. A. (2024, October 10). The AI playbook: What other sectors can learn from the creative industry's fight against AI. RAND Corporation. https://www.rand.org/pubs/commentary/2024/10/the-ai-playbook-what-other-sectors-can-learn-from-the.html

Zeng, J., Abidin, C., & Schäfer, M. S. (2021). Research perspectives on TikTok & its legacy appsl research perspectives on TikTok and its legacy apps—Introduction. *International Journal of Communication*, *15*, 12–12.

Zhang, D., Pee, L. G., & Cui, L. (2021). Artificial intelligence in E-commerce fulfillment: A case study of resource orchestration at Alibaba's Smart Warehouse. *International Journal of Information Management*, *57*, 102304. DOI: 10.1016/j.ijinfomgt.2020.102304

Zhang, W., Mei, J., Song, W., Evans, R., & Xiang, Y. (2021). Why do citizens engage with the TikTok accounts of public hospitals in China? *SAGE Open*, *11*(4), 1–11. DOI: 10.1177/21582440211061568

Zhang, X., Xie, H., Yang, H., Shao, H., & Zhu, M. (2020). A General Framework to Understand Vulnerabilities in Information Systems. *IEEE Access : Practical Innovations, Open Solutions*, *8*, 121858–121873. DOI: 10.1109/ACCESS.2020.3006361

Zhu, F., & Song, Z. (2022). Systematic Regulation of Personal Information Rights in the Era of Big Data. *SAGE Open*, *12*(1), 215824402110675. DOI: 10.1177/21582440211067529

Zitnik, M., Agrawal, M., & Leskovec, J. (2018). Modeling polypharmacy side effects with graph convolutional networks. *Bioinformatics (Oxford, England)*, *34*(13), i457–i466. DOI: 10.1093/bioinformatics/bty294 PMID: 29949996

Zou, W., Li, J., Yang, Y., & Tang, L. (2023). Exploring the Early Adoption of Open AI among Laypeople and Technical Professionals: An Analysis of Twitter Conversations on #ChatGPT and #GPT3. *International Journal of Human-Computer Interaction*, 1–12. PMID: 39989507

Zuiderwijk, A., Chen, Y. C., & Salem, F. (2021). Implications of the Use of Artificial Intelligence in Public Governance: A Systematic Literature Review and a Research Agenda. *Government Information Quarterly*, *38*(3), 101577. Advance online publication. DOI: 10.1016/j.giq.2021.101577

Zwitter, A., & Boisse-Despiaux, M. (2022). Blockchain for humanitarian action and development: Ethical considerations. *Nature Communications*, *13*(1), 355. DOI: 10.1038/s41467-022-27985-9 PMID: 35039495

Zwitter, A., & Gstrein, O. J. (2022). Decentralized digital identity: Essential infrastructure for the future. *Big Data & Society*, *9*(1). Advance online publication.

About the Contributors

Poshan (Sam) Yu is a Chinese-Australian professional with an astonishingly diverse background and an extensive collection of remarkable accomplishments. In the academic sphere, he currently teaches at the University of Victoria Program at Soochow University. Previously, he held the position of an external professor of financial technology and finance at SKEMA Business School. He also serves as a visiting researcher at the Australian Studies Centre of Shanghai University, a guest professor at emlyon business school China, and an adjunct faculty member at George Washington University's master (finance) program at Renmin University of China. As a visiting professor at Krirk University in Thailand and an EU-Asia research coordinator at the European Business Institute in Luxembourg, he brings a truly global perspective to his teaching and research activities. He is a member of the IP & Innovation Researchers of Asia (IPIRA) Network and the Global South Dialogue on Economic Crime (GSDEC), further enriching his academic engagement. His prior role as a casual academic at the University of South Australia Business School has significantly enhanced and diversified his academic portfolio. From September 2020 to February 2021, he held the position of visiting scholar at The Chinese University of Hong Kong on two occasions. His scholarly pursuits are extensive and multifaceted. As an editor for the European Academic Journal (Special Issue) - European Business Institute in Luxembourg, an associate editor for Humanities & Social Sciences Communications (an SSCI-indexed journal), and an associate editor for the International Journal of Interdisciplinary Research in Social Science, he plays a crucial role in the dissemination of knowledge. He is an ad hoc reviewer for a number of prestigious journals and academic presses, including Applied Economics, Applied Artificial Intelligence, Cogent Economics & Finance, Humanities and Social Sciences Communications, Journal of Chinese Economic and Business Studies, Journal of Digital Learning and Education, The Singapore Economic Review, ILMA Journal of Social Sciences & Economics, Journal of Environmental Issues and Climate Change, Advances in Global Economic and Business Journal, The

Interdisciplinary Social Sciences Journal Collection, Common Ground Research Networks, Imprint - Xi'an Jiaotong-Liverpool University, Innovative Teaching and Learning, IBT Journal of Business Studies, Journal of Infrastructure Policy and Development, and Innovative Informatics and Artificial Intelligence Research. Moreover, he is an international editorial advisory board member for several edited books such as Advancing STEM Education and Innovation in a Time of Distance Learning (Scopus-indexed), Multidisciplinary Perspectives Towards Building a Digitally Competent Society (Scopus-indexed), Handbook of Research on SDGs for Economic Development, Social Development, and Environmental Protection, and Implications for Entrepreneurship and Enterprise Development in the Blue Economy.

James Mulli is the Founder, Program Director and Academic Dean of the European Business Institute, Luxembourg. He lectures in Applied Blockchain Technologies and Artificial Intelligence integration, post-graduate Corporate Finance & Business Statistics, and Managerial Accounting. He has held lectureships at in Luxembourg, China, the USA, Kenya and Germany. Dr. Mulli is a former director of two successful New York based multinational conglomerates and has valuable experience in the field as an NASD Series 7 compliance officer for the broker dealer Heyka Capital Management. He worked as a Portfolio Manager for the Soros Fund Management's Quantum Industrial Partners and holds a Doctorate and Masters degree from Saint Johns University, New York, and an MBA from New Hampshire Plymouth State University with specialization in Investment and Finance. Dr. Mulli is an international educator, and educational technology consultant with a keen interest in looking over the horizon at how technology is shaping the world. His search for trends in emerging technologies allows him to be an educator who applies skills to help students thrive. He is the founder of Technology Enabled Learning - Africa, and European Business Institute, Luxembourg, both non-for-profit educational providers whose initiative is to support existing Institutions in developing e-learning . His belief is that education is the silver bullet for development and through existing expertise and support infrastructure, EBI can provide the needed “jump-start” to institutions in Europe, Africa, Asia and beyond. Dr. Mulli is a Rockefeller Foundation Grant Recipient, and a Prince Fellow Teaching Excellence recipient. He is a textbook author and has published numerous newspaper and magazine articles.

Zain Ali Shah Syed is currently affiliated with Soochow University.

Umm-e-Laila is a Pakistani academic, HEC Approved Supervisor, and registered engineer with nearly two decades of experience in teaching, research, and academic leadership. She currently serves as Associate Professor and Program Head of Software Engineering at the Institute of Business Management (IoBM),

Karachi, and is concurrently pursuing a Postdoctoral Fellowship at Multimedia University (MMU), CyberJaya, Malaysia (2025–onwards), focusing on machine learning-integrated AHP multicriteria analysis for open-source software adoption in mission-critical IT infrastructure. She holds a Ph.D. in Computer Science and Information Technology from NED University of Engineering and Technology, Karachi (2016–2021), an MEng in Computer Engineering (Software Engineering), and a BSc in Computer Engineering, both from Sir Syed University of Engineering and Technology (SSUET). She holds international certifications including SCJP, SCWCD, and Microsoft Certified Professional (MCP). Her research spans Open-Source Software, Software Engineering, IoT, Big Data, Deep Learning, and Artificial Intelligence. She has authored and co-authored more than 40 publications in internationally indexed journals including *Sensors (MDPI), Applied Sciences (MDPI),* and *multiple Springer and Elsevier journals.* She is co-editor of two IGI Global books — *Facilitating Global Collaboration and Knowledge Sharing in Higher Education With Generative AI* and *The Impact and Integration of AI: From Research to Realization* — and has two further book projects in progress. Dr. Umm-e-Laila serves as Editor of PJETS (HEC Y Category) and has served as Guest Editor for special issues in *Computers in Human Behavior (IF: 9.9), Cognition Technology & Work (IF: 2.8)*, and several other Springer journals. She has chaired sessions at more than eight international conferences, including IEEC, ICETBEST 2026, and IMTIC 2025, and is an Expert Member of the Board of Faculty at Begum Nusrat Bhutto Women's University, Sukkur. She was named Winner of the Future Leader Award 2026 (Torchbearers of Knowledge) at ITCN Asia and the Global CISO Forum. She is a Fellow Member of the Institute of Engineers Pakistan (IEP), a Registered Engineer with PEC, and an Elected Local Council Member of IEP Karachi (2023–2025). She delivered a Keynote Speech on women's technology empowerment at the IEP Center on International Women's Day 2024 and is a focal person for ERASMUS (CBHE) funding applications with national and international partners.

***

Daniel Opotamutale Ashipala (MN Sc) is a senior lecturer and Head of Department for the School of nursing and Public Health (SoNPH), General Nursing Science department in the Faculty of Health Sciences and Veterinary Medicine (FHSVM) at the University of Namibia, Namibia. His research interests include Primary health Care, Health care services and Nursing education, and Mental health. He has published a quite number of peer reviewed articles on the delivery of teaching and learning in higher education studies. He is a doctoral candidate at Stellenbosch University in Cape Town, South Africa. His doctoral work focuses on developing a framework to facilitate the implementation of a task shifting approach for nurses in the Namibia PHC services.

Vinay Bamil is an AI researcher specializing in enterprise-scale agentic and responsible AI, with a focus on the evaluation, governance, and trustworthiness of large language model systems. He currently works as a Principal AI Engineer at Salesforce. He has authored preprints and articles on trustworthy AI and LLM agents, has a forthcoming book on building trustworthy AI systcms, and is an inventor on a pending US provisional patent application for multi-agent AI verification. He is currently pursuing a Doctorate in Generative AI.

Frank Billingsley has had a diverse and extensive career spanning multiple domains, including education, social services, and administration. He has had extensive international experience, and qualifications in higher education administration and teaching are prominent, with notable roles as a program director, head of department, and faculty member at various institutions. He has contributed to academic publications and presented at conferences on topics related to higher education and gender studies. Beyond academia, the candidate has a social work and counseling background, demonstrating a versatile skill set. His precise community service and involvement in various organizations further underscore their commitment to positively impacting.

Neslihan Bulur graduated from Anadolu University, Faculty of Communication, Department of Communication Design and Management in 2017. She graduated from the Sociology Department, which is a double major program at Anadolu University, and the International Relations undergraduate program at Anadolu University, which she studied as a second university, in 2018. In 2020, she graduated from Galatasaray University Strategic Communication Management master's program. She completed her PhD at Marmara University Communication Sciences program in 2025.

Zindan Çakıcı graduated from Kadir Has University, Faculty of Communication, Department of Public Relations and Publicity in 2018 as the top student of the faculty. Having completed his master's degree at Galatasaray University, Graduate School of Social Sciences, Department of Strategic Communication Management in 2020, Çakıcı was awarded the title of Doctor in 2024 at Galatasaray University, Graduate School of Social Sciences, Department of Media and Communication Studies with his doctoral thesis titled “Visual Representation of Irregular Migration in the Turkish Press: Afghan Migration After Taliban Rule”.

Debarshi Das is a seasoned technology program leader with over a decade of experience driving digital transformation initiatives across the life sciences, healthcare, and pharmaceutical sectors. His work spans enterprise platforms, AI-enabled clinical trial optimization, and regulatory technology solutions. Currently, Debarshi is pursuing a Doctorate in Generative AI from Golden Gate University, where his research focuses on the ethical integration of AI in clinical trials—exploring applications in patient recruitment, data governance, and adaptive trial designs. Blending hands-on industry expertise with academic rigor, Debarshi is passionate about building responsible, human-centric AI systems that enhance both operational efficiency and patient outcomes in clinical research. He is also an active mentor and speaker, contributing to thought leadership at the intersection of AI, healthcare, and digital innovation.

Pranjal Khare is a distinguished corporate lawyer and is currently working as a law faculty at O.P. Jindal Global University, India. He specializes in Property Law, Corporate laws, Intellectual Property Rights, and International law. Mr. Pranjal is a highly regarded scholar and has published on numerous issues related to international affairs, economy, politics, data & privacy, artificial intelligence and IP disputes. He is also a frequent speaker at legal conferences and has served as a consultant to numerous law firms and corporations.

Le Hoang Kiet is a PhD candidate at the Faculty of International Studies, University of Social Sciences and Humanities, Vietnam National University, Hanoi. He has research experience in international relations focusing on security and global geopolitics. He has published in numerous international Scopus/Wos scientific publications and is regularly invited to be a reviewer for specialized journals.

Medusalem Hangula Joel holds a Bachelor of Nursing Science (Cli) (Hon) and a Postgraduate Diploma in Nursing Science (Nursing Education) from the University of Namibia, in Namibia. He is a registered nurse by profession and a researcher with an interests focused Health Care Services and Nursing Education. He has a research article publication Teaching and Learning in higher education studies.

Fransisco Chindongo Ntjamba holds a Bachelor of Nursing Science (Clinical) (Honours) from the University of Namibia, School of Nursing and Public Health, General Nursing Department in the Faculty of Health Sciences and Veterinary Medicine (FHSVM). He is also a NIMART-trained nurse and a Provider Initiated Testing and Counseling (PITC) nurse, specializing in sexual reproductive health and focusing on HIV prevention, treatment initiation, and ongoing care. In addition to his qualifications, Fransisco is an academic researcher who has published research studies in nursing education, mental health, and interdisciplinary innovations, including Artificial Intelligence on human healthcare. His career is defined by a commitment to advancing healthcare through both direct patient care and research, with key interests in Primary Health Care, healthcare services, nursing education, and mental health. He is dedicated to improving healthcare through evidence-based practices and community outreach initiatives, particularly in HIV prevention and treatment.

Miroslav Svítek graduated in radioelectronic from Czech Technical University in Prague. In 1996, he received the Ph.D. degree in radioelectronic at Faculty of Electrical Engineering, Czech Technical University in Prague. Since 2008, he has been Full professor in engineering informatics. In 2010 – 2018 he was Dean of Faculty of Transportation Sciences, Czech Technical University in Prague. Since 2018, he has been Visiting professor in smart cities at University of Texas at El Paso, USA. He is the author or co-author of more than 200 scientific papers and 10 books.

Yan Jingwen, an independent researcher, has research interests in economics, regtech, and fintech.

Ladislav Žák is interested in complex socio-economic systems in which the human factor plays a crucial role. He used the knowledge he gained in his practice as a crisis manager. He has published several works on the topic of ethics and the search for balance in complex systems. He regularly comments on current economic, social and political issues.

Index

A

B

C

D

E

F

G

H

I

K

L

M

N

O

P

R

S

T

U

V

www.ingramcontent.com/pod-product-compliance
Lightning Source LLC
LaVergne TN
LVHW062327300926
849161LV00007B/1605